A COMMENTARY ON
THE JCT
INTERMEDIATE
FORM OF
BUILDING CONTRACT

A COMMENTARY ON THE JCT INTERMEDIATE FORM OF BUILDING CONTRACT

Second Edition

NEIL F. JONES

LLB (Hons), ACIArb, Solicitor
of the Supreme Court

AND

DAVID BERGMAN

FRICS, ACIArb

OXFORD

BSP PROFESSIONAL BOOKS

LONDON EDINBURGH BOSTON

MELBOURNE PARIS BERLIN VIENNA

First Edition published by Collins
    Professional and Technical Books 1985
Second Edition published by BSP
    Professional Books 1990

British Library
Cataloging in Publication Data
Jones, Neil F.
    A commentary on the JCT Intermediate
    form of building contract.—2nd ed.
    1. Great Britain. Buildings. Construction.
    Contracts.
    Standard conditions: Joint Contracts
    Tribunal.
    Intermediate form of building contract
    I. Title   II. Bergman, David
    692.8

ISBN 0–632–02274–4

BSP Professional Books
A division of Blackwell Scientific
    Publications Ltd
Editorial Offices:
Osney Mead, Oxford OX2 0EL
    (Orders: Tel. 0865 240201)
25 John Street, London WC1N 2BL
23 Ainslie Place, Edinburgh EH3 6AJ
3 Cambridge Center, Suite 208, Cambridge
    MA 02142, USA
54 University Street, Carlton, Victoria 3053,
    Australia

Set by DP Photosetting, Aylesbury, Bucks
Printed and bound in Great Britain by
Billing & Sons, Worcester

Extracts from the Intermediate Form of Building Contract and copies of NAM/T, ESA/1
and Practice Note IN/1 appear with kind permission of the copyright holder, RIBA
Publications Ltd.

Contents

Contents

Preface to First Edition

The arrival of a completely new form of building contract from the Joint Contracts Tribunal is inevitably an important landmark for all those concerned with or interested in the construction industry. The Joint Contracts Tribunal Intermediate Form of Building Contract is no exception. It is likely to be very widely used and those using it will need to become familiar with its provisions as quickly as possible in order to do both those whom they represent, and themselves, justice.

We have both been fortunate in having been involved at all the relevant stages in the formulation and drafting of this new form of contract. This involvement, together with our interest and involvement in the building industry, led us to the writing of this book which we hope will assist those who will, in whatever capacity, be using or advising on the Intermediate Form of Building Contract. We would like to make it perfectly clear, however, that the commentaries, views and opinions expressed in this book are ours and are not intended in any way to represent those of the Joint Contracts Tribunal or any of its Committees.

This book is intended primarily to be of value to quantity surveyors, architects, contractors and employers, and it is hoped that it may be of some use to lawyers who are not already familiar with building contract law.

In order to put our commentary on the new contract into a proper legal context, a summary of the relevant general law has been included in each chapter. Clearly, however, this book is not meant to be a detailed legal text book. Accordingly, where what we consider to be uncontroversial points of law have been stated, these have not been supported by quoting cases. Cases have been used principally where we have dealt with matters which are controversial or are particularly germane to the building industry. As far as possible we have referred to recent cases.

Our commentary on the clauses is generally thorough and that on the clauses relating to named sub-contractors has been dealt with in considerable detail. The intention throughout has been to produce a book of practical value to the practitioners in the building industry.

The writing of a text book is almost inevitably a very time-consuming exercise. In the event, in order to provide what we hope is a detailed commentary on the new contract at the time when it is needed, it has been necessary to move apace. In this regard we are greatly indebted to Mary, wife of the first named co-author, who not only devoted an enormous amount of her time and energy to typing and checking, but in so doing also managed to get the better of a decidedly naughty word processor.

25 October 1984

Neil F. Jones   *David Bergman*
Neil F. Jones & Co., Solicitors   *Bucknall Austin*
Birmingham   *Quantity Surveyors*
                                                              Birmingham

Preface to Second Edition

The JCT Intermediate Form of Building Contract (IFC 84) has proved highly popular in use and this second edition has, we hope, benefited from experiences gained in the six years that the contract has been in use.

This edition also reflects a number of significant changes since publication of the first edition in 1985.

There have been five amendments to the contract issued by the Tribunal:

— *Amendment 1 1986.* This Amendment followed a wholesale review of the insurance and indemnity provisions in Section 6 of the IFC 84 Conditions. The amendments made to the provisions regarding insurance of the works were of a fundamental nature. Chapter 9, which deals with insurance of the works, has therefore been fully revised and rewritten.

— *Amendment 2 September 1987* introduced some minor amendments to the arbitration provisions and also to the terminology of the determination provisions to take into account the Insolvency Act 1986.

— *Amendment 3 July 1988* contained nine miscellaneous amendments, many of which are fundamental in nature, including relocating within the contract the arbitration provisions and significant amendments to these; the setting of a stricter and more positive time-table for the issue of the final certificate; an extension in the conclusiveness of the final certificate; and the deletion of the fair wages provision.

— *Amendment 4 July 1988* introduced the Standard Method of Measurement 7th Edition where bills of quantities are a contract document in place of the 6th Edition.

— *Amendment 5 April 1989* introduced changes to the value added tax provisions to take into account changes in legislation.

All of these changes have been dealt with in the text and the commentary on the arbitration provisions has been significantly expanded.

There have been many decisions of the courts arising out of construction contracts which are of relevance and all of the more important of these have been introduced together with comments upon their effects and likely implications where appropriate.

The opportunity has also been taken to deal specifically with the question of partial possession which can be introduced through the Appendix to Practice Note IN/1 as an additional clause 2.11 and also with the Sectional Completion Supplement published in 1985.

As with the first edition, the citing of legal cases has been used principally where we have dealt with matters of a controversial nature or where the decision is particularly germane to some aspect of the contract. We have throughout, as far as possible, referred to recent cases.

We are both greatly indebted to Mary, wife of the first named co-author, for bearing the burden of the typing, as well as much of the checking.

July 1990

Neil F. Jones,	*David Bergman,*
Neil F. Jones & Co., Solicitors,	*Bucknall Austin plc.,*
Number 3, Broadway,	*Quantity Surveyors*
Broad Street,	*and Construction*
Birmingham	*Cost Consultants,*
B15 1BQ	*5 Scotland Street,*
	Birmingham
	B1 2RR.

Chapter 1

Background to the Intermediate Form

The principal parties involved in drafting the earliest forms of standard building contracts were the Royal Institute of British Architects (RIBA) and the National Federation of Building Trades Employers (NFBTE), now the Building Employers Confederation (BEC). Originally, the contract was known as the 'RIBA Form', and even today it is not unusual to hear that name erroneously given to the present Standard Form produced by the Joint Contracts Tribunal (the Tribunal), notwithstanding that the RIBA is now only one of the constituent bodies of the Tribunal.

The Tribunal is not a tribunal in the more usual sense of the word, in so far as it does not sit in judgement on others. It is in fact a panel of representatives drawn from parties (the constituent bodies) which have a direct interest in the building industry. The constituent bodies at present are:

The Royal Institute of British Architects
The Building Employers Confederation
The Royal Institution of Chartered Surveyors
The Association of County Councils
The Association of Metropolitan Authorities
The Association of District Councils
The British Property Federation
The Committee of Associations of Specialist Engineering
    Contractors
The Federation of Associations of Specialists and Sub-Contractors
The Association of Consulting Engineers
The Scottish Building Contract Committee

The number of constituent bodies has grown steadily over the years, the Royal Institution of Chartered Surveyors becoming an endorsing body of the Standard Form in 1952 and being later joined on the Tribunal by the local authority associations. In 1963 the reconstituted Tribunal published a new Standard Form of Building Contract in four different versions, i.e. for use with and without bills of quantities, both with a local authority and private edition.

It was not until 1966 that the Federation of Associations of Specialists and Sub-Contractors (FASS) and the Committee of Associations of Specialist Engineering Contractors (CASEC), the two associations representing specialist sub-contractors, joined the Tribunal, and over the following years consideration was given to the possibility of radically revising the nominated sub-contract provisions included in the 1963 edition of the Standard Form. During the period of this review it became apparent that it would not be satisfactory to introduce such a radical change in the nomination provisions and, at the same time, to accommodate other revisions which were then under consideration without producing an entirely new form. Having taken that decision, it was further decided to completely revise the format and to publish at the same time the various related documents and sub-contracts, and this new form, together with the related documents, was finally published in 1980 and is known as the Joint Contracts Tribunal Standard Form of Building Contract 1980 Edition (JCT 80). It had a mixed reception. Whilst it was considered by many to encourage good practice in the administration of contracts, the procedural matters which were included inevitably tended to increase the complexity of the form. Largely because of this, it has been slow to receive acceptance, with some continued use of the previously much maligned Joint Contracts Tribunal Standard Form of Building Contract 1963 edition (JCT 63).

Whilst the procedural matters included in JCT 80 were not necessarily considered to be disadvantageous where the size and complexity of the contract works warranted them, many considered that a more simple form should be produced for the less complex type of works. The only alternative JCT form available was the Agreement for Minor Building Works (Minor Works Form) which was clearly unsuitable for other than simple contracts of low value. Responding to these pressures for a form of contract which would fill the gap between JCT 80 and the Minor Works Form the Tribunal, in 1982, set up a working party to consider the content of such a form and to make recommendations. This form was finally published by the Tribunal in September 1984 and is called the JCT

Intermediate Form of Building Contract 1984 (IFC 84). The Tribunal has also issued the following documents for use with IFC 84:

— JCT Fluctuations Clauses for use with JCT Intermediate Form of Building Contract IFC 84 (excluding Formula Rules);
— JCT Consolidated Main Contract Formula Rules October 1987;
— Form of Sub-contract Tender and Agreement for persons named under clause 3.3 (NAM/T[1]);
— Sub-Contract Conditions (including fluctuations clauses), referred to in NAM/T (NAM/SC);
— Sub-Contract Formula Rules, referred to in NAM/T and NAM/SC;
— Clause for insertion in the Intermediate Form as '2.11 – Partial Possession by Employer', if required.
— IFC 84 Sectional Completion Supplement July 1985.

In addition, a very useful Practice Note IN/1 has been issued. Furthermore, although not a Tribunal document, a separate agreement to be entered into between employer and a named person as sub-contractor dealing with the matter of design has been prepared by the RIBA and CASEC for use with IFC 84. It is known as ESA/1.

IFC 84 is designed to be suitable for building works:

(1) of a simple content involving the normally recognised basic trades and skills of the industry; and
(2) without any building service installations of a complex nature, or other specialist work of a similar nature; and
(3) adequately specified, or specified and billed, as appropriate prior to the invitation of tenders.

Finally, though again not a Tribunal document, the BEC, CASEC and FASS have produced standard Articles of Agreement and Sub-Contract Conditions for use between the Main Contractor and Domestic Sub-Contractors, known as IN/SC.

IFC 84 follows the general layout of the Minor Works Form, the clauses being grouped under very similar section headings, but there the similarity ends. The clauses are far more detailed and compre-

---

1. Printed in full in the Appendix to this book.

hensive although, by its very name, obviously less so than JCT 80.

IFC 84 is published in one edition only for both private and local authority use, and the text allows for a number of alternative documents to be included as contract documents. In addition to the drawings, the supporting documents can comprise bills of quantities, specification or schedule of works and one or more of these will be priced by the contractor. Where the drawings are supported by a specification only, which is not to be priced, then the contractor must supply either a schedule of rates or a contract sum analysis in support of his tender. It should be noted that in this contract, when there are no bills of quantities, the employer is responsible for the accuracy of any quantities given in the specification or schedule of works and in this respect it differs from JCT 80 (Without Quantities).

Practice Note 20 issued by the Tribunal (revised July 1988) gives guidance on the appropriate form of JCT main contract to use. In dealing with the use of IFC 84 it states in paragraph 13 as follows:

'The Intermediate Form has been prepared so as to be suitable for contracts for which the more detailed provisions of the Standard Form (1980) Edition) are considered by the Employer or by his professional consultants to be unnecessary in the light of the foregoing criteria. The Form would normally be the most suitable form for use, subject to these criteria, where the contract period is not more than twelve months and the value of works is not more than £280,000 (1987 prices), but this must be read together with paragraph 14 on the money limits within which the use of the Minor Works Form may be appropriate.

The Intermediate Form may however be suitable for somewhat larger or longer contracts, provided the three criteria referred to in the endorsement are met, but Employers and their professional consultants should bear in mind that the provisions of the Intermediate Form are less detailed than in the Standard Form (1980 Edition) and that circumstances may arise, if it is used for unsuitable works, which could prejudice the equitable treatment of the parties.'

From a consideration of the three criteria set out above regarding the nature of the building works for which IFC 84 is designed (see page 3) in conjunction with these two paragraphs of Practice Note 20, it is clear that the three criteria should govern the situation. It is not therefore the contract value or contract period which is the determining factor. It is the nature of the work which is undertaken. So for instance, there seems no reason at all why IFC 84 should not be used for the construction of a large estate of conventional dwelling houses whatever the total value of the works.

The second quoted paragraph above from Practice Note 20 refers to the fact that circumstances may arise, if IFC 84 is used for unsuitable works, which could prejudice the equitable treatment of the parties. When the form is scrutinised, it is almost certain that the reference to unsuitable works is a reference to the use of named sub-contractors to carry out specialist work of a complex nature. If such specialist works, e.g. complex service installations, form a significant proportion of the total cost of the contract and take up a significant part of the construction period, it might be argued by some that the increased risks which a main contractor takes under IFC 84 for named sub-contractors when compared with the responsibility of the main contractor under JCT 80 for nominated sub-contractors, is such that it is an unfair burden on the contractor in such circumstances. Others may well disagree. Under IFC 84 it is of course the main contractor who takes the risk of delay and disruption caused by named sub-contractors. This is in some respects different from the case of nominated sub-contractors under JCT 80.

Furthermore, in the event of a named sub-contractor's employment being determined, one of the options for the architect is to require the contractor to finish off the outstanding work of the named sub-contractor. This may be regarded as unfair in relation to highly specialist complex work. There is no equivalent provision in relation to nominated sub-contractors under JCT 80.

There are other risks of inequitable treatment of the parties where specialist or complex work is concerned. For example, the provision in JCT 80 clause 8.1.1 which provides relief for the contractor where materials or goods of the kinds and standards described in the contract documents are not procurable, is absent in IFC 84. JCT 80 in clause 11 provides for the architect to have access to off-site workshops or other places of the contractor and sub-contractors where work is being prepared. Again there is no such provision in IFC 84. A final example is in relation to off-site goods and materials. Both JCT 80 and IFC 84 provide for the architect in his discretion to include the value of off-site goods and materials in interim payment certificates. However, JCT 80, in clause 30.3, expressly provides for precautions to be taken, such as the goods being separately stored off-site and identified as belonging to the employer etc. Such precautions are clearly sensible and become particularly important in respect of high value goods or materials stored off-site. There is no equivalent provision in IFC 84. No doubt, in practice, if the architect is minded to exercise his discretion to include the value of off-site goods or materials in interim payment certificates, he will ensure that, as a condition of exercising his discretion in favour of the contractor, the type of precautions described in clause 30.3 of JCT 80 must be taken.

Probably the main difference between JCT 80 and IFC 84 is that there is no provision for nomination. In its place IFC 84 allows the employer to 'name' a sub-contractor to execute work. The work may be specified in the main contract tender document, the tender submitted by the named sub-contractor being furnished to tendering main contractors in order that they can themselves price the work. This means that tenders from prospective 'named' sub-contractors must be obtained prior to inviting main contract tenders. Alternatively, if it is intended to 'name' a sub-contractor after the main contract has been placed, this can be done by including a provisional sum for the work in the main contract tender document. There is no provision for the use of prime cost sums in IFC 84. As with nomination under JCT 80, the Tribunal has published related documents which have to be used in the naming procedure.

The basic concept for the first method of naming is simple. The contractors tendering are given detailed information as to the named sub-contractor's price and the conditions, programme and attendances upon which he has based his offer. The contractor then takes responsibility for the named sub-contractor as with any other domestic sub-contractor. However, if the sub-contractor named in the tender documents fails to enter into a sub-contract or if, following his appointment, he fails to complete the sub-contract works, then certain safeguards for the contractor have been built into the system which tend to make named sub-contracting similar in a number of respects to nomination under JCT 80.

It appears to be the case that IFC 84 has been well received by the industry. It is one of the hazards of drafting a form which attempts to achieve a reasonable balance of responsibility between the parties that at the end, neither side feels entirely happy with the outcome. However, it is proving to be a very useful contract. Certainly it is in very frequent use.

The comments on the clauses in IFC 84 which are contained in the following pages are dealt with in the same order as they appear in the contract. The contract places each clause within a section and each section is dealt with as a separate chapter in this book except for section 3 which, due to the extensive treatment given to naming, has been divided into two parts. Each chapter commences with a summary of the general law appertaining to the matters covered by the particular section of the contract. This is followed by the full text of each clause included in the section, together with a general commentary on the clause and notes on specific extracts. For convenience the term 'the architect' has been used in place of 'the architect/the contract administrator' and should therefore be read as including the contract administrator unless the context otherwise requires.

In spite of the simplicity of the building works for which it is intended IFC 84 should be handled with care and the respective responsibilities of the parties should be clearly understood. In this respect it is hoped that the summaries of the relevant law, together with the commentaries on the various clauses, followed by the notes, will assist towards a better understanding of the contract.

Chapter 2

The agreement,
recitals and articles

CONTENT

This chapter deals with the form of agreement, recitals and articles.

SUMMARY OF GENERAL LAW

The recitals, if included in an agreement, usually begin with the word 'Whereas'. The recitals are not within the operative part of a document and they cannot generally be used as an aid to the construction of that part. However, if there is a doubt about the construction of the operative part of the contract, the recitals may then be looked at in order to see if they assist in determining the true construction. The recitals really introduce what it is that the parties are intending to achieve by their contract. They are in the nature of an introduction and background.

The articles of agreement will often commence with words such as 'Now it is hereby agreed as follows:' and these words introduce the operative part of the agreement.

This Agreement

is made the _____ day of _____ 19_____

between _____

_____

_____

of (or whose registered office is at) _____

_____

_____

(hereinafter called the 'Employer') of the one part AND

_____

_____

_____

of (or whose registered office is at) _____

_____

_____

(hereinafter called the 'Contractor') of the other part.

Whereas

The Works

1st   The Employer wishes the following work _____

_____

_____

_____

(hereinafter called 'the Works')

[a] to be carried out under the direction of the Architect/the Contract Administrator named in Article 3 hereunder and has caused the following documents showing and describing the Works to be prepared:

the Contract Drawings numbered _____
[b] the Specification,
[b] the Schedules of Work,
[b] Bills of Quantities,

[c] and, in respect of any work described and set out therein for pricing by the Contractor and for the execution of which the Contractor is required to employ a named person as sub-contractor in accordance with clause 3·3·1 of the Conditions annexed hereto, has provided all of the particulars of the tender of the named person for that work in a Form of Tender and Agreement NAM/T with Sections I and II completed together with the Numbered Documents referred to therein.

[d] **Pricing by the Contractor**

2nd   A. The Contractor has priced the Specification/priced the Schedules of Work/priced the Bills
[b] of Quantities (as priced, called 'the Contract Bills') and the total of such pricing is the Contract Sum as mentioned in Article 2 hereof;

and such priced document and the Contract Drawings, both signed by or on behalf of the parties (together with, where applicable, the particulars, referred to in the 1st recital, of the tender of any named person in a certified copy of a Form of Tender and Agreement NAM/T with Sections I and II completed, also signed by or on behalf of the parties hereto), the Agreement and the Conditions annexed hereto are hereinafter called the 'Contract Documents'.

2nd   B. The Contractor has –

stated the sum he will require for carrying out the Works shown on the Contract Drawings and described in the Specification and these documents, both signed by or on behalf of the parties (together with, where applicable, the particulars, referred to in the 1st recital, of the tender of any named person in a certified copy of a Form of Tender and Agreement NAM/T with Sections I and II completed, also signed by or on behalf of the parties hereto), the Agreement and the Conditions annexed hereto are hereinafter called the 'Contract Documents',

and
supplied to the Employer either a Contract Sum Analysis or a Schedule of Rates on which the Contract Sum is based.

---

[a] See clause 8·4 and Article 3. Delete 'the Architect' or 'the Contract Administrator'.

[b] Delete as appropriate.

[c] Delete if no items specifying a named person are included in the documents.

[d] Delete alternative A or alternative B as appropriate.

[e] Strike out the words in italics in Article 3, when the Architect/the Contract Administrator, or in Article 4, when the Quantity Surveyor, is an official of the Local Authority.

[f] If the Architect/the Contract Administrator is to exercise the functions ascribed by the Conditions to the Quantity Surveyor, his name should be inserted in Article 4.

Now it is hereby agreed as follows

Article 1

Contractor's
obligation

For the consideration mentioned in Article 2 the Contractor will upon and subject to the Contract Documents carry out and complete the Works.

Article 2

Contract Sum

The Employer will pay to the Contractor the sum of _____

_____

_____ (£ _____ . _____)
exclusive of VAT (hereinafter called 'the Contract Sum') or such other sum as shall become payable hereunder at the times and in the manner specified in the Conditions.

Article 3

The Architect/
The Contract
Administrator

[a] The term 'the Architect'/'the Contract Administrator' in the Conditions shall mean the person referred to in the 1st recital namely:

_____

of _____

_____

or in the event of his death or ceasing to be so appointed for the purpose of this Contract such other person as the Employer shall within 14 days of the death or cessation nominate for that purpose *not being a person to whom the Contractor shall object for reasons thought*
[e] *to be sufficient by an Arbitrator appointed in accordance with Article 5.* Provided that no person subsequently so appointed under this Contract shall be entitled to disregard or overrule any certificate or instruction given by any person for the time previously appointed.

[f] **Article 4**

The Quantity
Surveyor

The term 'the Quantity Surveyor' in the Conditions shall mean

_____

of _____

_____

or, in the event of his death or ceasing to be the Quantity Surveyor for the purpose of this Contract, such other person as the Employer shall nominate for that purpose, *not being a person to whom the Contractor shall object for reasons considered to be sufficient by an*
[e] *Arbitrator appointed in accordance with Article 5.*

Article 5

Settlement of
disputes –
Arbitration

5·1 If any dispute or difference as to the construction of this Contract or any matter or thing of whatsoever nature arising thereunder or in connection therewith shall arise between the Employer and the Architect/the Contract Administrator on his behalf and the Contractor either during the progress or after the completion or abandonment of the Works or the determination of the employment of the Contractor except under Supplemental Condition **A7** *(Value Added Tax)* or Supplemental Condition **B8** *(Statutory tax deduction scheme)* it shall be and is hereby referred to arbitration in accordance with clause 9.

[g] Not used.

[†] As witness the hands of the parties hereto

[†] Signed by or on behalf of the Employer _____

    in the presence of: _____ _____

[†] Signed by or on behalf of the Contractor _____

    in the presence of: _____

[∗] Signed sealed and delivered by/[∗∗] The common seal of: _____

    in the presence of/[∗∗] was hereunto affixed in the presence of _____

[∗] Signed, sealed and delivered by/[∗∗] The common seal of: _____

    [∗] in the presence of/[∗∗] was hereunto affixed in the presence of: _____

[†] For Agreement executed under hand.

[∗] For Agreement executed under seal by an individual or unincorporated body.

[∗∗] For Agreement executed under seal by a local authority, a company or other body corporate.

Recitals

COMMENTARY

The recitals include a brief description of the works to be constructed and indicate the documents which are to be included as contract documents. Unlike the JCT Standard Forms of Building Contract, where contracts based on differing documentation are published separately, for example with and without quantities, IFC 84 has been designed to allow the maximum degree of flexibility which can be accommodated in one document. Hence the various types of alternative documentation referred to.

The first recital allows for the insertion of a brief description of 'the Works' and this term appears throughout the conditions of IFC 84 and particularly in clause 1.1 covering the contractor's obligation to '... carry out and complete the Works in accordance with the Contract Documents ...'. The first recital also makes reference to the person under whose direction the works are to be carried out (that person being named in article 3) and, depending on whether the person so named is an architect, i.e. a person registered under the Architects Registration Acts 1931 and 1969, either the words 'the Architect' of the words 'the Contract Administrator' should be deleted. The deletion made will then be deemed to have been made throughout the conditions (clause 8.4).

The first recital also includes provision for identifying the documents prepared by or on behalf of the employer and '... showing and describing the Works ...'. These will include the contract drawings and one or more of the three documents listed. The contractor then uses the contract drawings and the other documents in order to compile his tender bid.

The contractor must either price one of the three documents mentioned in the first recital i.e. the specification, the schedules of work or the bills of quantities (see A of the second recital), following which it will become one of the contract documents; or alternatively, where neither schedules of work nor bills of quantities are used and the specification is unpriced, then the contractor must provide the necessary pricing information in a contract sum analysis or schedule of rates (see B of the second recital). Once the contract sum analysis or schedule of rates has been supplied, the unpriced specification will become one of the contract documents. In practice it might occasionally be convenient to have, on the one hand, part of the works covered by a priced specification or schedules of works or bills when alternative A of the second recital will apply, and on the other, part of the work covered by an unpriced specification or contract sum analysis or schedule of rates when alternative B of the second recital will apply. Unfortunately footnote (d) to the second recital (see page 10) only contemplates the use of either alternative A or B for the whole of the works.

The drawings must be identified by listing the drawing numbers. Where any work is to be carried out by a named person who has been named as a sub-contractor in the main contract documentation (dealt with in detail in Chapter 5), the employer must also provide all of the particulars of the tender of the named person in respect of that work in a form of tender and agreement known as NAM/T with sections I and II completed together with the numbered documents referred to therein. Accordingly, the main contract documents will include the agreement, the conditions (including the appendix and supplemental conditions), the contract drawings, a priced bill of quantities (which then become known as 'the Contract Bills'), and/ or a priced specification, and/or a priced schedule of work, or an unpriced specification together with, where a named person is being appointed as a sub-contractor, the form of tender and agreement NAM/T.

A valuation of a variation is based upon the 'priced document' (see clause 3.7). Where alternative A in the second recital is applicable, the priced document will be the priced specification and/or schedule of works and/or the contract bills. Where alternative B in the second recital is applicable, the priced document will be the contract sum analysis or the schedule of rates (see clause 3.7.1).

Further, where alternative A is applicable, the total of the pricing equals the contract sum referred to in article 2. Where alternative B is applicable, then the contract sum in article 2 is based on the contract sum analysis or schedule of rates. The contract sum analysis means an analysis of the contract sum provided by the contractor in accordance with the stated requirements of the employer (see clause 8.3). In practice it may be little more than a breakdown of the contract sum into elements, in which case it would have only limited use in the valuation of variations but may be of greater use for the purpose of interim valuations. Alternatively it may take the form of an elemental breakdown with separately priced items showing in detail the breakdown of costs within each element. The greater the refinement of the document, the greater will be its value both in terms of evaluating the tender and subsequently as a means of valuing variations. JCT Practice Note 23 gives guidance on the use and structure of contract sum analyses. This Practice Note, at page 1, refers to the purposes for which a contract sum analysis is required as:

> '1. the valuation of variation and provisional sum work insofar as reference to the Contract Sum Analysis for this purpose is required by:
>
>> clause 13.5 in the Standard Form Without Quantities, and
>> clause 3.7 in the Intermediate Form;
>
> 2. in the Standard Form Without Quantities where clause 40 (use of the price adjustment formulae) is included in the Conditions, to enable the operation of this clause in accordance with the Formulae Rules.

'A further purpose, which is not specifically required by the Conditions, is to facilitate the determination of the amounts to be stated in Interim Certificates under clause 30 in the Standard Form Without Quantities and for Interim Payments under clauses 4.2 and 4.3 in the Intermediate Form'.

It can be seen that so far as JCT 80 without quantities is concerned, the contract sum analysis enables the use of price adjustment formulae in accordance with the Tribunal's Consolidated Main Contract Formula Rules where no bills of quantity have been used. This is specifically catered for by Amendment 3 to JCT 80 Without Quantities Edition (issued March 1987).

A similar amendment to IFC 84 could have been made but wasn't. Accordingly as presently drafted IFC 84 does not provide for the use of the Consolidated Main Contract Formula Rules unless priced bills of quantities have been used (see paragraph 2c of Section 1 of the Consolidated Main Contract Formula Rules dated October 1987).

However, if it is desired to use IFC 84 without priced bills but with the price adjustment formula then it should be quite possible with careful drafting to incorporate similar amendments into IFC 84. This should make IFC 84 even more flexible.

An alternative to the use of a contract sum analysis, where alternative B is applicable, is the use of a schedule of rates which should not be confused with a schedule of work. The latter will generally include quantities, whereas the former will not and will usually take the form of a list of priced items described in a similar fashion to work included in a bill of quantities, and the unit of measure on which the prices are based will be stated. It is likely to have only limited use in the evaluation of the contractor's tender. Furthermore, unless it is possible to have access to the detailed figures building up to the tender sum, it will also be very difficult to say with any certainty whether the rates are those on which the tender was based. That may not necessarily be important providing one can be satisfied that the rates are, in any case, reasonable and that they would form a fair basis for the pricing of variations. It could of course be argued that there are advantages in including a schedule of rates priced much higher than those on which the contract sum is based. The argument would be that it would discourage variations which can have such a disastrous effect on the contractor's efforts to organise the contract works efficiently. That view will clearly not be shared by everyone and it will therefore be prudent to give careful thought to the level of rates included in a schedule of rates when this type of document is to be used in preference to a contract sum analysis. For the reasons stated above however, a priced document with quantities with the rates extended and totalled to equal the contract sum will generally be preferable to a schedule of rates.

The contract drawings will always be provided but the extent of detailing included in those drawings will depend on the choice of the other documents. Where contract bills are used it is likely that only a limited number of drawings will be included since the contractor will look to the contract bills to indicate the quality and quantity of work (clause 1.2) and will have priced accordingly. As the work proceeds, the architect will then issue '. . . such further drawings or details as are reasonably necessary to enable the Contractor to complete the Works . . .' (clause 1.7).

It can be seen therefore that the contract allows considerable flexibility in the type of document on which the contract sum is to be based. Where the contract sum is based upon drawings and a specification or schedule of works then a more comprehensive set of drawings will be required than would be the case if bills of quantities were incorporated as a contract document. The briefer the specification or schedule of works, the greater the reliance which must be placed on the drawings. Further drawings and details will still be issued during the course of the contract but, in case of dispute, it may be more difficult to decide whether these further drawings and details simply amplify the contract drawings or whether they represent a departure which would qualify as a variation. The same problem would not normally present itself where bills of quantities were used since the quality and quantity of work will be defined in the bills and not on the drawings.

Articles

COMMENTARY

Article 1 states the contractor's obligation to carry out and complete the works on, and subject to, the contract documents for the consideration mentioned in article 2, i.e. the contract sum.

Article 2 requires the insertion by the contractor of a sum which becomes 'the Contract Sum' which is capable of adjustment and is payable at the times, and in the manner specified, in the conditions of contract. The contract is a lump sum contract in the sense that it is a contract to complete the whole work for a lump sum. The contractor must carry out and complete the works in accordance with the contract documents (see clause 1.1). Whilst the contractor is entitled to interim payments in accordance with the conditions, these payments are on account of the finally adjusted contract sum. An interim payment is not therefore a final payment in respect of the work to which its value relates.

Article 3 requires the name of the architect or contract administrator to be inserted, or in the event of his death or his ceasing to be so appointed, such other person as the employer nominates within fourteen days of the death or cessation. In relation to subsequent appointments the contractor has a right to object for reasons thought to be sufficient by an arbitrator appointed in accordance with article 5 and clause 9. The new nominee is not entitled to disregard or overrule any certificate or instruction given by his predecessor.

The use of the name 'architect' is restricted by the Architects Registration Acts 1931 to 1969. No one may practise or carry on business under any name, style or title containing the word 'Architect' unless he is a person registered by the Architects Registration Council. However, a body corporate, firm or partnership can carry on the business under the style or title of 'Architect' provided that certain conditions are fulfilled as to the business being under the control and management of a registered architect. It is permissible therefore for the name of an architectural practice to be inserted in article 3 as well as that of an individual ('person' is defined in clause 8.3 to include a partnership or body corporate). It is submitted that this is so, despite the evidential difficulties which may arise in any dispute where the architect's opinion is called into question and which he is required to defend or explain. In practice it does not appear to impose insuperable problems.

It is quite common for local authority employers to insert the name of a chief officer as either architect or contract administrator, although actual decisions and opinions are given and expressed by others acting in the name of the chief officer concerned. The chief officer in person may subsequently be called upon to justify a decision or opinion which in truth he may not hold. This can be embarrassing.

If the person appointed to administer the contract is not entitled to describe himself as an 'Architect' the reference to 'the Architect' in article 3 should be deleted, and in accordance with clause 8.4, the term 'the Architect' shall be deemed to have been deleted throughout the contract, and of course the reverse applies where the words 'the Contract Administrator' is deleted in article 3.

Article 4 requires completion with the name and address of the quantity surveyor. Similar provisions apply as in article 3 in the event of death or a cessation of the appointment. The architect or supervising officer can act as the quantity surveyor and if this is the case then his name should be inserted in article 4 – see footnote (f) on page 2 of IFC 84.

ARBITRATION

Article 5 deals with arbitration. Any dispute or difference concerning the contract which arises between the employer, or the architect on his behalf, and the contractor must be referred to arbitration and not to the courts for determination. Two types of dispute or difference are excluded from arbitration under IFC 84, namely:

— concerning any employer's challenge in regard to Value Added Tax claimed by the contractor in which case the matter is dealt with in accordance with clause A7 of Supplemental Condition A (Value Added Tax) incorporated into IFC 84 by virtue of clause 5.5 of the Conditions.

— concerning the Statutory Tax Deduction Scheme under the Finance (No 2) Act 1975 and the Income Tax (Sub-Contractors in the Construction Industry) Regulations 1975 SI No 1960 to the extent that the Act or Regulations or any other Act of Parliament of statutory instrument rule or order provides for some other method of resolving such dispute or difference – see clause 5.6 and Supplemental Condition B8;

Article 5 provides that the dispute or difference 'shall be and is hereby referred to arbitration in accordance with clause 9.' Reference should be made to chapter 12 dealing with clause 9 – Settlement of disputes – Arbitration (see page 377).

In addition, it should be borne in mind that the issue of the final certificate can restrict the matters which can be referred to arbitration (see clause 4.7).

Chapter 3

Intentions of the parties

CONTENT

This chapter looks at section 1 which is an assemblage of topics under the general heading 'Intentions of the parties'. They comprise the following:

Contractor's obligations
Quality and quantity of work
Priority of contract documents
Instructions as to inconsistencies, errors or omissions
Contract bills and standard method of measurement
Custody and copies of contract documents
Further drawings and details
Limits to use of documents
Issue of certificates by the architect
Unfixed materials or goods: passing of property etc.
Off-site materials and goods: passing of property etc.

With such a mixture of topics, it is difficult to give them a suitable composite heading. The heading given, namely, 'Intentions of the parties' could apply equally to each and every clause throughout the IFC 84. However, the precedent for this particular grouping was set with the JCT Agreement for Minor Building Works and it has been repeated in IFC 84.

SUMMARY OF GENERAL LAW

(A) Contractor's obligations

The obligation to complete
In any building contract, the contractor's prime obligation is to carry

out and complete the contract works in accordance with the contract documents. This usually takes the form of an express term in the contract to this effect. However, even without such an express term, this obligation would be implied.

The contractor's obligation to complete can have quite extreme consequences unless the contract concerned in some way limits or qualifies the obligation. In its unqualified form, the obligation will mean that if the contract works are damaged or destroyed, even on the last day before completion, the contractor will generally be responsible for reinstating or rebuilding them at his own cost. For the contractual limitations placed upon this general obligation in IFC 84 – see Note [1] on page 31. However, even apart from qualifications in the contract itself, there are certain legal excuses for a failure by the contractor to perform this prime obligation. Briefly they are as follows:

(i) FRUSTRATION

It is not possible in a book of this kind to consider in detail the law relating to the frustration of contracts. Suffice it to say that there must be some intervening event of a fundamental nature which renders continued performance of the contract impossible or of a totally different nature to that which was envisaged when the contract was entered into. Only very rarely will a contractor be excused performance of his obligation to complete the contract works by reason of the contract becoming frustrated.

The mere fact that performance of the contract turns out to be more difficult or expensive than envisaged by the contractor at the outset will not be sufficient to frustrate the contract. Furthermore, if the contract itself expressly caters for the eventuality concerned, then it cannot be a frustrating event. A modern example of a frustrating event in relation to a building contract occurred in the case of *Wong Lai Ying and Others* v. *Chinachem Investment Co. Ltd* (1979), heard by the Judicial Committee of the Privy Council on appeal from the Court of Appeal of Hong Kong. It involved a landslip which took with it a block of flats of 13 storeys together with hundreds of tons of earth which landed on the site of partly completed buildings completely obliterating them. It was accepted for the purposes of argument that the landslip was an unforeseeable natural disaster. Following the landslip it was uncertain whether the partly completed contract could ever be completed and even if it could as to when it could be completed. This was held to be a frustrating event.

In the case of *Davis Contractors Ltd* v. *Fareham UDC* (1956), a fixed price contract to build 78 houses in eight months was held not to be frustrated when, owing partly to severe ,and unforeseeable shortages of labour and materials, completion took 22 months. Lord Radcliffe in this case said as follows:

'. . . it is not hardship or inconvenience or material loss itself which calls the principle of frustration into play. There must be as well such a change in the significance of the obligation that the thing undertaken would, if performed, be a different thing from that contracted for'.

It has been said that 'Frustration is a doctrine only too often invoked by a party to a contract who finds performance difficult or unprofitable, but it is very rarely relied on with success. It is in fact a kind of last ditch' : *per* Harman LJ in *Tsakiroglou & Co. Ltd* v. *Noblee Thorl GMB* (1961).

A contract may become frustrated if the government prohibits or restricts the work contracted for: *Metropolitan Water Board* v. *Dick, Kerr & Co. Ltd* (1918).

In the event that the contract is frustrated, the financial position between the parties will generally be governed by sections 1 and 2 of the Law Reform (Frustrated Contracts) Act 1943, which gives the court discretion to award reasonable sums for work done or benefits conferred by the parties to the contract.

(ii) EXCLUSION CLAUSES

The contract between the parties may seek to relieve the contractor from liability under the contract or from liability for a tort connected with the contract. It may well be that in many instances, particularly where the parties are of equal bargaining power, such exclusions or restrictions do no more than apportion the risk between the parties which will, in turn, be reflected in the make-up of the tender sum. The more the contractor is able to exclude, limit or define a risk, the more likely it is that a lower tender sum will result. It is not possible in this book to deal with exclusion clauses in any detail and reference should be made to text books which deal with the topic, e.g. *Chitty on Contracts*, 26th edition, chapter 14.

What can be said at this point is that, apart from statutory controls (referred to in the next paragraph), it is possible, provided the exclusion clause is appropriately worded in all the circumstances, to exclude liability even for fundamental breaches of contract. There is no rule of law that a fundamental breach committed by one party to a contract inevitably deprives that party of the right to rely upon an exclusion clause: *Photo Production Ltd* v. *Securicor Transport Ltd* (1980), a House of Lords case disapproving the Court of Appeal's decision in *Harbutt's Plasticine Ltd* v. *Wayne Tank and Pump Co. Ltd* (1970). The House of Lords unanimously rejected the view that a breach of contract by one party, accepted by the other as discharging him from further performance of his obligations under the contract, brought the contract to an end and, together with it, any exclusion clause. It all depends upon the construction of the particular contract as to whether or not an exclusion clause is adequate to limit what would otherwise be the liability of the party at fault.

There is some statutory control of exclusion clauses by virtue of the Unfair Contract Terms Act 1977. The Act provides some control over contract terms which exclude or restrict liability for breach of certain terms implied by statute or at common law into building contracts. It also controls some contract terms which purport to entitle one of the parties to render a contractual performance substantially different from that reasonably expected of him. Generally speaking, where a consumer transaction takes place, i.e. one of the parties is a consumer not purchasing goods or materials or work in the course of a business and the other contracting party is selling or providing the work in the course of a business, then the exclusion or restriction of liability is rendered absolutely ineffective. On the other hand, if it is a business transaction, then the exclusion or restriction clause is likely to be effective only in so far as the exclusion clause satisfies the requirement of reasonableness contained in the Act.

(iii) BREACH OF CONTRACT BY EMPLOYER

Where the employer is guilty of a sufficiently serious breach of contract, the contractor is entitled to treat the employer's breach as a repudiation which releases the contractor from any further obligation to perform the contract. It is not every breach that will have this effect. It must be of a very serious nature. If it is not, it will entitle the contractor to claim damages for breach of contract but will not entitle him to decline further performance of his contractual obligations.

(iv) MISREPRESENTATION

A misrepresentation is an untrue statement of fact made by one contracting party to the other at, or before, the time of entering into the contract and which acts as an inducement to that other party to enter into the contract.

Where such a statement made by one party is relied upon by the other party to his detriment, then that other party may, in appropriate circumstances, be able to obtain a rescission of the contract and also, depending upon the circumstances, to claim damages under the Misrepresentation Act 1967. If the statement concerned also becomes a term of the contract then the contractor will be entitled to sue for breach of contract. For a detailed discussion of misrepresentation the reader is referred to the appropriate text books covering this topic, e.g. *Chitty on Contracts*, 26th edition, chapter 6.

(v) ACCORD AND SATISFACTION

There is nothing to stop the employer and contractor agreeing to excuse one another from any further performance of their contractual obligations upon agreed terms, e.g. the employer may be running short of funds and the contractor may have entered into a very unprofitable contract and they may well both be happy to terminate their obligations. The parties are of course always free to agree whatever terms they wish to end the contract before completion of the contract works. This is known as an accord and satisfaction.

In most building contracts, the obligation upon the contractor to carry out and complete the contract works is linked to a greater or lesser degree to the obligation to insure them. So far as the IFC 84 is concerned, this is dealt with later under Note (1) page 31.

Implied terms

It has been stated above that, in the absence of an express term requiring the contractor to carry out and complete the contract works in accordance with the contract documents, such a term would be implied. Building contracts, as with all other contracts, are likely to have incorporated within them certain implied terms apart from their express terms.

Certain terms are implied as matters of law, e.g. that in a contract for work and materials the goods will be of merchantable quality; that to the extent that the builder's skill and judgment is being relied on, the work will be reasonably fit for its purpose; and that reasonable skill and care will be used in relation to the workmanship employed. These implied terms were the product of the common law. They are now however embodied in statute law – see sections 4 and 13 of the Supply of Goods and Services Act 1982. This Act also incorporates other implied terms where the contract is silent e.g. the time for performance – see section 14 of the Act. The ability of a contracting party to exclude or restrict these implied terms is limited by the provisions of the Unfair Contract Terms Act 1977 (see section 7). This can have particular significance where, despite carrying out the work in accordance with the contract documents, the contractor is in breach of building regulations e.g. as to the adequacy of the foundations. It has been held that in this situation a contractor may, despite having observed the strict terms of the contract, nevertheless be liable to the building owner – see the case of *Street and Another* v. *Sibbabridge Ltd and Another* (1980) in which it was held that there was an implied term in a building contract that the contractor would comply with building regulations even if his failure to do so was because he did no more than comply with the architect's design. (This particular issue is dealt with in more detail when clause 5.3 is considered – see page 273.)

There are other implied terms which are not implied as a matter of law but are implied as a matter of fact. Where the express terms of a contract do not cover a particular situation, the court may imply a term based upon the imputed intentions of the parties gleaned from the actual circumstances of the case, where this is required to give the contract what is called 'business efficacy'. The extent and nature of such implied terms will clearly vary from case to case.

Statutory obligations

The contractor's obligation to carry out and complete the works is bound to be subject to a greater or lesser degree to statutory control – for example, in relation to the safety aspects of equipment and methods of working, the Health and Safety at Work etc Act 1974.

Furthermore, the quality or standard of work to be achieved may be affected by statute e.g. the building regulations under the Public Health Act 1961; the Defective Premises Act 1972. Obligations under the building regulations and under the Defective Premises Act 1972 are dealt with in more detail under the heading 'Summary of general law' in chapter 8 when dealing with clause 5 of IFC 84 (statutory obligations etc – see page 264).

(B) The passing of ownership in goods and materials

For the sake of the contractor's cash flow it is important that he receives as much as possible as soon as possible. It is therefore of considerable benefit to him to be paid for materials and goods which are intended to be incorporated into the works before actual incorporation e.g. as soon as they are delivered to the site or even off-site if they have been allocated to the particular contract concerned. Many building contacts provide for this.

On the other hand, the employer wishes to be sure that if he pays for materials or goods before they are incorporated into the works, the ownership will vest in him. These are often conflicting interests and difficulties can arise.

Once materials or goods are physically incorporated into the building there will be no problem as the property will pass to the landowner. This legal rule carries the Latin tag *quicquid plantatur solo, solo cedit* (the property in all materials and fittings, once incorporated in or affixed to a building, will pass to the freeholder). The employer can then safely pay for them.

Before such incorporation there is a risk. Simply to state in a contract that property is to pass from contractor to employer is not a full protection for the employer – for instance, where at the time that the contract says the property is to pass, e.g. at the time of payment for the goods or materials, the contractor does not own them. There is a general legal rule that no-one can transfer a better title than he himself possesses. The Latin tag is *nemo dat quod non habet*. There are exceptions to this rule, most of which relate to contracts for the sale of goods rather than a building contract. A building contract is a contract for work and materials which is not governed by the law relating to the sale of goods. However, these exceptions will nevertheless be relevant to the position between the contractor and his suppliers and this in turn can determine the ability of the contractor to pass title to the employer even though he himself may not own the materials or goods. It is not possible in a book of this kind to deal with this rule and its exceptions in detail. Reference should be made to the appropriate text books, e.g. *Chitty on Contracts* (specific contracts, 26th edition, paragraphs 4807 to 4832).

If a contractor accepts payment from the employer under a contract which expressly provides that the property in materials or goods is to pass to the employer upon payment, when the contractor does not own such materials or goods at that time, clearly he is in breach of contract. If there is no express term in the contract relating to title, there will nevertheless be an implied term as to title by virtue of section 2 of the Supply of Goods and Services Act 1982 which applies, *inter alia*, to building contracts. Section 2 provides that:

'In a contract for the transfer of goods . . . there is an implied condition on the part of the transferor that in the case of a transfer of the property in the goods he has the right to transfer the property and in the case of an agreement to transfer the property in the goods he will have such right at the time when the property is to be transferred.'

There is also an implied warranty that:

(a) the goods are free and will remain free from any charge or incumbrance not disclosed or known to the transferee before the contract is made; and

(b) the transferee will enjoy quiet possession of the goods except so far as it may be disturbed by the owner or other person entitled to the benefit of any charge or incumbrance disclosed or known.

There are statutory restrictions on the ability of a contracting party to exclude or limit liability for breach of this implied condition or these implied warranties (see Unfair Contract Terms Act 1977 section 7(3a)).

While it is clear therefore that an employer will have a right of redress against a contractor who is in breach of an express or implied term concerning the transfer of title, such a remedy is of little value if the contractor becomes bankrupt or goes into liquidation.

The most common reason why a contractor does not own the materials or goods at the time of their delivery to site is that the contract between the contractor and his supplier provides that title is not to pass to the contractor until payment in full has been received by the supplier. Such clauses are of various types and degrees of elaboration e.g. purporting to deal with the situation where the materials or goods are mixed with others or become part of other materials or goods; making ownership dependent upon the discharge of all debts from that contractor to that supplier, and so on. These retention of title clauses are sometimes called Romalpa clauses after the leading case of that name, *Aluminium Industrie Vaassen BV* v. *Romalpa Aluminium Ltd* (1976).

A successful retention of title clause will mean that if the contractor becomes bankrupt or goes into liquidation or a receiver is appointed, without the supplier having been paid, the supplier's title will hold good as against the trustee in bankruptcy, liquidator or receiver and also as against the employer even though he may have paid the contractor for them. However, such clauses, to be effective, require very careful drafting.

Where the sub-contract is not for the mere supply of materials or goods under a sale of goods contract but is a sub-contract for works and materials, the position can be even more precarious from the employer's point of view, as the sub-contract may well not expressly state when title in the materials or goods is to pass to the contractor, if at all. It could well be that on a true interpretation of the subcontract the title may never be intended to pass at all to the contractor. For instance, if the sub-contract assumes that the materials or goods will be incorporated into a building before payment is made to the sub-contractor by the contractor, then the title will transfer directly from the sub-contractor to the landowner. The sub-contract may however provide for payment upon delivery to the site, and for the title to pass when delivery or payment is made. If there is no express contractual term dealing with the passing of title, then it will be a question of determining from the circumstances when the parties intended the title in the property to pass: see also the case of *Dawber Williamson Roofing Co. Ltd* v. *Humberside County Council* (1979) page 120.

CONSIDERATION OF THE RELEVANT CLAUSES OF IFC 84

Clause 1.1

Contractor's obligations

1.1   The Contractor shall carry out and complete[1] the Works in accordance with the Contract Documents[2] identified in the 2nd recital: provided that where and to the extent that approval of the quality of materials or of the standards of workmanship is a matter for the opinion of the Architect/the Contract Administrator such quality and standards shall be to the reasonable satisfaction of the Architect/the Contract Administrator.[3]

COMMENTARY ON CLAUSE 1.1

This clause covers the basic obligation of the contractor to carry out and complete the contract works in accordance with the contract documents and with materials of the quality and workmanship to the standards specified in those documents. As between the employer and the contractor, the employer is generally responsible for design. There is therefore no reference to any design obligation falling upon the contractor so that he is not responsible for the suitability of the materials and goods specified. The contract simply requires him to carry out the works as designed. While there is probably no implied contractual or tortious duty upon the contractor to inform the employer of any design defect, the contractor would be well advised to notify the architect in writing if he considers that either the design or the specification is in any way deficient.

Furthermore, if the contractor discovers that there is a divergence between any statutory requirements in relation to the works and the contract documents or any instruction of the architect, he is required immediately to give the architect written notice specifying the divergence (clause 5.2).

The architect's function in inspecting the works is to ensure that the contractor carries them out in accordance with the contract documents and he cannot insist upon a higher standard of workmanship or a higher quality of materials than is expressly stated in those documents, unless he issues an instruction under clause 3.6 requiring a variation, in which case there will be an adjustment to the contract sum.

Where the contract documents provide that the approval of the quality of materials or the standards of workmanship is to be a matter for the opinion of the architect, then there is more discretion, as the quality or standards must then be to the reasonable satisfaction of the architect. This particular topic is dealt with in more detail later under Note [3], page 32.

NOTES TO CLAUSE 1.1

[1] '. . . carry out and complete . . .'
For a general discussion of this obligation see earlier, page 21.

For a general discussion on the time for completion see chapter 4 dealing with possession and completion.

This obligation to complete generally requires the contractor to carry out and complete the works for the contract sum so that damage or loss, including total destruction of the works, is the contractor's risk. However, this will not be the case in relation to those risks to be covered by 'All Risks Insurance' (see later page 322) where clause 6.3B or 6.3C.2 applies or those risks known as ''Specified Perils' where clause 6.3C.1 applies as in these cases the restoration, replacement or repair of loss or damage is treated as if it were a variation with the contractor being paid accordingly. This aspect is dealt with in more detail in chapter 9 dealing with injury to persons and property, insurance and indemnity, and insurance of the works. It is also possible for either party to seek determination of the contractor's employment under the contract following such loss or damage where it is just and equitable to do so where clause 6.3C applies (see clause 6.3C.4.3) or, where the loss or damage is due to a specified peril, if it causes a prolonged suspension of the works (see clause 7.8.1(b)).

[2] '. . . in accordance with the Contract Documents . . .'
The contract documents will be the contract drawings together with either a priced specification, and/or priced schedules of work and/or the contract bills, or alternatively the contract drawings together with an unpriced specification (see second recital IFC 84).

The contract documents will usually specify the materials or goods. They may also specify the standard of workmanship. Where they do not so specify, there will be an implied term that the materials or goods will be of merchantable quality and that the workmanship will be carried out with reasonable care and skill (see section 4 and 13 of the Supply of Goods and Services Act 1982).

The Unfair Contract Terms Act 1977 places restrictions on the ability of a contracting party to exclude or limit the operation of the implied terms as to quality. Where the contract falls within the definition of 'business transaction' any such exclusion or restriction must satisfy 'the requirement of reasonableness,' see sections 7 and 11 and schedule 2 to the Act.

So far as a contract involving one party 'dealing as consumer' is concerned, the implied terms as to quality cannot be excluded or restricted at all: see section 12 of the Act. The majority of contracts let under IFC 84 will form part of a business transaction. In most contracts where the contract documents specifically require a lesser standard than that which would be implied under the Supply of Goods and Services Act 1982, this will be the requirement of the employer, and there is no exclusion or restriction upon which the Unfair Contract Terms Act 1977 can operate.

[3] '... where and to the extent that approval of the quality of materials or of the standards of workmanship is a matter for the opinion of the Architect/the Contract Administrator such quality and standards shall be to the reasonable satisfaction of the Architect/the Contract Administrator'

To fully appreciate the relevance of these words, reference must also be made to clause 4.7 which provides that where this proviso applies, the quality of materials or standards of workmanship will be treated conclusively as having been to the reasonable satisfaction of the architect. This means that where the contract documents or an instruction requiring a variation require that approval of the quality of materials or standards of workmanship is to be reserved for the opinion of the architect, then after the expiry of 28 days from the date of the final certificate for payment, it shall be conclusive in any legal proceedings (which includes arbitration) that the quality or standards are to his satisfaction. The contractor would therefore be well advised, where this approval is a stated requirement, to obtain that approval in writing, remembering always the possibility that such approval may not be conclusive until the issue of the final certificate.

Although the use of these words is no doubt intended to avoid the conclusive effect of the final certificate in relation to materials or workmanship where the quality and standards are not specifically made subject to the opinion of the architect, the effect of the conclusiveness of the certificate in relation to materials or workmanship covered by the proviso still renders the position, in certain situations, unfair to the employer. If there is a defect concerning quality or standards which has escaped the eye of the architect, then the employer could be without a remedy in respect of defective work as the expression of satisfaction by the architect is likely to be binding on the employer. No-one can reasonably expect the architect to be constantly on the site ensuring correct quality and standards. It can be argued therefore that these words assume infallibility in the architect, and to the extent that this is not achieved the employer can be prejudiced when in fairness he should not be.

On the other hand, one must appreciate the contractor's point of view. Instead of having to meet an objective criterion in relation to quality of materials or standards of workmanship, he has to try to satisfy the subjective standards of the particular architect albeit the satisfaction is qualified by the word 'reasonable'. Having satisfied the architect, the contractor does not want the employer to be able to argue, perhaps years later, that the quality or standards were unsatisfactory when there is no independent specified criterion to which the contractor can then turn.

Furthermore, it is not clear from the quoted wording and the wording in clause 4.7 that it is only those matters where approval is expressly reserved by the contract documents for the opinion of the architect which are caught by the conclusiveness of the final certificate. It has not been made clear beyond doubt that the approval referred to is in relation to a subjective standard rather than approval of an objective criterion. After all, in a sense all questions of quality of materials and standards of workmanship are a matter for the opinion (and possibly therefore the approval) of the architect as he must be satisfied in the first instance that the requirements of the contract have been met before the value of materials and workmanship are included in an interim certificate for payment. A contractor might therefore still argue that, by including the value of materials and work in an interim certificate for payment, which is not subsequently adjusted, the architect has in effect expressed an approval so that all materials and workmanship, whether judged against an objective criterion or the subjective opinion of the architect, are affected by the conclusiveness of the final certificate issued under clause 4.6. The distinction between those cases involving an objective criterion and those involving purely an approval by the architect should be made clearer in this clause and in clause 4.7.

Presumably the reference to materials in this connection includes goods.

As the quality or standards are to be to the reasonable satisfaction of the architect, the expression of satisfaction by the architect can no doubt be challenged by the employer as well as by the contractor, provided the appropriate steps stated in clause 4.7. are taken.

If the architect's expression of satisfaction or his lack of it is challenged, it is a matter for argument as to whether the price contained in the contract for the materials or work is a relevant factor. In reliance on the case of *Cotton* v. *Wallis* (1955) it is possible to argue that the architect can take the price into account as being a relevant factor in determining whether or not he should express his satisfaction. However, the architect's first point of reference is the description in the contract documents and if this assists in determining the quality or standards then it would be wrong for him to take the price or rate for the work into account. If however the contract documents leave the architect genuinely in doubt, he may bring into his consideration the price or rate.

The Supply of Goods and Services Act 1982, which applies (*inter alia*) to building contracts, states in relation to the implied condition that goods supplied under such a contract are to be of merchantable quality, that they are of merchantable quality if they are as fit for the purpose or purposes for which goods of that kind are commonly supplied, as it is reasonable to expect having regard to any description applied to them, the price (if relevant) and all the relevant circumstances (see section 4(9)). This provision to some extent, therefore, begs the question. If the price is a relevant factor i.e. where it is clearly intended to have a direct bearing upon quality or standards then the architect is, it is submitted, obliged to take the price into account.

There is no specified limit as to the time within which the architect must express any dissatisfaction. Clearly to express it belatedly could result in much good work becoming abortive. It might be argued that there is an implied term that the architect will express any dissatisfaction within a reasonable time of the work being carried out. There is no provision in IFC 84 equivalent to clause 8.2.2 of JCT 80 which provides for the architect to 'express any dissatisfaction within a reasonable time from the execution of the unsatisfactory work'.

Clauses 1.2 and 1.3

Quality and quantity of work

1.2   Where or to the extent that quantities are not contained in the Specification/Schedules of Work and there are no Contract Bills, the quality and quantity of the work included in the Contract Sum (stated in Article 2) shall be deemed to be that in the Contract Documents taken together; provided that if work stated or shown on the Contract Drawings is inconsistent with the description, if any, of that work in the Specification/Schedules of Work then that which is stated or shown on the Contract Drawings shall prevail for the purpose of this clause.

Where and to the extent that quantities are contained in the Specification/Schedules of Work, and there are no Contract Bills, the quality and quantity of the work included in the Contract Sum for the relevant items shall be deemed to be that which is set out in the Specification/Schedules of Work.

Where there are Contract Bills, the quality and quantity of the work included in the Contract Sum shall be deemed to be that which is set out in the Contract Bills.

Priority of Contract Documents

1.3   Nothing contained in the Specification/Schedules of Work/Contract Bills shall override or modify[1] the application or interpretation of that which is contained in the Articles, Conditions, Supplemental Conditions or Appendix.

COMMENTARY ON CLAUSES 1.2 AND 1.3

These two clauses deal with priority within the contract documents in relation to the quality and quantity of work. The position can be summarised as follows:

(a) if contract bills are used then the quality and quantity of work is deemed to be that set out in the bills; but if there are no contract bills, then;

(b) if there is a specification/schedules of work which contain quantities then the quality and quantity of work shall be deemed to be that set out therein for the relevant item; but where and to the extent that there are no quantities contained therein, then;

(c) the quality and quantity of work is deemed to be that in the contract documents taken together, but if work shown on the contract drawings is inconsistent with any description in the specification/schedules of work then the contract drawings prevail.

NOTES TO CLAUSES 1.2 AND 1.3

[1] '. . . override or modify . . .'
It appears on the face of the contract that, except where the quality and quantity of work is concerned, no specification/schedule of works or contract bills can override or modify the conditions of contract. Great care must therefore be taken if it is intended in any way to alter the conditions of contract. This should not be done other than by inserting the amendment into the contract itself or possibly in a separate note of amendments which must itself be made a contract document in its own right.

The courts have considered the effect of these or similar words in a number of cases:

Gold v. *Patman and Fotheringham Ltd* (1958)
Gleeson v. *Hillingdon LBC* (1970)
North-West Metropolitan Regional Hospital Board v. *T. A. Bickerton & Sons Ltd* (1970)
English Industrial Estates Corporation v. *George Wimpey & Co. Ltd* (1972)
Henry Boot Construction Ltd v. *Central Lancashire New Town Development Corporation* (1980)
E. Turner & Sons Ltd v. *Mathind Ltd* (1986).

It may be said that, whilst lip service is paid to the dominant effect of words such as these in preventing any special provisions in the contract bills (or specification/schedule of works) affecting the contract conditions, nevertheless in practice the courts seem, even if by dint of semantic gymnastics, to take heed of what these other documents say even though they may be inconsistent with the contract conditions themselves.

Whether it is a sound policy to give standard printed conditions priority over specially formulated documents is doubted by some. However, it does help prevent those specially formulated documents accidentally affecting the conditions of contract. If amendments are necessary, then provided those using the contract are familiar with its terms, the conditions themselves may be amended in the proper way.

Clauses 1.4 and 1.5

Instructions as to inconsistencies, errors or omissions

1.4 The Architect/The Contract Administrator shall issue instructions in regard to the correction:
- of any inconsistency in or between the Contract Documents or drawings and documents issued under clause 1.7 *(Further drawings)* or 3.9 *(Levels)*; or
- of any error in description or in quantity or any omission of items in the Contract Documents or in any one of such Documents; or
- of any error or omission in the particulars provided by the Employer of the tender of a person named in accordance with clause 3.3.1 *(Named sub-contractors)*; or
- of any departure from the method of preparation of bills of quantities[1] included in the Contract Documents referred to in clause 1.5 (including any error in or omission of information in any item which is the subject of a provisional sum for defined work*);

and no such inconsistency, error or departure shall vitiate this Contract. Where the Contract Documents include bills of quantities and the description therein of a provisional sum for defined work* does not provide the information required by General Rule 10.3 in the Standard Method of Measurement the correction shall be made by correcting the description so that it does provide such information.

If any such instruction changes the quality or quantity of work deemed to be included in the Contract Sum as referred to in clause 1.2 *(Quality and quantity of work)* or changes any obligations or restrictions imposed by the Employer, the correction shall be valued under clause 3.7 *(Valuation of Variations)*.

Bills of quantities and SMM

1.5 Where the Contract Documents include bills of quantities, those bills, unless otherwise expressly stated therein in respect of any specified item or items, are to have been prepared in accordance with the Standard Method of Measurement of Building Works, 7th Edition, published by the Royal Institution of Chartered Surveyors and the Building Employers Confederation.

---

* See footnote to clause 8.3 (Definitions).

COMMENTARY ON CLAUSES 1.4 AND 1.5

These clauses can be summarised as follows:

(1)   Where the contract documents include bills of quantities then, unless they expressly state otherwise, it is assumed that the Standard Method of Measurement of Building Works, 7th edition, published by the Royal Institution of Chartered Surveyors and the Building Employers Confederation (SMM 7) has governed their preparation.

(2)   If the contract bills depart in any particular SMM 7, such departure is to be regarded as an error and must be corrected by an instruction from the architect unless the bills expressly deal with the departure in relation to any specified item or items. Such instruction, if it changes the quality or quantity of the work or any obligations or restrictions imposed by the employer shall be valued as a variation under clause 3.7. Even where, therefore, there has been an intentional departure from SMM 7 in respect of an item, if this is achieved by necessary implication rather than expressly, it nevertheless appears to be the case that the contractor may be able to disregard it so far as his rates are concerned and then require a correction which will be valued, if appropriate. This seems unfair if no-one was misled. Perhaps in an appropriate case, e.g. where it is established that the contractor had included in his rates for an omitted item, or where the departure has no practical effect on the work to be carried out, the valuation of the instruction would be nil.

(3)   The following shall be corrected by an instruction and will, if it changes the quality or quantity of work or changes obligations or restrictions imposed by the employer, be valued under clause 3.7 as a variation:

(i)   inconsistencies in or between the contract documents or further drawings or details issued under clause 1.7 or drawings or other information under clause 3.9 in relation to setting out; or

(ii)   errors in description or in the quantities or omissions of items in the contract drawings; or

(iii) errors or omissions in the particulars provided by the employer in relation to the tenders of named sub-contractors (see clause 3.3.1).

(iv)   departures from SMM 7 in the preparation of bills of quantities included in the contract documents including any error in or omission of information in any item which is the subject of a provisional sum for defined work.

Where there is an error in or omission of information in any item which is the subject of a provisional sum for defined work, the description of the item is to be corrected so that the mistake or omission is made good.

The phrase 'provisional sum' is defined in clause 8.3 (see page 374) to include defined and undefined work. SMM 7 in General Rules 10.1 to 10.6 includes a description of what amounts to a provisional sum for defined or undefined work (see footnote to clause 8.3 page 375).

If work is to be classified as defined work certain minimum information is required in accordance with General Rule 10.3 namely:

(a) The nature and construction of the work.
(b) A statement of how and where the work is fixed to the building and what other work is to be fixed thereto.
(c) A quantity or quantities which indicate the scope and extent of the work.
(d) Any specific limitations and the like identified in Section A35 of SMM 7.

If such information cannot be given the work will be treated as a provisional sum for undefined work. The significance of this is in relation to the contractor's programming, planning and pricing of preliminaries. If the work is classified as a provisional sum for defined work the contractor will be deemed to have made due allowance for all of these matters; whereas if the work is classified as undefined the contractor will be deemed not to have made any such allowance.

The importance of classification is clearly relevant in relation to the valuation of preliminary items (see clause 3.7.6), extensions of time (see clause 2.4.5) and loss and expense (see clause 4.12.7).

Clause 1.4 attempts to make it clear that if the description in the bills amounts to an attempt to provide the information required by General Rule 10.3 of SMM 7 but fails to do so as a result of an error in or omission of information, it is nevertheless still a provisional sum for defined work and will be corrected accordingly. The item is not to be treated, as a result of the inaccurate or omitted information, as a provisional sum for undefined work giving rise to a valuation of preliminaries, an extension of time or reimbursement of loss or expense on that basis.

Even so there may be situations where not all of the information required under General Rule 10.3 can be given, rather than where it can but has been omitted by accident. In such cases, e.g. where there is an item for a reception desk where all the information has been given other than that as to how and where it is to be fixed, the contractor will nevertheless to a large extent be able to make an accurate allowance in his programming, planning and pricing preliminaries and is very likely to reflect this in his actual tender. It seems absurd if he were then able to demand that the item be treated as a provisional sum for undefined work resulting in his tender being 'deemed' to exclude any such allowance.

(4)   No such inconsistencies, errors or departures will vitiate the contract. However, if the inconsistency etc. is of a truly fundamental nature, the correction of which would render performance of the contract radically different from what the contractor could have reasonably envisaged, then it may vitiate the contract and entitle the contractor to rescind the contract and to be paid for work done on a *quantum meruit* basis, that is, as much as it is worth, rather than in accordance with the contract documents. Such situations will be very rare indeed.

NOTES TO CLAUSES 1.4 AND 1.5

[1] '... bills of quantities ...'
Amendment No 4 to IFC 84 issued in July 1988 substituted these words for the former words 'Contract Bills'. This is to ensure that clauses 1.4 and 1.5 apply also to bills of quantities forming part of any named sub-contract documents which includes bills of quantities. Such bills of quantities are not within the meaning of 'Contract Bills' as described in the second recital to the agreement recitals and articles for IFC 84.

Clause 1.6

Custody and copies of Contract Documents

1.6   The Contract Documents shall remain in the custody of the Employer so as to be available at all reasonable times for the inspection of the Contractor.

The Architect/The Contract Administrator without charge to the Contractor shall provide the Contractor with one copy of the Contract Documents certified on behalf of the Employer and two further copies of the Contract Drawings and the Specification/Schedules of Work/Contract Bills.

COMMENTARY ON CLAUSE 1.6

This clause calls for no comment.

Clause 1.7

Further drawings and details

1.7   The Architect/The Contract Administrator without charge to the Contractor shall provide him with 2 copies of such further drawings or details as are reasonably necessary to enable the Contractor to carry out and complete the Works in accordance with the Conditions.

COMMENTARY ON CLAUSE 1.7

This clause provides for what is almost invariably the case, that is, that the contract documents as defined must be supplemented with additional drawings or details to enable the contractor to carry out and complete the works.

The drawings included in the contract documents are often small scale drawings sufficient for the contractor to assess the scope of the works for tendering purposes. In order to construct the works it is necessary for the architect to supply the contractor with construction details, reinforcement bending schedules and finishing schedules etc. If these depart in any way from the scope of the works envisaged by the contract documents taken as a whole, then they should be accompanied by an instruction varying the works under clause 3.6, specifying the extent to which the further drawings or details depart from those provided in the contract documents.

There is no time limit laid down as to when such further drawings or details should be made available to the contractor but the architect would be well advised to have regard to the obligation placed on the contractor to complete the works by the date for completion stated in the appendix. It is the duty of the architect on behalf of the employer to provide the necessary further drawings or details at an appropriate time, and failure to do so is likely to amount to a breach of contract by the employer, whether or not the contractor has requested the architect to issue them. However, if the architect fails to issue the further drawings or details sufficiently early to prevent the progress of the works being delayed, such that the contractor is unable to complete by the date for completion or any extended time for completion, then an extension of time can only be made if the contractor has specifically applied for the drawings or details in writing at an appropriate time. For the difficulties in making an extension of time in the absence of the contractor's written application see Note [17] to clause 2.4, page 91.

The contractor should be careful to assess his need for further information sufficiently in advance so as to be able to inform the architect of that need.

It was held in the case of *H. Fairweather & Co. Ltd* v. *London Borough of Wandsworth* (1987), a case on the JCT 63 form of contract, that where a nominated sub-contractor was late in supplying installation drawings (held in this case not to be design drawings) to the main contractor for onward transmission to engineers on behalf of the architect, who in turn approved them and formally issued them back to the main contractor, this was not a breach by the employer of clause 3(4) of JCT 63 (which is worded similarly to clause 1.7 of IFC 84). This was so despite the fact that the main contractor had dealt with them promptly. The only delay was on the part of the nominated sub-contractor. Accordingly the contractor could not claim for an extension of time or reimbursement of any direct loss or expense based upon late receipt of information under clauses 23(f) and 24(1)(a) of JCT 63 (similar to clauses 2.4.7 and 4.12.1 of IFC 84).

The reasoning is to be found in the following extract from the judgment of His Honour Judge Fox-Andrews QC

'Whatever means of conveying and approving the drawings may have been agreed upon Conduits (the nominated sub-contractor) had a contractual obligation to Fairweather to supply installation drawings. It necessarily follows that Fairweather had a similar obligation to Wandsworth.

'If without faults on the part of the architect or engineer Fairweather were delayed in their work by reason of Conduits' failure to supply installation drawings in good time, it would be surprising if Wandsworth were to be liable to Fairweather for such delay. It may be that if and to the extent that delay was due to the architect or engineer Wandsworth would be liable for breach of an implied term to consider and approve such installation drawings with reasonable expedition (see for example *London Borough or Merton* v. *Stanley Hugh Leach Ltd* (1985) 32 BLR 51) or for breach of a collateral contract.

'I am satisfied, however, that the arbitrator was correct in rejecting a claim under clause 3(4) of the JCT contract.'

Bearing in mind that under the main contract it was for the architect to furnish the contractor with such drawings or details as were necessary to enable the contractor to carry out and complete the works in accordance with the contract conditions rather than requiring the contractor to obtain such information directly from a nominated sub-contractor, the decision is in some respects surprising, particularly as under JCT 63 there will generally be a 'Grey Form' of warranty between the employer and the nominated sub-contractor in force under which the sub-contractor promises to provide information to the architect in such manner as not to give the contractor an entitlement to an extension of time or reimbursement of loss and expense on the basis of late receipt of information from the architect.

Turning to IFC 84, if the contract is set up on the basis that installation drawings are to be issued by the architect to the main contractor, the fact that for administrative convenience the flow of information from a named sub-contractor to the architect is channelled through the main contractor should not, without more, relieve the employer of responsibility if the information is delivered late due to the named sub-contractor's default. The situation might be different if there is a clear contractual obligation in the sub-contract requiring the sub-contractor to provide the information to the main contractor. The lesson for the contractor is clear: any information which has to find its way from the named sub-contractor to the architect should at best by-pass the contractor and at worst flow through the main contractor on the explicit understanding that it is for administrative convenience only and is not the subject of any express or implied sub-contractual obligation. It should be borne in mind that clause 3.2 of Agreement ESA/1 (for use between the employer and a named sub-contractor under IFC 84 – see appendix 3 page 414) places an obligation on the sub-contractor to provide the architect with such further information as is reasonably necessary to enable the architect to provide the main contractor with sufficient information to enable the main contractor to complete the main contract works in accordance with the IFC 84 Conditions, which of course includes the sub-contract works themselves.

Clause 1.8

Limits to use of documents

1.8   None of the documents mentioned in clauses 1.3 *(Priority of Contract Documents)* or 1.7 *(Further drawings)* shall be used by the Contractor for any purpose other than this Contract, and neither the Employer, the Architect/the Contract Administrator nor the Quantity Surveyor shall divulge or use except for the purposes of this Contract any of the rates or prices in the Contract Documents or, where the 2nd Recital, alternative B, applies, in the Contract Sum Analysis or in the Schedule of Rates.

COMMENTARY ON CLAUSE 1.8

This clause calls for no comment.

Clause 1.9

Issue of certificates by Architect/Contract Administrator

1.9   Except where provided otherwise any payment or other certificate to be issued by the Architect/the Contract Administrator shall be issued to the Employer and a duplicate shall at the same time be sent to the Contractor.

COMMENTARY ON CLAUSE 1.9

This clause calls for no comment.

Clause 1.10

Unfixed materials or goods: passing of property, etc.

1.10   Unfixed materials and goods delivered to, placed on or adjacent to the Works and intended therefor shall not be removed[1] except for use upon the Works unless the Architect/the Contract Administrator has consented in writing to such removal which consent shall not be unreasonably withheld.

Where the value of any such materials or goods has in
accordance with clause 4.2.1(b)[2] *(Interim payments)*
been included in any payment certificate under which the
amount properly due to the Contractor has been discharged[3] by the Employer, such materials and goods
shall become the property[4] of the Employer, but subject
to clause 6.3B and 6.3C.2 to .4[5] *(Insurance by Employer)*
(if applicable), the Contractor shall remain responsible for
loss or damage to the same.[6]

COMMENTARY ON CLAUSE 1.10

For a summary of the general law in relation to the passing of
property see page 27.

This clause seeks, *inter alia*, to pass the ownership in materials
and goods delivered to the site, but not yet incorporated into the
works, from the contractor to the employer once their value has
been included in any payment certificate under which the amount
properly due to the contractor has been paid or otherwise
discharged by the employer. Problems can arise where the employer
pays but the contractor does not own the materials or goods. This is
a breach of contract by the contractor but if the contractor becomes
bankrupt or goes into liquidation, the employer's remedy for breach
of contract, i.e. generally to sue for damages, may be worthless.

The contractor naturally enough wishes to be paid as early as
possible. The employer needs to be as certain as he reasonably can
be that, having paid for unfixed materials or goods, they will
become his property; otherwise, between the period of payment for
the materials or goods and either ownership passing to the
contractor and thence to the employer, or incorporation into the
works (whichever is the sooner), there is a risk that the true owner
might recover the materials or goods and the employer will then
receive no value for the money paid.

If the architect is aware that the contractor is not in a position to
transfer the title in the materials or goods, their value ought not to
be included in the payment certificate. This could of course result in
the contractor having to pay his supplier or his sub-contractor
before being paid by the employer and this in turn could involve the
contractor in cash flow problems. Should this occur to any
considerable extent in the industry, it could have an effect on tender
prices.

In very many instances, it will not be practicable for the architect
or, where appointed, the quantity surveyor, to check whether or not

the contractor is in a position to transfer the title in materials or goods to the employer. In many cases even detailed investigation would not necessarily clarify the position and even if it did, the administrative load would be out of proportion to the level of risk being taken. However, where there are deliveries to the site of materials or goods having a particularly high value then the lengths to which the architect, or the quantity surveyor as the case may be, should go to check the contractor's ability to transfer the title will depend upon a number of factors e.g.:

(1) the value of such materials or goods;
(2) his knowledge of the contractor and his methods of conducting business, e.g. does he usually pay for materials or goods before being paid?
(3) the terms of the supply contract or the sub-contract;
(4) the financial state and status of the contractor.

For high value items it is suggested that the least that the architect should do is to check, so far as he reasonably can, that the contractor's own contract for the supply of the materials or goods does not purport to retain title.

Even this limited step is of course of little use if the title still vests in a sub-supplier or sub-sub-contractor. Where a named sub-contractor is engaged under NAM/SC, by virtue of clause 19.5.2 the named sub-contractor will be stopped from denying the title of the main contractor to the materials or goods delivered to site by the named sub-contractor, provided the main contractor has been paid under the payment certificate including the value of the materials or goods; see also on this the Commentary to clause 3.2 page 117. This is of some limited protection to the employer but does not of course protect him where the named sub-contractor does not own the materials or goods and is not in a position to transfer title to them.

The question of title remains a difficult area and it is necessary to strike a reasonable balance between, on the one hand, going to inordinate lengths in trying to fully protect the employer and, on the other, to accepting that the practical difficulties of establishing title to all materials and goods before their value is included in a payment certificate is such that some element of risk must be taken. It is in the end a matter of using professional judgment in all the circumstances of the situation: see also Commentary to clause 3.2 under the heading 'Materials and goods' page 119.

NOTES TO CLAUSE 1.10

[1] '. . . *shall not be removed* . . .'
This restriction against removal is a necessary safeguard to the employer's interests. Once materials or goods have been paid for, it would be rarely proper for the architect to unconditionally allow their removal. Although this clause, by its terms, prevents the contractor from removing materials and goods delivered to the site, even if delivered prematurely, as they will not have been paid for, the architect would have less reason to withhold his consent to their removal, though there may still be circumstances in which he could reasonably object e.g. consider the case where the type of materials or goods concerned are in short supply and the contractor wishes to remove them for use on another contract thus causing potential delays to completion.

Furthermore, if too large a quantity of materials or goods is brought on to the site so that some deliveries are premature but others not, and a proportion is included in the valuation for a payment certificate, it may not be possible to differentiate between those in which the property has passed to the employer from those in which the property has not e.g. a stock pile of bricks. It may be totally impracticable to separately store or identify the two categories and in such circumstances the architect will be acting reasonably in refusing consent to any of them being removed from the site.

[2] '. . . *in accordance with clause 4.2.1(b)* . . .'
By clause 4.2.1(b), in order to qualify for payment in a payment certificate, the materials and goods must have been reasonably and properly and not prematurely delivered and must be adequately protected against weather and other casualties.

[3] '. . . *has been discharged* . . .'
It is clear that such materials and goods are, so far as the contract can ensure it, to become the employer's property once the certificate has been duly discharged – not necessarily that its full amount has been paid. It is the amount properly due which must have been paid or otherwise discharged. This need not equal the amount shown as due in the certificate e.g. where the employer deducts liquidated damages.

[4] '. . . *shall become the property* . . .'
The transfer of title is dealt with earlier on page 45.

[5]   '. . . subject to clause 6.3B and 6.3C.2 to .4 . . .'
By virtue of clause 6.3B and 6.3C.2 to .4 the employer is required to
take out 'All Risks Insurance' (see clause 6.3.2 for definition) for the
benefit of the employer and contractor in a 'Joint Names Policy' (see
clause 8.3 for definition). This will cover the risk of physical loss or
damage to 'Site Materials' (see clause 6.3.2 for definition). The
payment structure within clauses 6.3B and 6.3C in relation to
replacement or repair of site materials which have been lost or
damaged as a result of one or more of the insured risks is such that
the cost falls upon the employer rather than the contractor (see
clauses 6.3B.3.2 and 6.3B.3.5 and 6.3C.4.1 and 6.3C.4.4).

*[6]   '. . . the Contractor shall remain responsible for loss or damage to
the same.'*
Subject to what was said under Note (5) above, the contractor is
liable to replace lost or damaged materials or goods even though
they belong to the employer. If clause 6.3A applies, the contractor
must insure against the risk of loss or damage by taking out all risks
insurance in a joint names policy for the benefit of both contractor
and employer. Any shortfall of insurance monies will not relieve the
contractor of his obligation to replace or repair the lost or damaged
site materials.

Clause 1.11

Off-site materials and goods: passing of property, etc.

1.11   Where as provided in clause 4.2.1(c)[1] (*Interim payments: Architect's discretion*) the value of any materials
or goods intended for the Works and stored off-site has
been included in any payment certificate under which
the amount properly due to the Contractor has been
discharged[2] by the Employer, such materials and goods
shall become the property[3] of the Employer and
thereafter the Contractor shall not, except for use on the
Works, remove or cause or permit the same to be
moved or removed[4] from the premises where they are,
but the Contractor shall nevertheless be responsible for
any loss thereof or damage thereto and for the cost of
storage, handling and insurance[5] of the same until
such time as they are delivered to and placed on or
adjacent to the Works whereupon the provisions of
clause 1.10 (except the words 'Where the value' to the
words 'the property of the Employer, but') shall apply
thereto.

COMMENTARY ON CLAUSE 1.11

For a general summary of the law relating to the transfer of title, see earlier page 27.

This clause deals with payment for materials or goods intended for the works which are stored off-site. Whereas in clause 1.10, provided the conditions therein set out are satisfied, the architect must include the value of such materials and goods in a payment certificate, under this clause it is a matter for the discretion of the architect whether or not to include in a payment certificate the value of any such off-site goods or materials. This discretion is absolute and a refusal to exercise it in favour of the contractor cannot, it is submitted, be challenged in arbitration proceedings despite the wide scope of article 5 and clause 9. However, if the architect, having agreed to exercise his discretion to include the value of off-site materials or goods in a payment certificate, he cannot, if the contractor has relied upon it, e.g. by ordering or paying for off-site materials or goods, or even by programming the works on this basis, subsequently refuse to exercise it. Therefore if the architect intends to exercise his discretion on a limited basis, e.g. for specific materials or goods only or for a limited quantity only, he should take care to make any such limitations clear to the contractor.

There are varying views as to whether, from the employer's point of view, it is good practice to pay for materials and goods stored off-site. Unless the architect is particularly careful, difficulties can arise in the event of the bankruptcy or liquidation of the contractor. Where materials or goods are stored at some considerable distance from the site, the checking of these for the purpose of preparing a valuation may also present difficulties.

Although this type of provision is perhaps more suitable for inclusion in contracts for larger projects e.g JCT 80, it may still be necessary for certain materials or goods, which are difficult to store on site, to be obtained in order that the contractor can meet his programme. In such circumstances, the arrangement can be of benefit to both sides provided that proper precautions are taken in separating and identifying the materials or goods as belonging to the employer and not the contractor. This question of separation and identification is dealt with again under Note [1] below.

NOTES TO CLAUSE 1.11

[1]   '. . . clause 4.2.1(c) . . .'
By this clause, the architect has complete discretion to include the value of such materials or goods in a payment certificate. The exercise of that discretion will depend upon many considerations e.g.:

(i)   the total value concerned;
(ii)   the financial position of the contractor – he may need payment to stay solvent;
(iii)   the ease with which it can be ascertained that ownership will pass to the employer;
(iv)   the nature of the materials or goods and the storage facilities available on site.

In any event it will be essential, if the value of such materials or goods is included in a payment certificate, to ensure that the materials or goods are separately stored and clearly marked and identified as belonging to the employer.

The architect would be taking a grave risk of attracting personal liability in including the value of such materials or goods in a payment certificate without first obtaining the consent of the employer after fully explaining to him, if necessary, the risks involved, especially the risk of a retention of title clause applying. The type of safeguards for the employer which the architect should consider stipulating as a condition of the exercise of his discretion can be found in other JCT contracts, e.g. see JCT 80 clause 30.3 which can readily be adapted for use with IFC 84.

[2]   '. . . has been discharged . . .'
See Note [3] to clause 1.10 earlier on page 47.

[3]   '. . . shall become the property . . .'
The transfer of title is dealt with earlier on page 45.

[4]   '. . . the same to be moved or removed . . .'
Whilst under both clause 1.10 and clause 1.11 the materials or goods are not to be removed, under clause 1.11 they may not even be moved within the premises in which they are stored. This is sensible from the employer's point of view as he needs to ensure that such

materials and goods are stored separately and are thereafter kept separate from other materials and goods.

There is no provision, as in clause 1.10, for the materials or goods to be removed with the consent of the architect. Such consent could still be requested and given either directly by the employer or by the architect on his behalf if so authorised. This could be necessary e.g. in the event of damage to the premises in which the materials or goods are stored.

[5]   '. . . insurance . . .'
This should be carefully checked by the architect to ensure that the cover is adequate and that any conditions required by the insurance policy in connection with storage and protection have been complied with.

Chapter 4

Possession and completion

CONTENT

Section 2 deals in 10 clauses with matters relating to possession and completion. It covers such important topics as:

The giving of possession
Liquidated damages and extensions of time
Practical completion
Defects liability.

In dealing with section 2, clauses 2.3 to 2.8 inclusive have been grouped together. They all relate to the question of liquidated damages and extensions of time and must be considered as a whole. The commentary on this group of clauses therefore follows on after the print of clause 2.8 (see page 71). Similarly, clause 2.1 and 2.2 dealing with possession are grouped together, as are clause 2.9 dealing with practical completion and clause 2.10 dealing with defects liability.

SUMMARY OF GENERAL LAW

(A) Possession

The obligation of the employer to give to the contractor possession of the site is of course fundamental. The degree of possession to be given by the employer will vary, c.f. the case of a new building on a green field site with additions to an existing building which remains occupied.

Most forms of building contract restrict the degree of possession to be given to the contractor both in terms of physical extent and duration. The possession given must however be such as to reasonably enable the contractor to complete by the contractual completion date.

If a building contract failed expressly to provide for possession of the site to be given to the contractor, such a provision would be implied to the extent that possession was required to enable the contractor to complete by any agreed completion date. It would be necessary to imply such a term to give the contract business efficacy.

If the employer fails to give the appropriate degree of possession to enable the contractor to complete on time or probably to carry out the work in accordance with any agreed programme, this will be a breach of contract by the employer. This is so even though the employer's failure was due to circumstances completely outside his control.

In practice, although the employer may have entered into a contract with the very best of intentions, some difficulty in giving possession may remain, e.g. a tenant who refuses to vacate without a court order. It is advisable therefore for the building contract to deal specifically with such problems.

The failure by the employer to hand over possession by an agreed date may, depending on its degree and duration, amount to a serious breach of contract entitling the contractor to treat the contract as at an end and to sue the employer for damages. Apart from this, unless there is an express provision for an extension of time for delay in giving possession, such delay, even if not amounting to a fundamental breach, will invalidate any liquidated damages clause – see *Rapid Building Group Ltd* v. *Ealing Family Housing Association Ltd* (1984)). However, the employer will still retain a right to claim unliquidated damages at common law should the contractor fail to complete within a reasonable time, subject to the possibility that any such claim cannot in any event exceed the sum stated as liquidated damages.

Once in possession, it is a question of some debate whether or not the contractor's licence to remain in possession is revocable or irrevocable whilst the contract period is still running. In other words, if the employer purports to determine the contractor's employment and this is disputed by the contractor can he remain on site or is he compelled to leave?

In the case of *London Borough of Hounslow* v. *Twickenham Garden Developments Ltd* (1970) it was held that the contractor's licence to remain on site during the contract period was irrevocable. This decision was however criticised and not followed in the New Zealand case of *Mayfield Holdings Ltd* v. *Moana Reef Ltd* (1973). It is suggested that whilst the English decision may be the stronger authority which courts, at any rate of first instance, may feel compelled to follow, the New Zealand decision is the more logical. It appears that the *Hounslow* decision which concerned a JCT 63 contract has been followed in a case under the JCT 80 contract, namely *Vonlynn Holdings* v. *T. Flaherty* (1988). In a case under significantly different wording in the ICE 5th Edition Form of Contract, namely *Tara Civil Engineering Ltd* v. *Moorfield Developments Ltd* (1989), it was held that the contractor was bound to leave the site despite contesting the employer's right to determine. It should be noted that under clause 63 of the ICE Conditions, once the engineer has certified in writing to the employer that in his opinion the contractor has defaulted in one of the stipulated ways, the employer, after giving seven days notice to the contractor, is expressly entitled to enter upon the site of the works and expel the contractor therefrom. This is significantly different to the wording under JCT 80.

However, IFC 84 in clause 7.4(a), expressly requires the contractor to give up possession of the site in the event of the employer determining the contractor's employment under the contract. It is submitted that this requires the contractor to give up possession of the site even if the validity of the determination is challenged. This interpretation is also supported by the provisions of clause 6.3 in relation to the insurance of the works which provides that the obligation to insure the works ceases upon a determination of the contractor's employment even if it is validly challenged. (For further comments see pages 321 and 323.)

To allow the contractor to remain on site when there has been a complete breakdown of relationships seems a nonsense. Even if the employer is in the wrong, he cannot easily be compelled to make interim payments and the architect cannot be compelled to issue necessary drawings and other information etc. so that there will be little or no progress made. So too, the employer will be prevented, by the continued presence of the contractor, from employing others to finish off the work. It is submitted that the contractor should be required to leave the site and to pursue his claim for damages for breach of contract if he is wrongly removed from the site. Whilst it may be contended that for the contractor to be compelled to leave, even though he may be the innocent party, could result in damage to his reputation, nevertheless, to permit him to stay will cause a hopeless impasse in many situations. The *Hounslow* case is discussed again when dealing with determination of the contractor's employment – see page 333.

The site

Under the general common law, an employer does not impliedly warrant the condition of the site. This is of course subject to the express terms of the contract made between the parties. Much may depend upon whether use is made of a specification or a bill of quantities. If a specification is used and nothing at all is stated so that the contractor is to satisfy himself as to the site conditions, then the risk of unforeseen difficulties will generally have to be borne by him e.g. unforeseen ground conditions. On the other hand, if a bill of quantities is used then the express terms of the contract will have to be examined to decide the status of the bill of quantities. For example, the contract may say that the bills have been prepared in accordance with a particular standard method of measurement, in which case, to a greater or lesser extent, a warranty as to site conditions is given either by reason of what is stated in the bills by way of site investigation results, or assumptions to be made in the absence of such information. The precise extent of the warranties given in this way is difficult to assess e.g. to what extent does information obtained from a trial pit or a bore hole amount to a warranty by the employer that the contractor can freely assume in pricing that such information is typical of the whole of the site?

If the site conditions are such that it is necessary for the employer to vary the design, then this will almost invariably amount to a variation to the contract works for which the contractor will generally be compensated by express terms of the contract.

On particular occasions, e.g. where the employer states the assumptions as to ground conditions on which the contractor is asked to design the works himself, there may be an implied warranty on the part of the employer as to site conditions. See *Bacal Construction (Midlands) Ltd* v. *Northampton Development Corporation* (1975).

(B) Completion

A building contract may or may not have a fixed date for completion. If it does not, then there will be an implied term to the effect that the contractor will be obliged to complete within a reasonable time.

If the contract contains a fixed completion date, it is a matter of interpretation, in the absence of express words, as to whether the time stated for completion is of the essence of the contract so that failure to achieve it puts the contractor into breach of a fundamental term entitling the employer to treat the failure as a repudiation of the contract and as releasing him from his obligations under the contract. Historically, at common law, if the contract gave a completion date, this was then regarded as of the essence of the contract. However, courts of equity gave relief and did not generally regard time as of the essence and prevented the innocent party from treating the contract as at an end, though the right to claim damages for breach of contract remained.

Nowadays, unless it is otherwise clear from the terms of the contract or the surrounding circumstances, a fixed completion date will not make time of the essence of a building contract. The standard forms of construction contract in common use in the United Kingdom at the present time, whilst providing for a definite completion date, do not, by their terms, make time of the essence. This is demonstrated by the fact that such contracts envisage practical completion later than the contractual or extended contractual completion date. However, it is possible for the parties to agree to the completion date being of the essence despite the contract also having a liquidated damages clause: see *Peak Construction (Liverpool) Ltd* v. *McKinney Foundations Ltd* (1970).

(C) Liquidated damages

Most standard forms of building contract provide for the payment of a sum by the contractor to the employer by way of liquidated damages for delay in completion of the contract works.

There is little doubt that the topic of liquidated damages and extensions of time is one of the more emotive aspects of building contracts and has given rise over the years to many disputes and difficulties. Depending upon whether you are an employer, an architect or a contractor, the extension of time provisions in a building contract will be viewed in a different light.

By means of such a provision an agreed sum is payable as damages for delay in completion. In building contracts the sum may be expressed simply as so much per week of delay, or alternatively, in a suitable case, e.g. a housing development, and where the contractual framework permits it, as so much per week per incomplete dwelling.

The main advantage to the employer of such a provision is that he is not required to prove his actual loss as a result of the delay.

The existence of a valid liquidated damages clause can be of particular benefit to an employer where the cost of delay cannot easily be measured e.g. a fire station or a church. In such a case the employer would have difficulty in actually proving a loss beyond perhaps the loss of the use of capital money invested in the project with a delayed return.

Provided that the stipulated sum represents, at the time of entering into the contract, a genuine attempt at a pre-estimate of the likely loss, it is not a penalty and will, other things being equal, be upheld by the courts: *Dunlop Pneumatic Tyre Company Ltd* v. *New Garage and Motor Co. Ltd* (1915). This is so even if, in the result, no actual loss is suffered at all – see *BFI Group of Companies Ltd* v. *DCB Integration Systems Ltd* (1987). The courts will construe liquidated damages and extension of time clauses in printed forms of contract strictly *contra proferentem* (against the person seeking to rely upon it). As these clauses are inserted primarily for the benefit of the employer they will be construed strictly against him: *Peak Construction (Liverpool) Ltd* v. *McKinney Foundations Ltd* (1970).

An employer may lose the right to deduct liquidated damages in a number of ways:

(i) where the sum stipulated for is in truth a penalty. A penalty clause is unenforceable in English Law. For instance, if the amount stipulated for goes beyond what could conceivably be a genuine pre-estimate of the likely loss due to delay; or where the same sum is stipulated for in differing circumstances where it is clear that the actual loss must vary so that the pre-estimate is not genuine, then it will amount to a penalty: *Ford Motor Co.* v. *Armstrong* (1915).

(ii) by the employer so delaying the contractor as to prevent the contractor completing by the agreed completion date. If, in such circumstances, the contract contains no provision for an extension of time covering such delay, then the employer cannot insist upon completion by the agreed date and will lose his right to liquidated damages: see *Dodd* v. *Churton* (1897); *Peak Construction (Liverpool) Ltd* v. *McKinney Foundations Ltd* (1970); *Rapid Building Group Ltd* v. *Ealing Family*

Housing Association Ltd (1984). In such a case, the fixed completion date will be lost and the contractor will have a reasonable time within which to complete the contract works. The employer may still have a right to claim unliquidated damages (see the text immediately following the reference to the *Rapid Building Group* case on page 53).

(iii) by the breakdown of the contractual machinery for calculating liquidated damages. A liquidated damages clause needs to be operated from one specific date until another. If therefore the starting date has not been specified or the finishing date is lost, it can lead to difficulties in calculation and thereby also in the operation of the liquidated damages provision.

(iv) failure of the employer to comply with a condition precedent. For example, the employer may first require a certificate from the architect that the contractual completion date has passed and the works remain uncompleted: *Token Construction Co. Ltd* v. *Charlton Estates Ltd* (1973).

(v) wilful failure by the architect or other supervising officer under the contract to properly administer the machinery of an extension of time clause: *Peak Construction (Liverpool) Ltd* v. *McKinney Foundations Ltd* (1970). The mere failure even if due to incompetence and even possibly negligence on the part of the architect in administering the extension of time machinery under the contract may not be enough to invalidate the liquidated damages provision. The contractor can generally seek immediate arbitration to put matters right: see *Temloc Ltd* v. *Errill Properties Ltd* (1987); *Lubenham Fidelities and Investment Co. Ltd* v. *South Pembrokeshire District Council and Another* (1986); *Pacific Associates, Inc. and Another* v. *Baxter and Others* (1988).

Should the liquidated damages clause become invalidated the employer will still be able to claim damages at common law if any delay in completion by the contractor amounts to a breach of contract. See the reference to the *Rapid Building Group* case at page 53.

(D) Extensions of time

Many building contracts make provision for extensions of time to the contractual completion date in certain events. Such clauses are directly related to any liquidated damages provision in the contract.

Extension of time provisions have the effect of extending the original contractual completion date where the delaying event falls within its terms. Where the delay is what might be called a neutral delay i.e. not the direct fault of either the employer or the contractor, such as exceptionally adverse weather conditions, civil commotion, local combination of workmen, strike or lockout etc. (being, it is submitted, events which might otherwise be at the contractual risk of the contractor – see for example *Percy Bilton Ltd* v. *Greater London Council* (1982) 20 BLR 8 per Lord Fraser at pages 13 and 14), the extension of time clause benefits the contractor by relieving him of any liability to pay liquidated damages. On the other hand, if the event which causes the delay is the fault of the employer or someone acting on his behalf e.g. the architect issuing late instructions, then the effect of extending the contractual completion date to a new later date will keep intact the liquidated damages clause for the benefit of the employer, which would not be the case if the employer's delay was not covered by the extension of time provision.

(E) Practical completion

Most standard forms of construction contract by their express terms treat completion as something less than entire completion. They do not require the contract works to be completed in every detail before contractual completion is achieved. They permit small details to be completed during a maintenance or defects liability period. Completion is qualified by reference to its being practical.

Practical completion probably means that the contract works have been completed to the extent that nothing important or which significantly affects their use for their intended purposes remains outstanding.

Once practical completion has been achieved and an appropriate certificate under the contract issued to that effect, it cannot later be cancelled by reason of the discovery of a defect which renders the works unusable for their intended purpose: *The Lord Mayor Alderman and Citizens of the City of Westminster* v. *J. Jarvis & Sons Ltd and Another* (1970).

(F) Defects liability period

A defects liability clause will usually refer to a stated period for which it is to run commencing with the date of practical completion. Often it is six or 12 months.

Often under building contracts such provisions require the contractor to remedy defects appearing within the period but do not expressly refer to the completion of any known outstanding or defective items which existed at the time when practical completion was achieved. This can lead to a strict interpretation of what amounts to practical completion, namely, that nothing at all must remain to be done however minor in nature it may be.

Unless a contract by its terms expressly and unequivocally states to the contrary, the contractor's obligation and right to attend to defects etc. discovered within the defects liability period will not exclude or limit the employer's legal right to recover damages for losses suffered, if any, as a result of defective work. It simply means that the employer can call for the physical presence on site of the contractor to carry out the remedial work and that the contractor has a contractual right to remedy the defects. This can be of considerable importance to a contractor. Firstly, in the absence of such a right the employer could get the defects remedied by another contractor and the reasonable cost of so doing would be payable as damages by the contractor for breach of contract. The opportunity therefore for the contractor himself to attend to the defects can save him money. Secondly, it helps to minimise the risk to the contractor of his reputation being tarnished if he can himself attend to the defects rather than having a third party examine his work.

CONSIDERATION OF THE RELEVANT CLAUSES OF IFC 84

Clauses 2.1 and 2.2

Possession and Completion Dates

2.1   Possession of the site[1] shall be given to the Contractor on the Date of Possession stated in the Appendix.[2] The Contractor shall thereupon begin and regularly and diligently proceed with the Works[3] and shall complete the same on or before the Date for Completion stated in the Appendix,[4] subject nevertheless to the provisions for extension of time in clause 2.3.

Possession by Contractor – use or occupation by Employer
For the purposes of the Works insurances the Contractor

shall retain possession of the site and the Works up to and including the date of issue of the certificate of Practical Completion, and the Employer shall not be entitled to take possession of any part or parts of the Works until that date.

Notwithstanding the provisions of the immediately preceding paragraph the Employer may, with the consent in writing of the Contractor, use or occupy the site or the Works or part thereof whether for the purposes of storage of his goods or otherwise[5] before the date of issue of the certificate of Practical Completion by the Architect/the Contract Administrator. Before the Contractor shall give his consent to such use or occupation the Contractor or the Employer shall notify the insurers under clause 6.3A or clause 6.3B or clause 6.3C.2 to .4 whichever may be applicable and obtain confirmation that such use or occupation will not prejudice the insurance. Subject to such confirmation the consent of the Contractor shall not be unreasonably withheld.

Where clause 6.3A.2 or clause 6.3A.3 applies and the insurers in giving the confirmation referred to in the immediately preceding paragraph have made it a condition of such confirmation that an additional premium is required the Contractor shall notify the Employer of the amount of the additional premium. If the Employer continues to require use or occupation under clause 2.1 the additional premium required shall be added to the Contract Sum and the Contract shall provide the Employer, if so requested, with the receipt of the insurers for that additional premium.

Deferment of possession

2.2　Where this clause is stated in the Appendix to apply the Employer may defer the giving of possession for a time not exceeding the period stated in the Appendix calculated from the Date of Possession, which should not exceed six weeks.[6]

COMMENTARY ON CLAUSES 2.1 AND 2.2

The giving of possession of the site by the employer to the contractor and the contractor's commencement of work effectively heralds the beginning of the post-contract period by which time the building team should have reached agreement on procedures and should have established relationships that will allow the works to proceed smoothly to completion. The first steps in that direction should have been taken at the preliminary or briefing meeting. Although much has been written about pre-contract planning, and of course it is very important, the satisfactory progress of the works and running of the contract can to some extent depend on attitudes and relationships during the contract period. Even the effects of inadequate pre-contract planning can to some extent be mitigated if that fact is acknowledged by all concerned, if the contractor is aware of it when he tenders and if the building team work together to solve the problems as and when they occur, giving full recognition to the difficulties and additional costs which stem from this situation.

Contracts which are operated on this basis are more likely to be completed successfully than an adequately pre-planned scheme which is bedevilled by poor site performance by the contractor or an unco-operative attitude on the part of the building team generally.

IFC 84 requires the employer to give to the contractor possession of the site on the date of possession stated in the appendix to the contract. Clause 2.1 expressly limits the extent of possession to be given by the employer to the contractor by allowing the employer to use or occupy part of the site 'for the purposes of storage of goods or otherwise . . .'. Quite apart from this express restriction, the IFC 84 Conditions elsewhere also make it clear that the contractor's right to possession is not absolute, see for example clauses 2.2, 2.4.8, 2.4.14 and 3.11. The issue of late possession is considered in more detail in the commentary below on clause 2.2 and in the discussion of clause 2.4.14 on page 83.

The second paragraph of clause 2.1 was inserted as part of Amendment 2 of 1984. It provides that, for the purposes of the works insurances, the contractor shall retain possession of the site and of the works up to and including the date of issue of the certificate of practical completion and that the employer shall not be entitled to take possession of any part of the works until that date. This express right is, it is submitted, to be read subject to clause 7.4(a) and to other clauses such as clause 2.2, 2.4.8, 2.4.14 and 3.11 as well as being qualified expressly by the paragraph of clause 2.1 which follows it. In addition the possession is expressly stated as being for 'the purposes of the Works insurances'.

The third paragraph gives the employer certain limited rights to occupy the site or the works themselves or a part thereof for the storage of goods or otherwise prior to practical completion, but requires the consent in writing of the contractor. However, this consent must not be unreasonably withheld. The employer's use or occupation for storage or otherwise must be cleared by the works insurer and provided it is confirmed that the cover is not prejudiced the contractor must not unreasonably withhold his consent.

If the work's insurers are prepared to maintain existing cover only on the payment of an additional premium, the employer, if he still wishes to take advantage of this provision, must meet the cost of this additional premium. If the relevant works insurance clause is 6.3A (contractor to insure in joint names) any additional premium will be added to the contract sum. This facility is clearly very useful for employers, e.g. where the contract is over-running and the employer has to take delivery of furniture or equipment for the building.

The contract conditions do not provide for insurance against loss or damage to any stored goods etc. unless perhaps they are stored in a building which already existed when the contract was formed – see clause 6.3C.1. Otherwise the risk of loss or damage will be governed, if at all, by clause 6.1.2.

If the employer makes use of this provision he should be very careful to ensure that its use or occupation will not impede the contractor. There is no event to cover such a situation within clause 2.4 dealing with extensions of time, so that the employer could find that the liquidated damages provision is invalidated should the contractor be unable to complete by the completion date. Furthermore, there is no relevant matter in clause 4.12 entitling the contractor to reimbursement of direct loss or expense if the progress of the works is materially affected by the employer's use or occupation. As the contractor will have consented to the employer's use or occupation, it can hardly be a breach of contract by the employer, at any rate to the extent to which any disruption can be seen to be inevitable. In such circumstances the contractor could consider requiring suitable safeguards for recompense as a condition of giving his consent.

The degree of use or occupation required by the employer may prompt the contractor to suggest that the employer ought to operate the provisions of optional clause 2.11 (partial possession) so that there will be a deemed practical completion of the relevant part of the works. The situation in which clause 2.1 rather than clause 2.11 should be used will depend upon the circumstances. It is submitted that the contractor's withholding of consent under clause 2.1 may on occasions be reasonable if at the same time he indicates that he would consent to a partial possession under clause 2.11.

Clause 2.2, where stated in the appendix to apply, gives the employer the power to defer the giving of possession of the site. It should be noted that it is not the architect that has this power but the employer. The maximum period recommended for which the deferment can be given is six weeks. Bearing in mind that IFC 84 is intended for contract periods of no more than 12 months, a period of six weeks is perhaps more than fair to the employer. In the event of such a deferment being made, the contractor may be entitled to an extension of time for completion and to reimbursement of loss and expense. Should the deferment exceed the period stated in the appendix, the employer will be in breach of contract entitling the contractor to damages and possibly, depending on the seriousness of the breach, entitling the contractor also to treat the employer's breach as a repudiation of the contract thereby excusing the contractor from further performance of his own contractual obligations.

NOTES TO CLAUSES 2.1 AND 2.2

[1]   '. . . the site . . .'
Warranties in connection with site conditions will usually depend on whether on the one hand a specification is used or on the other hand a bill of quantities. If IFC 84 is used in conjunction with a specification, without a bill of quantities, then the position depends on what is stated in the specification, if anything, as to conditions of the site. If nothing at all is stated, so that the contractor is to satisfy himself as to the site conditions, then the risk of unforeseen difficulties will generally have to be borne by him e.g. unforeseen ground conditions.

On the other hand, if a bill of quantities is used then, by virtue of clause 1.5, the contract bills will be treated as having been prepared in accordance with the Standard Method of Measurement of Building Works, 7th edition (SMM) published by the Royal Institution of Chartered Surveyors and the Building Employers Confederation. Accordingly, particulars will have been given of soil and ground conditions such as water level, trial pits or boreholes and over or underground services. If that information is not available then a description of the ground and strata which is to be assumed shall be stated (see Section D20 of SMM 7). To this extent therefore a warranty as to site conditions is given. There are other examples within SMM. If the actual information given turns out to be inaccurate or misleading the contractor will be able to make a claim for extra costs incurred in overcoming the unforeseen problems.

[2] '. . . *Date of Possession stated in the Appendix* . . .'
The appendix should contain a specified date. However, the insertion of a specified date may well be impracticable from the employer's point of view, bearing in mind such matters as the tendering procedures involved or last minute difficulties in obtaining possession of land by the employer. It may be appreciated that for one reason or another there may be difficulty in accepting a tender in sufficient time to give possession of the site to the contractor by a stipulated date. Often, therefore, the date will be stated in terms of some days or weeks from the date of acceptance of the contractor's tender. It is submitted that the insertion of such words as 'on a date to be agreed' at tender stage is fraught with danger for the employer. If such a phrase is used, it is vital that before the contractor's tender is accepted, the date for possession is agreed and inserted in the appendix. If this is not done at the time of acceptance of the contractor's tender and no date can subsequently be agreed, then presumably possession must be given within a reasonable time and this could arguably invalidate the liquidated damages provision in the contract. Further, the absence of agreement as to the date for possession of the site would deprive the employer of any power he might have had to defer possession under clause 2.2 even if stated in the appendix to apply.

[3] '. . . *regularly and diligently proceed with the Works*. . . .'
Even without these express words it has been argued that such a term would be implied in a building contract: see *Hudson's Building and Engineering Contracts*, 10th edition, page 611 *et seq* and also the supplement to the 10th edition. Failure by the contractor to so proceed can lead to a determination of the contractor's employment under this contract (see clause 7.1 (b)). This requirement to regularly and diligently proceed with the works should be of much more practical use to the employer than the obligation on the contractor to complete by a certain date. In the latter case, generally no action can be taken by an employer until the completion date is past even though the contractor is clearly falling behind; whereas, in the former case, it is submitted that the appropriate notices etc. can be served in accordance with the contract and ultimately the contractor's employment can be determined if necessary.

In the case of *Greater London Council* v. *Cleveland Bridge and Engineering Co. Ltd and Another* (1984) at first instance Mr Justice Staughton held that a failure by the contractor to execute the works with due diligence and expedition entitling the employer to discharge the contractor for such failure did not itself render the contractor in breach of contract and liable for damages. It was only in the event of a failure to use such diligence and expedition as would reasonably be required to meet the contractual deadlines, which would amount to a breach of contract by the contractor. The result is, while the employer's express right to determine the contractor's employment may be relied on, always assuming that the express contractual remedy of determination of the contractor's employment is drafted clearly enough to enable its use even where it may still be physically possible to complete on time, the employer is likely to be taking a considerable risk in treating such a failure by the contractor before the completion date has expired as a repudiation and as a ground for determining the contract itself. He may have to wait and see if the contractual completion date is met, or at least until it becomes absolutely clear that it will not be.

If this is the correct position following the *Cleveland Bridge* case this will deprive the employer of one of the main benefits sought to be achieved by the inclusion of such a term i.e. the ability to take early positive action under the common law when serious delays are taking place. The Court of Appeal in dismissing the appeal of the Greater London Council did not deal with this issue in detail.

Although there is no contractual requirement for a programme, if one is produced by the contractor, his failure to keep to it may be some evidence of a failure to proceed 'regularly and diligently'.

These words were considered by Mr Justice Megarry in the *Hounslow* case in relation to the determination provisions of JCT 63 – see page 347.

[4] '*. . . on or before the Date for Completion stated in the Appendix . . .*'
The contractor is fully entitled to complete before the date for completion stated in the appendix and the architect will be obliged to issue a certificate of practical completion pursuant to clause 2.9. If the employer does not want practical completion until the date for completion is reached, then the proper method is to amend the contract conditions themselves rather than to insert such a provision elsewhere e.g. in the bills of quantities or specification (see clause 1.3).

A date for completion is to be stated in the appendix to the contract. The contract is therefore designed for a fixed date for completion. The fact that a fixed date for completion is provided for does not, under this contract, make the time for completion of the essence of the contract so that mere failure by the contractor to complete on time is not a sufficiently serious breach to entitle the employer to treat the breach as a repudiation of the contract by the contractor. Subject to the extension of time provisions, it does however entitle the employer to liquidated damages under the contract.

[5] '... for the purposes of storage of his goods or otherwise ...'
The reference to 'or otherwise' seems unnecessarily wide and vague. Coupled with the earlier reference to the use or occupation of the site or the works themselves it offers the possibility of extensive occupation by the employer while work is continuing and with works insurance remaining in place. Presumably, if the employer pushes too far in this direction, he will be faced with works insurers refusing to confirm that cover will be maintained in place, and with the contractor claiming that in truth there has been a partial possession if the option in clause 2.11 has been adopted for the contract.

[6] '... should not exceed six weeks.'
These words make it clear that, while six weeks may be the recommended maximum, it is quite possible to insert a longer period in the appendix and this is likely to be valid even without any amendment to clause 2.2 or to the words '(period not to exceed six weeks)' contained in the appendix item.

Clauses 2.3 to 2.8

Extension of time

2.3 Upon it becoming reasonably apparent that the progress of the Works is being or is likely[1] to be delayed, the Contractor shall forthwith[2] give written notice of the cause of the delay to the Architect/the Contract Administrator, and if in the opinion of the Architect/the Contract Administrator the completion of the Works is likely[3] to be or has been delayed beyond the Date for Completion stated in the Appendix or beyond any extended time previously fixed under this clause, by any of the events[4] in clause 2.4 then the Architect/the Contract Administrator shall[5] so soon as he is able to estimate the length of delay beyond that date or time make in writing[6] a fair and reasonable[7] extension of time for completion of the Works.

If an event referred to in clauses 2.4.5, 2.4.6, 2.4.7, 2.4.8, 2.4.9, 2.4.12 or 2.4.15[8] occurs after the Date for Completion (or after the expiry of any extended time previously fixed under this clause) but before Practical Completion is achieved the Architect/the Contract Administrator shall so soon as he is able to estimate the length of the delay, if any, to the Works resulting from that event make in writing a fair and reasonable extension of the time for completion of the Works.

At any time up to 12 weeks[9] after the date of Practical Completion, the Architect/the Contract Administrator may[10] make an extension of time in accordance with the provisions of this clause 2.3, whether upon reviewing a previous decision or otherwise[11] and whether or not the Contractor has given notice as referred to in the first paragraph hereof. Such an extension of time shall not reduce any previously made.

Provided always that the Contractor shall use constantly his best endeavours[13] to prevent delay and shall do all that may be reasonably required to the satisfaction of the Architect/the Contract Administrator to proceed with the Works.[12]

The Contractor shall provide such information required by the Architect/the Contract Administrator as is reasonably necessary for the purposes of clause 2.3.[14]

Events referred to in 2.3

2.4   The following are the events referred to in clause 2.3:

2.4.1   force majeure;

2.4.2   exceptionally adverse[15] weather conditions;

2.4.3   loss or damage caused by any one or more of the Specified Perils;

2.4.4   civil commotion, local combination of workmen, strike or lock-out affecting any of the trades employed upon the Works or any trade engaged in the preparation, manufacture or transportation of any of the goods or materials required for the Works;

2.4.5   compliance with the Architect's/the Contract Administrator's instructions under clauses

1.4   *(Inconsistencies)*, or
3.6   *(Variations)*, or
3.8   *(Provisional sums)*
except, where the Contract Documents include bills

of quantities, for the expenditure of a provisional
sum for defined work[16]* included in such bills, or
3.15 *(Postponement)*
or, to the extent provided therein, under clause 3.3
(Named sub-contractors);

2.4.6 compliance with the Architect's/the Contract Administrator's instructions requiring the opening up or the testing
of any of the work, materials or goods in accordance with
clauses 3.12 or 3.13.1 (including making good in consequence of such opening up or testing), unless the inspection or test showed that the work, materials or goods
were not in accordance with this Contract;

2.4.7 the Contractor not having received in due time necessary
instructions (including those for or in regard to the expenditure of provisional sums) drawings, details or levels
from the Architect/the Contractor Administrator for
which he specifically applied in writing[17] provided that
such application was made on a date which having
regard to the Date for Completion stated in the Appendix
or any extended time then fixed was neither unreasonably distant from nor unreasonably close[18] to the date on
which it was necessary for him to receive the same;

2.4.8 the execution of work not forming part of this Contract by
the Employer himself or by persons employed or otherwise engaged by the Employer as referred to in clause
3.11 or the failure to execute such work;

2.4.9 the supply by the Employer of materials and goods which
the Employer has agreed to supply for the Works or the
failure so to supply;

2.4.10 where this clause is stated in the Appendix to apply, the
Contractor's inability for reasons beyond his control and
which he could not reasonably have foreseen at the Base
Date to secure such labour as is essential to the proper
carrying out of the Works;

2.4.11 where this clause is stated in the Appendix to apply, the
Contractor's inability for reasons beyond his control and
which he could not have foreseen at the Base Date to
secure such goods or materials as are essential to the
proper carrying out of the Works;

2.4.12 failure of the Employer to give in due time[19] ingress to or
egress from the site of the Works or any part thereof
through or over any land, buildings, way or passage
adjoining or connected with the site and in the possession and control[20] of the Employer, in accordance with
the Contract Documents, after receipt by the Architect/
the Contract Administrator of such notice, if any, as the

Contractor is required to give, or failure of the Employer to give such ingress or egress as otherwise agreed[21] between the Architect/the Contract Administrator and the Contractor;

2.4.13 the carrying out by a local authority or statutory undertaker of work in pursuance of its statutory obligations in relation to the Works[22] or the failure to carry out such work;

2.4.14 where clause 2.2 is stated in the Appendix to apply, the deferment[23] of the Employer giving possession of the site[24] under that clause;

2.4.15 by reason of the execution of work for which an Approximate Quantity is included in the Contract Documents which is not a reasonably accurate forecast of the quantity of work required.

* See footnote to clause 8.3 (Definitions).

Further delay or extension of time

2.5   In clauses 2.3, 2.6 and 2.8 any references to delay, notice, extension of time or certificate include further delay, further notice, further extension of time, or further certificate as appropriate.

Certificate of non-completion

2.6   If the Contractor fails to complete the Works by the Date for Completion or within any extended time fixed under clause 2.3 then the Architect/the Contract Administrator shall issue a certificate to that effect.

In the event of an extension of time being made after the issue of such a certificate the Architect/the Contract Administrator shall issue a written cancellation[25] of that certificate and shall issue such further certificate under this clause as may be necessary.[26]

Liquidated damages for non-completion

2.7   Subject to the issue of a certificate under clause 2.6[27] the Contractor shall, as the Employer may require in writing[28] not later than the date of the final certificate[29] for payment, pay or allow the Employer liquidated damages at the rate stated in the Appendix[30] for the period during which the Works shall remain or have remained incomplete, and the Employer may deduct the same from any monies due or to become due to the Contractor under this Contract[31] (including any balance stated as due to the Contractor in the final certificate for payment) or may recover the same from the Contractor as a debt.

Repayment of liquidated damages

2.8   If after the operation of clause 2.7 the relevant certificate under clause 2.6 is cancelled the Employer shall pay or repay to the Contractor any amounts deducted or recovered under clause 2.7 but taking into account the effect of a further certificate, if any, issued under clause 2.6.

COMMENTARY ON CLAUSES 2.3 TO 2.8

Clauses 2.3, 2.4, 2.5, 2.6, 2.7 and 2.8 deal with the making of extensions of time by the architect and the deduction or recovery of liquidated damages by the employer from the contractor.

IFC 84 provides for the date for completion to be extended where completion of the works is likely to be or has been delayed by one or more of the events referred to in clause 2.4.

Extensions of time clauses probably share with variation clauses the doubtful honour of having caused more difficulty in building contracts over the years than any other factor. They cause difficulty not only between contractor and architect but also between the architect and employer since few employers can understand why so many events are built into the contract allowing the contractor to claim an extension of time. They also find it hard to understand the advantage of granting the contractor an extension of time for so called 'neutral' events which may be included in an extension of time clause, regarding such events as being matters which should be accepted as the contractor's responsibility and his 'commercial risk'. However, such a concept may not always work to the advantage of the employer as the contractor may cover such a risk in his tender prices only to find that they may not always occur to the degree which he had assumed.

Much difficulty experienced in relation to extension of time clauses has arisen firstly because contractors have failed to notify the architect when it has become reasonably apparent that the progress of the works has been or is likely to be delayed, thus depriving the architect of the opportunity to take, or to instruct the contractor to take, measures to alleviate the consequences; and, secondly, because the architect has often failed to make an extension of time sufficiently early to allow the contractor to plan completion of the works by the revised completion date.

The drafting of JCT 80 attempted to improve this position by laying down a timetable for response by the architect and at the same time making the contractor responsible for providing information on the anticipated effects and extent of the delay in order to assist the architect in making his assessment.

In the drafting of IFC 84 the JCT has abandoned the procedural matters contained in JCT 80, reverting to the general concept to be found in JCT 63, but with some significant departures which are referred to later in this chapter. The abandonment of the procedural matters, including a strict timetable for response by the architect, was clearly an attempt to simplify the conditions rather than any recognition that these matters would be less important in a form of contract designed for less complex projects. Experience indicates that it is in fact just as important whatever the size of project and architects would be well advised to respond as rapidly as possible to a contractor's written notice of the cause of delay, notwithstanding the fact that the only reference to the timing of the architect's response in IFC 84 is that he should act 'so soon as he is able to estimate the length of delay'.

As already stated, the clause is more akin to JCT 63 than JCT 80. Clause 25 of JCT 80 requires the contractor to provide specific information concerning the expected effects and an estimate of the delay or likely delay attributable to each and every relevant event together with updates. IFC 84 is less specific. It requires the contractor to provide such information required by the architect as is reasonably necessary for the purposes of operating clause 2.3. Whilst this may be an improvement over JCT 63, it may nevertheless be a disappointment to those who would prefer to see the more specific requirements of JCT 80 repeated in ICF 84.

An extension of time made under clause 2.3 will have the effect in relation to those 'neutral' events which are a contractor's risk of absolving the contractor from liability to pay or allow liquidated damages to the extent that the time for completion is thereby extended. Where the extension of time is made in respect of events which are the employer's responsibility, the effect of the extension is to keep the employer's entitlement to claim liquidated damages intact.

It had been thought that JCT 63 had a potentially serious defect in its extension of time provisions, namely, that by its terms (clause 23 of that contract), it did not allow the architect to make an extension of time after the date for completion had gone by. Where the ground for an extension of time was what might be termed a neutral event, e.g. exceptionally adverse weather conditions, this would in any event often not have attracted an extension of time as the contractor may well not have been delayed by the event but for the fact that he was in culpable delay at the time in question. Had he not been at fault, he would not have suffered delay from the neutral event. However, this would not inevitably be so, e.g. discovery of a defective water main which needs to be replaced by a statutory undertaker.

If the delay was caused by the act of the employer or someone acting on his behalf e.g. the issue of a variation instruction requiring extra work, the employer would have prevented the contractor from completing as soon as he would otherwise have done, and as there was no apparent power to make an extension of time, the liquidated damages provision would be invalidated. Under JCT 80 the position may arguably be otherwise as it envisages the making of an extension of time after practical completion, presumably, even where practical completion takes place after the expiry of the contractual completion date i.e. after a period of culpable delay during which the contractor is liable to liquidated and ascertained damages. JCT 80 also provides for the repayment of liquidated damages upon the fixing of a later completion date (see clauses 25.3.3 and 24.2.2 of that contract). However, the position is not, it is submitted, clarified beyond any doubt in JCT 80 (see particularly the words '. . . delayed thereby . . .' in JCT 80 clause 25.3.1.2). This is so despite an apparent attempt to resolve the debate by an amendment introduced as part of Amendment 4 to JCT 80 in July 1987.

In IFC 84, the second paragraph of clause 2.3 attempts to put the issue beyond question by expressly empowering the architect to make an extension of time upon the occurrence, after the expiry of the date for completion or any extended time previously fixed under the clause, of an event referred to in clauses 2.4.5, 2.4.6, 2.4.7, 2.4.8, 2.4.9 or 2.4.12, all of which may be regarded as being in one way or another the responsibility of the employer. Thus, despite such an event and the timing of it, the liquidated damages provision should remain effective.

So far as the length of the extension is concerned, the words in the second paragraph, namely '. . . the length of the delay . . . resulting from that event . . .' lend support to the contention that the extension given should be the net effect on the actual completion date which the event causes and not the gross period between the pre-existing contractual or extended completion date and the date when the delaying event has expended itself. On the other hand, the gross extension argument seems more logical than the net extension argu-

ment when considering the operation of clause 2.6. Where there has
been a clause 2.6 certificate followed by a further extension of time
involving the cancellation of the original clause 2.6 certificate and
requiring the issue of a second clause 2.6 certificate to satisfy the
condition precedent of an architect's certificate of non-completion
before the employer can deduct any liquidated damages at all, the
second certificate issued under clause 2.6 would, if the net extension
argument is to prevail, have to contain a statement by the architect
that the contractor has failed to complete by an extended completion
date. This, by definition (as the contractor's culpable delay will not
have been included within the extension of time given), must be a
date earlier than that on which the delay for which the extension of
time is granted expired, and possibly even before it commenced.
When the employer then exercises his rights under clause 2.7 to
deduct liquidated damages this will in fact cover a period of time
during part of which the contractor was delayed by the employer.
Even so, on balance it appears that the probable intention behind
clauses 2.3 to 2.8 taken together favours the net extension approach.

A possible flaw in the second paragraph of clause 2.3 is the
absence of any reference to clause 2.4.15. It may have been that
when clause 2.4.15 was introduced in Amendment 4 to IFC 84 in
July 1988, the consequential drafting amendment to this paragraph
was overlooked.

The third paragraph of clause 2.3 provides for the architect, at any
time up to 12 weeks from practical completion, to make an extension
of time whether on reviewing a previous decision or otherwise, and
whether or not the contractor has given the appropriate notice under
the first paragraph of clause 2.3. This appears to give the architect
considerable discretion where the contractor has failed to give the
required notice. The architect would be prudent in seeking the
employer's authority before exercising the discretion in appropriate
circumstances. Any such extension must not reduce an extension of
time previously made. Any review can therefore only be in an
upwards direction so far as the period of extension of time is
concerned. No doubt, however, if there are grounds for making an
additional extension of time, but also factors which fairly reduce the
extension required by the contractor e.g. some variations requiring
additional work and others omitting work, the architect can look at
the net effect in order to determine the appropriate period for the
extension, provided of course it does not have the effect of reducing
the total of the extensions of time already given. If the architect fails

to make a further extension of time when one is warranted, this will not invalidate the liquidated damages provisions in the contract, though of course the contractor could seek arbitration to secure an extension of time and recover any liquidated damages deducted – see *Temloc Ltd* v. *Errill Properties Ltd* (1987).

It must be at least arguable, though not as strongly as under JCT 80, that once the 12 weeks has expired the architect no longer has any power under the contract to make further extensions of time. He appears to be ex-officio in this regard.

Clauses 2.3 and 2.4 are basically traditional in character and scope and are for the most part consistent with other standard forms of contract issued by the Tribunal. They do, however, contain some additional features and these will be dealt with under the next heading and also under the *NOTES* which follow on page 85.

In relation to the grounds for making an extension of time, the contract adopts a listing approach rather than stating a general ground for an extension of time e.g. 'for reasons beyond the control of the contractor'. This listing approach is to be preferred as the courts may be unwilling to construe a general provision for an extension of time as covering delays which are the responsibility of the employer e.g. late instructions. This could be disastrous for the employer as his entitlement to liquidated damages would be lost: *Peak Construction (Liverpool) Ltd* v. *McKinney Foundations Ltd* (1970).

If IFC 84 is compared with JCT 80, it can be seen that one of the relevant events in JCT 80 has been omitted, namely delay caused by the UK Government exercising any statutory power which directly affects the carrying out of the works by restricting availability of essential labour, goods or materials, fuel or energy in connection with the proper carrying out of the works (clause 25.4.9 of JCT 80). This provision was inserted following the 'three-day week' in the early 1970s. In certain circumstances such an event may be a *force majeure* falling within clause 2.4.1. All the other JCT 80 events are included, some of them with amendments.

THE EVENTS IN RESPECT OF WHICH AN EXTENSION OF TIME CAN BE MADE (CLAUSE 2.4)

Introduction

There are 15 clause 2.4 events listed although these break down into many more items. Those events in clauses 2.4.10, 2.4.11 and 2.4.14 do not apply unless stated to apply in the appendix.

If the employer intends to amend any of the events without this affecting the freezing of fluctuations it will be necessary to amend also, by deletion or otherwise, clause C4.7 and C4.8 of supplemental condition C or clause D13 of supplemental condition D as appropriate.

Some of the clause 2.4 events, namely those in clauses 2.4.5, 2.4.6, 2.4.7, 2.4.8, 2.4.9, 2.4.12, 2.4.14 and 2.4.15 have words which are similar or identical to some of the sub-clauses of clause 4.12 (or clause 4.11(a) in the case of 2.4.14) entitling the contractor to claim for loss and expense incurred by him as a result of one or more of those matters.

Although the wording may therefore be similar or identical, the clauses are not directly linked. If the architect makes an extension of time in respect of one of the clause 2.4 events, which has a similar or identical wording to any of the matters referred to in the sub-clauses of clause 4.12, this is evidence that the contractor has suffered delay as a result of a relevant event. This does not automatically mean that the delay has caused loss and expense. This has to be separately established – see *H. Fairweather & Co. Ltd* v. *London Borough of Wandsworth* (1987). On the other hand, the contractor may incur loss and expense by reason of one of these matters even though it does not delay the completion of the works e.g. where plant or labour is idle or under-productive in relation to a non-critical part of the works, the delay and disruption to which does not affect the overall completion date.

The Events Themselves

Clause 2.4.1

Force majeure has been described as having reference to all circumstances independent of the will of man, and which it is not in his power to control. It has a meaning wider than *vis major* (act of God). Its interpretation will be conditioned to some degree by the nature and type of the other clause 2.4 events. It may well not cover an event, however catastrophic, which has been brought about by the negligence of the contractor: see *J. Lauritzen A/S* v. *Wijsmuller BV (the Super Servant Two)* (1989) (*Times*, 17 October 1989).

Clause 2.4.2

The weather must not be just adverse, it must be exceptionally adverse. This should render it a truly unusual, if not rare occurrence. It is to be wondered, therefore, why architects appear so often to grant an extension of time under this head. Architects will often require evidence of exceptionally adverse weather conditions in the form of detailed meteorological records demonstrating the exceptional nature of the weather.

There are many debating points concerning this event. Examples are:

(a) should the contractor have to provide meteorological evidence for the previous five, ten or fifteen years in order to demonstrate that the weather conditions were exceptionally adverse?

(b) to what extent, if any, should the architect have regard not only to the period of exceptionally adverse weather conditions but any exceptionally beneficial weather conditions occurring shortly before or after which may mean that *overall* progress has not been affected by such weather? e.g. an exceptionally mild February which is particularly beneficial to the progress of the contractor's external works, during which month however there were two days of unprecedented snowfall which prevented progress being made on those two particular days.

Clause 2.4.3

The specified perils are defined in clause 8.3 (Definitions). The list includes many of the insurable risks e.g. fire, flood. It may be possible for the contractor to obtain an extension of time under this clause where the loss or damage was brought about by his own negligence or that of someone for whom he is responsible, even if this amounts to a breach of an implied term of the contract to the effect that the contractor would take reasonable care in carrying out his contractual obligations. It might have been thought that as the contractor must not benefit from his own breach of contract he might lose his entitlement to any extension of time. A contracting party cannot be seen to profit from his own breach of contract. However, in the case of *Surrey Heath Borough Council* v. *Lovell*

Construction Ltd (1988) in which it was held under a JCT 81 With
Contractor's Design Contract that the negligent destruction by fire
by a sub-contractor of part of the works was a breach by the
contractor of an implied term of the kind mentioned above, it was
nevertheless assumed by His Honour Judge Fox-Andrews QC,
though without argument on the point, that an extension of time
granted by the architect under a similar clause to that now under
consideration was appropriate.

It is to be noted that the loss or damage is not restricted to the
works. Any loss or damage caused by a specified peril which causes a
delay to completion can fall within this event. This view is supported
by the fact that the definition of 'Specified Perils' is to be found in
clause 8.3 rather than in clause 6.3.2 dealing with insurance of the
works; and by the scope of the insurance cover for lost liquidated
damages set out in clause 6.3D.1 which includes loss or damage to
temporary buildings, plant and equipment.

An argument that, in such circumstances, no extension of time
need be made because the contractor has failed to use constantly his
best endeavours to prevent delay is, it is submitted, untenable. The
contractor's best endeavours relate to preventing *the delay* rather
than preventing *the cause* of the delay.

There is provision for the employer to require the contractor to
obtain insurance for the benefit of the employer for lost liquidated
damages due to an extension of time for this event. This is discussed
later in Chapter 9 (page 327).

Clause 2.4.4

It should be noted that the effect of such matters as civil commotion,
strikes or lockouts etc. need not be direct, so long as it affects any of
the trades employed on the works or any trade engaged in the
preparation, manufacture or transportation of any of the goods or
materials required for the works. It can therefore be of extremely
wide application and virtually removes from the contractor all risks
of delay (though not of course of costs which may result from any
disruption) caused by industrial disputes and civil disobedience.
However, it is submitted that the effect on any trade in the
preparation, manufacture or transportation of any goods or materials
must be of a national, or at any rate fairly general rather than local,
nature as otherwise this sub-clause would appear to overlap 2.4.10
and 2.4.11.

Clause 2.4.5

See the specific clauses referred to for the instructions to which this clause relates.

Clause 2.4.6

This relevant event has to be considered together with clauses 3.12 and 3.13 (instructions as to inspection – tests and instructions following failure of work etc.) and clause 4.12.2 (loss and expense). So far as the application of this sub-clause is concerned, it matters not whether the instruction for opening up or testing etc. was issued under clause 3.12 or clause 3.13.1. In either case an extension of time can only be given where the inspection or test showed that the work, materials or goods were in accordance with the contract. However, oddly enough (see clause 3.13.2), it appears that an arbitrator could award an extension of time in respect of an instruction given under clause 3.13.1 even though the result of the inspection or test etc. showed that the work, materials or goods were not in accordance with this contract.

Clause 2.4.7

The time when the architect must provide the necessary instructions etc. must generally relate to the contractor's need for the same in order that he is not thereby prevented from completing by what would otherwise be the date for completion (or other later date where the event occurs after the expiry of the date for completion or extended time for completion). However, the employer's own circumstances may have some relevance (see *Neodox Ltd* v. *Swinton and Pendlebury Borough Council* (1958). In the *Neodox* case it was held to be an implied term that the necessary information etc. would be '. . . given in a reasonable time . . .' and it was further held that such an expression meant reasonable from the employer's point of view as well as from the contractor's. However, the relevant wording in the *Neodox* case was different from the wording in this sub-clause and the closing words in particular – namely, 'nor unreasonably close to the date on which it was necessary for him to receive the same' – may make it difficult for an employer to introduce his particular difficulties or problems into the equation.

Where the instructions are for or in regard to the expenditure of provisional sums this event can apply to both defined and undefined provisional sums. In the case of undefined provisional sums it may be very difficult for the contractor to know in advance when he should apply in writing.

It could happen that through no fault of the employer or the contractor at the time of the contractor's written application the instructions etc. required are bound to be received too late to be 'received in due time' in the sense that some delay to the completion date is by then already inevitable e.g. where there is a change in statutory requirements during the contract period necessitating a further instruction from the architect. The contractor may in such a case immediately apply for instructions or information on the point but some delay may be inevitable even if the architect responds immediately. However, it is submitted that the contractor cannot argue that as it was never possible to obtain the instructions or information in time for him to complete by the contractual completion date, sub-clause 2.4.7 has no application and cannot be relied upon by the architect to cover any delay. In such a case the words 'in due time' will be construed as meaning in a reasonable time in all the circumstances: see *Percy Bilton Ltd* v. *Greater London Council* (1982).

It should be noted that the timing of the application is expressed to relate to the contract completion date or extended completion date and not to any earlier completion date shown on the contractor's programme: see *Glenlion Construction Ltd* v. *The Guinness Trust* (1987).

A schedule of dates by which the information is required provided by the contractor to the architect at the commencement of the contract or early on in the contract period, while useful to the architect, does not, it is submitted, satisfy the requirements of this sub-clause, having regard to the words 'provided that such application was made on a date which ... was neither unreasonably distant from ... the date on which it was necessary ...'. A finding to the contrary in the case of the *London Borough of Merton* v. *Stanley Hugh Leach Ltd* (1985) in relation to JCT 63 clause 23, was based upon different wording and was in any event a surprising finding on the wording of sub-clause 23(f) of JCT 63, even if it had the virtue of according with common sense.

Clause 2.4.8

This clause is to be read in conjunction with clause 3.11 (work not forming part of the contract). If it is intended at the outset that such other work shall be undertaken during the contract period, as much information as possible should be supplied to the contractor prior to his tendering. A certain amount of disturbance and disruption can then be allowed for by the contractor in his programming and tender price. In such a case it is only where the extent of the delay could not reasonably be foreseen by an experienced contractor that the architect should make an extension of time.

Sometimes local authorities and other public bodies employ their own direct labour organisation to carry out work. If that work is kept separate from the work for which the contractor is responsible under IFC 84 i.e. if the work does not form part of the contract, this sub-clause will apply. However, if their work forms part of the main contract work under IFC 84 (often the direct labour organisation is engaged by the contractor under a notional 'sub-contract'), this sub-clause has no application. In legal terms the public body concerned will be both the employer under the main contract for the works as a whole but also a sub-contractor to the contractor for the relevant sub-contracted part of the works.

If the public body, in breach of its sub-contract, delays the contractor, there appears to be no clause 2.4 event which readily covers the situation. Two possibilities arise. Firstly, if the contractor is thereby put in breach of the main contract and becomes liable for liquidated damages to the public body as employer, the contractor will have a right of recourse against the public body as sub-contractor. This will lead to what is known as circuity of action and could be used to defeat the employer's claim for liquidated damages. Secondly, and more probably, the result will be that the public body as employer will have prevented (even if in its sub-contract role) the contractor from completing by the contractual completion date and as no appropriate event exists in clause 2.4 to cover the situation the employer will lose his right to claim liquidated damages.

Clause 2.4.9

The effect of a failure by the employer to supply or of a delay by the employer in supplying materials and goods is likely to have a significant impact on the contractor's programme – more so than delays by the employer or others on his behalf in relation to the execution of work. Subject to this important point, similar comments can be made of this clause as have been made in respect of clause 2.4.8 except that if it is a public body employer's direct service organisation which has entered into a notional 'supply' contract with the main contractor (in law it will be a supply contract between the public body and the contractor), this sub-clause will apply as it relates to the supply of materials or goods for the works themselves. Hence if it is a supply only rather than a supply and fix situation, an extension of time can be made and the liquidated damages provision will be preserved.

Clause 2.4.10

This sub-clause applies only where it is stated in the appendix to apply. Some employers are of the view that the risk of delay due to such an event as this is properly a contractor's commercial risk. Others may be of the opinion that including such a provision is more likely to produce comparable tenders as the need to price the risk is removed from the contractor, and if the event does not materialise the employer has not had to pay against the risk, whereas this might happen if the clause were deleted. Much may depend upon the state of both the construction and labour markets.

Is the contractor expected to incur extra cost to overcome the problem e.g. paying exceptionally high wages to attract skilled labour where there is a shortage of it? No doubt it is a matter which the architect will take into account in determining whether or not the contractor has used his best endeavours to prevent delay. It is submitted that the contractor would not be expected to incur considerable expenditure in order to overcome the problem.

What if the contractor obtains specialist sub-contract labour which produces defective work which no amount of supervision by the contractor could have avoided? Could the contractor contend that he has been unable 'for reasons beyond his control . . . to secure such labour as is essential to the proper carrying out of the Works'? It would be surprising if such an argument was to succeed, as this sub-clause has generally been treated as relating to shortages of labour rather than to defective work. That said, it is by no means certain that the argument would fail: see the House of Lords case of *Scott Lithgow* v. *Secretary of State for Defence* (1989).

Clause 2.4.11

Similar comments can be made of this clause as have been made of clause 2.4.10.

Clause 2.4.12

It should be appreciated that this event has nothing to do with the employer giving possession of the site to the contractor as such. It is restricted to the means of access to, or egress from, the site over other connected or adjoining property or way in the possession and control of the employer.

Clause 2.4.13

Similar wording in JCT 63 has received some detailed judicial consideration: see *Henry Boot Construction Ltd* v. *Central Lancashire New Town Development Corporation* (1980) and Note [22] on page 93.

Clause 2.4.14

This sub-clause applies only where stated in the appendix to apply. It should be read in conjunction with clause 2.2. It can only operate if the date of possession in the appendix is correctly completed either with a specific date inserted or alternatively, but less satisfactorily, a stated number of days or weeks from the date of acceptance of the tender. If the appendix carries some such phrase as 'on a date to be agreed' then it is vital that the date is agreed before the tender is accepted. To leave agreement until afterwards runs the risk of failure to agree a specific date with a possible invalidation of the liquidated damages provision.

This event is inserted very much for the employer's benefit in order to prevent an employer's delay from invalidating the liquidated damages provision. If it is known before the tender is accepted that there is going to be a delay in giving possession of the site to the contractor, then a new date for possession can be reached by agreement and there will be no place for this provision to operate. It applies only where, after a binding contract has been formed, the employer is unable to give possession of the site in accordance with clause 2.1. As this sub-clause refers to the deferment in giving possession of the site, it may well not include a deferment by the employer in giving possession of a part only of the site rather than a deferment in relation to the whole of it. If this is the case, then a deferment in relation to part of the site would amount to a breach of contract by the employer, at any rate where that part was required by the contractor in order for him to commence work on site.

Clause 2.4.15

Where bills of quantities are a contract document, they are to be prepared in accordance with SMM 7 which recognises that bills drawn in accordance with the standard method can include approximate quantities which are the subject of a measure when the work is carried out. This is an exception to the lump sum principle of IFC 84, which does not in itself require the quantities upon which the original unadjusted contract sum is based to be actually measured as completed work. This is considered again in the commentary to clause 3.7 (page 176).

If upon measuring the work for which an approximate quantity is included in the bills, it is found that the approximate quantity stated is not a reasonably accurate forecast of the work as measured, this sub-clause applies. What is reasonably accurate will depend upon all the circumstances.

CERTIFICATE OF NON-COMPLETION: PAYMENT AND REPAYMENT OF LIQUIDATED DAMAGES

Clause 2.6

The first paragraph makes it clear that the architect has a duty, not a discretion, as to the issue of a certificate of non-completion once the date for completion or any extended time has expired.

However, this certificate is subject to cancellation where, after its issue, a further extension of time is made whether or not the event occurred before the date stated in the certificate (e.g. on a reconsideration of a previous extension of time based on an estimate) or afterwards in respect of those events referred to in the second paragraph of clause 2.3. There is no provision for fixing an earlier date for completion than an existing one even where a subsequent event reduces the overall delay e.g. a variation instruction requiring an omission. If there are subsequent events causing delays as well as variations by way of omission then the architect would be acting reasonably in making an extension of time in respect of the net delay only.

If an extension of time is made after the certificate has been issued and the architect fails to cancel the certificate and, if appropriate, issue a further updated certificate in its place, any deduction of liquidated damages by the employer will be wrongful, even if limited to the actual period of culpable delay between the revised completion date and practical completion: see *A. Bell and Son (Paddington) Ltd* v. *C.B.F. Residential Care and Housing Association* (1989).

Clause 2.7

The benefit to the employer of a liquidated damages clause has been discussed elsewhere (page 57). On occasions, where the figure inserted in the appendix in respect of liquidated damages is less than the actual loss suffered by the employer as a result of the contractor's breach of contract in failing to meet the completion date or any extended time for completion, the liquidated damages clause in effect operates as a limitation of liability in favour of the contractor. This is not always appreciated.

Clause 2.8

This clause provides for the repayment of liquidated damages where, subsequent to their deduction or recovery, the time for completion is extended or further extended. It is submitted that any claim by the contractor for interest on payment or repayment of liquidated damages by the employer would be likely to fail. The contract envisages the deduction followed by repayment where appropriate. The initial deduction cannot therefore subsequently be regarded as a breach of contract by the employer and the contract contains no express provision for interest in these circumstances. Whilst the decision of an Irish court in the case of *Department of Environment for Northern Ireland* v. *Farrans (Construction) Ltd* (1981) (a decision on JCT 63) may appear to indicate that interest is recoverable on repaid liquidated damages, JCT 63, unlike IFC 84, does not expressly provide for repayment of liquidated damages. Furthermore, the decision in this case has received some critical comment (see the commentary by the learned editors of *Building Law Reports* at 19 BLR page 3 *et seq*).

NOTES TO CLAUSES 2.3 TO 2.8

[1] '. . . *progress of the Works is being or is likely* . . .'.
The contractor is required to give notice in respect of the progress of the works being delayed or of the likelihood of their being delayed and from whatever cause. This gives the architect the opportunity of taking action to negate or reduce the effect of the delay. It need not relate to a relevant event. It should be noted that what is required is

notice of the cause of the delay and not an application for an extension of time by the contractor as such. Also, it is delay to the progress of the works rather than delay pushing the works beyond the date for completion which is the key to the giving of the notice.

[2] '. . . forthwith . . .'.
This must mean immediately progress of the works is or is likely to be delayed. Is the requirement that the contractor's notice should be given forthwith a condition precedent to the operation of the clause? It is suggested not – certainly, in relation to those clause 2.4 events which are inserted primarily for the benefit of the employer. To construe the requirement for notice as a condition precedent would, in the absence of due notice, prevent the architect from making an extension of time and thus invalidate the liquidated damages provision. This cannot be the intention of clause 2.3. Furthermore, the third paragraph of clause 2.3 appears to give the architect a discretion to extend time in the absence of the notice.

In the case of *London Borough of Merton* v. *Stanley Hugh Leach Ltd* (1985) it was held, under a JCT 63 contract, that the notice of delay required from the contractor by that contract's extension of time clause (clause 23) was not a condition precedent to the contractor's entitlement to an extension of time. However, the failure to give notice could well be a breach of contract by the contractor. If as a result either the architect is prejudiced in his ability to make an accurate assessment of the delay, e.g. if his opportunity to investigate the matter is now limited, or if the architect has lost the chance to issue appropriate instructions, or to take other measures which could have reduced the overall delay, then this will no doubt be taken into account by the architect when assessing what amounts to a fair and reasonable extension of time.

[3] '. . . is likely . . .'.
The architect must grant a fair and reasonable extension of time for completion of the works if he is of the opinion that completion has been or is likely to be delayed. Clearly, when the delay is merely anticipated rather than actual, the assessment of an appropriate extension of time may be somewhat speculative. Nevertheless, as soon as he is able to make a reasonable estimate of the length of delay he must make the extension of time. The estimate made can be reviewed under the provisions of the third paragraph of clause 2.3 during the progress of the works or at any time up to 12 weeks after the date of practical completion.

[4] '. . . *by any of the events* . . .'.
There is absolutely no power under this contract for the architect to
make an extension of time in respect of a cause of delay which does
not fall within the clause 2.4 events listed.

Where the same period of delay is attributable to both a clause 2.4
event and some non-qualifying event, then it is reasonable, it is
submitted, for the architect to grant an extension of time. The
reasoning for this is as follows: liquidated damages clauses are
generally construed strictly against the employer – see *Peak Construction (Liverpool) Ltd* v. *McKinney Foundations Ltd* (1970). It is for
the employer to claim his entitlement to liquidated damages. In
order to do so he must, it is submitted, be able to demonstrate that,
but for the contractor's culpable delay (i.e. a delay for which no
extension of time event is applicable), the employer would have got
his building on time. Now if there is a genuinely concurrent delaying
event for which the contract provides an extension of time, running
alongside the contractor's culpable delay, it follows that even if the
contractor had not been at fault the employer would still not have
got his building on time due to a delay for which the contractor is
entitled to an extension of time and thus to relief from liability to
pay liquidated damages. This ought clearly to be the position where
the concurrent delay is one of those events in clause 2.4 for which the
employer is responsible. It seems likely that it should also be the
position where the concurrent delay is of the 'neutral' kind, though
the argument here is not as strong as the employer will not have
contributed to the delay.

The situation could well be reversed in relation to claims by the
contractor for reimbursement of direct loss or expense where there
are concurrent causes one of which is a clause 4.12 matter and the
other a matter for which the contractor takes the commercial risk, as
it is the contractor in this instance who is in the role of claimant and
has to prove his case.

Having said all this, there is some authority for saying that neither
delay should dominate the other and that some sort of apportionment is required. See *H. Fairweather & Co. Ltd* v. *London Borough of
Wandsworth* (1987). In practice it will often be impossible for the
architect to do this on any rational or certain basis.

[5] '. . . *the Architect/the Contract Administrator shall* . . .'.
Failure to do so if wilful could well prevent the employer from being
able to operate clause 2.7 to deduct liquidated damages. Further, the
freezing effect of the fluctuations provisions at the date for comple-

tion or any extended time for completion will not operate (see clause 4.9 and supplemental condition C (clause C4.7 and C4.8) and supplemental condition D (clause D13)).

If the architect's failure is not wilful it is possible that the employer's right to deduct liquidated damages may still remain on the basis that the architect's failure to make the assessment '*so soon as he is able . . .*' relates to a matter of administration rather than of substance. It should be noted however that in the case of *Temloc Ltd* v. *Errill Properties Ltd* (1987), it was, exceptionally, the employer who was claiming that the right to liquidated damages was lost due to the architect's failure to comply with the time-table for determining extensions of time under a JCT 80 contract. If it had been the contractor so claiming the decision might just have gone the other way. Of course, if the contractor is aggrieved that an extension of time to which he is entitled has not been granted this can be the subject of a challenge in arbitration proceedings.

[6] '*. . . in writing . . .*'.
The architect should make it absolutely clear that he is making an extension of time and should state the extended time for completion of the works.

[7] '*. . . fair and reasonable . . .*'.
The decision of the architect is reviewable by an arbitrator. The architect must take into account the proviso in the fourth paragraph of clause 2.3, namely, that the contractor shall constantly use his best endeavours to prevent delay.

[8] '*. . . an event referred to in clause 2.4.5, 2.4.6, 2.4.7, 2.4.8, 2.4.9, 2.4.12 or 2.4.15 . . .*'
For the suggested reasoning behind this paragraph see earlier page 72.

[9] '*. . . At any time up to twelve weeks . . .*'.
Presumably, by implication, the architect has no power to operate this paragraph after the twelve-week period has expired, though it appears that under a similar though not identical clause in JCT 80, clause 25.3.3, the court found otherwise: see *Temloc Ltd* v. *Errill Properties Ltd* (1987).

[10] '... *may* ...'
The exercise by the architect of the power to make further extensions of time under this paragraph of clause 2.3 is thus not mandatory. This is to be compared with the somewhat similar provision in JCT 80 clause 25.3.3 which uses the word '*shall*'.

[11] '... *or otherwise* ...'.
These words do not mean that the contractor is entitled to an extension of time for delaying events which occur between the date for completion or any extended time for completion and the date of practical completion, unless such events fall within those listed in the second paragraph to clause 2.3. The events there listed are of course inserted for the benefit of the employer.

[12] '*Provided always that the Contractor shall use constantly his best endeavours to prevent delay and shall do all that may be reasonably required to the satisfaction of the Architect/the Contract Administrator to proceed with the Works*'.
This paragraph contains the normal type of proviso to be found in extension of time provisions. In the event of delay occurring, it is clear that the contractor is not permitted to sit back and let the delay take its natural course. He must take such positive steps as are available to prevent delay. In any event this will usually be in his own best interests, especially in those situations where any disruption costs will have to be borne by him.

The proviso has two strings to it. Firstly, the constant use by the contractor of his best endeavours: this is independent of any instructions of the architect. Secondly, the doing of all that may be reasonably required to the satisfaction of the architect to proceed with the works: this enables the architect to issue instructions as to how the delay should be overcome or reduced. This is qualified by the use of the word 'reasonably'.

[13] '... *best endeavours* ...'
This is often a difficult phrase to interpret in a given context. What is the position, for instance, if the contractor could take certain very expensive measures to prevent delay e.g. importing labour to avoid the effect of a widespread strike in a particular trade. Is he bound to do so? It is submitted that there is no requirement for the contractor to commit himself to significant expenditure even though the word

used is 'best' and it is nowhere qualified in terms of the cost involved.

If the contractor fails to use his best endeavours to prevent the delay then he will be liable to pay or allow to the employer liquidated damages.

If the clause 2.4 event was one which also materially affected the regular progress of the work and so could form the basis of a claim by the contractor for reimbursement of direct loss and expense under clauses 4.11 and 4.12, could any of the extra cost incurred by the contractor in using his best endeavours to prevent delay become part of that claim for reimbursement? There seems no reason in principle why not, if the steps taken were reasonable.

It is to be hoped that the words 'best endeavours' will be construed relatively. It would have been preferable from the contractor's point of view if the words 'reasonable endeavours' had been used.

[14] 'The Contractor shall provide such information required by the Architect/the Contract Administrator as is reasonably necessary for the purposes of clause 2.3.'
This requirement for the contractor to provide information is worded very generally. Clearly it will cover matters such as an assessment of the length of any delay already suffered, together with an estimate of the extent of any expected delay. It is submitted that the contractor will also have to give information as to the expected effects of the delay where this is reasonably required by the architect in order that he can determine what may be reasonably required of the contractor to proceed with the works. It may also be reasonably necessary for the architect to have this information in order to determine whether or not the contractor has used his best endeavours to prevent delay.

As to the extent and detail of the information required from the contractor, the case of *London Borough of Merton* v. *Stanley Hugh Leach* (1985) is instructive. Though considering the requirement for a contractor to provide information relating to a claim for reimbursement of direct loss and expense rather than an extension of time, the summary of Mr Justice Vinelott in this case is still apposite when he said at 32 BLR p. 97/98:

'But in considering whether the contractor has acted reasonably and with reasonable expedition it must be borne in mind that the architect is not a stranger to the work and may in some cases have a very detailed knowledge of the progress of the work and of the contractor's planning. Moreover it is always open to the architect to call for further information either before or in the course of investigating a claim. It is possible to imagine circumstances where the briefest and most uninformative notification of a claim would suffice: a case, for instance, where the architect was well aware of the contractor's plans and of a delay in progress caused by a requirement that the works be opened up for inspection but where a dispute whether the contractor has suffered direct loss or expense in consequence of the delay had already emerged. In such a case the contractor might give a purely formal notice solely in order to ensure that the issue would in due course be determined by an arbitrator when the discretion would be exercised by the arbitrator in place of the architect.'

[15] '. . . adverse . . .'.
This can include hot and dry as well as cold and wet weather conditions.

[16] '. . . except . . . for the expenditure of a provisional sum for defined work . . .'
This is because if the provisional sum is for defined work it is deemed to have been taken into account by the contractor in programming and planning the works so as to complete them by no later than the contractual completion date – see SMM 7 General Rule 10.4. General Rule 10.4 is set out as part of the footnote to clause 8.3 Definitions (see page 375). See also the discussion on this matter earlier on page 39.

[17] '. . . for which he, specifically applied in writing . . .'.
What if the contractor does not so apply? Is the architect empowered nevertheless to make an extension of time, remembering that this clause is principally for the employer's benefit to keep his entitlement to liquidated damages intact? If no extension of time can be made, does the fixed date for completion remain intact so that the employer can deduct liquidated damages or is completion to be within a reasonable time instead? It is tentatively submitted that unless and until there is a specific application in writing in accordance with the clause, any delay by the architect in relation to the matters stated in this event will, except in an extreme case, be at the risk of the

contractor who will not be relieved of his liability to pay liquidated damages.

[18] '... unreasonably close ...'
If the contractor delays in requesting instructions etc. then this could render his application unreasonably close to the date on which it was necessary for him to receive the same. If progress is thereby held up by the absence of such instructions etc. the contractor will be the author of his own misfortune (even if, it is submitted, the hold up would have occurred despite the written application owing to the inability of the architect to issue the instructions etc. at the time when it was necessary for the contractor to receive the same). See Note [17] above.

[19] '... in due time ...'
If a time is agreed or clearly indicated on an agreed contractor's programme this will no doubt be the due time, other things being equal. If no precise time is agreed in advance, then access or egress must be given in reasonable time taking into account the overall situation including, amongst other matters, the employer's position.

[20] '... in the possession and control ...'
It is submitted that the employer must have not only the right to possession but also the *de facto* control of the land, buildings, way or passage. Presumably some form of shared control is sufficient e.g. a joint tenancy.

[21] '... as otherwise agreed ...'
The question of access and egress may be dealt with in the contract documents or otherwise agreed. If dealt with in the contract documents, the agreement is clearly directly between the employer and the contractor. If otherwise agreed, the clause suggests that the agreement is between the architect and the contractor with the former clearly acting as the agent of the employer. The architect should therefore check his own authority to make sure that he is permitted to bind the employer in this fashion. As the contract clearly bestows such power on the architect, the contractor need not concern himself with checking the authority of the architect unless perhaps he has knowledge that the architect has limited actual authority.

[22] '... *its statutory obligations in relation to the Works* ...'.

In reality, the carrying out by a statutory body of work strictly in pursuance of its statutory obligations will often be very limited. Cases can arise where part of the work is carried out pursuant to statutory obligations and part not. If the statutory body carries out work otherwise than pursuant to its statutory obligations, this will sometimes take the work outside the contract so that clause 2.4.8 will apply. On other occasions the work will remain part of the contract works and will be carried out under a sub-contract between the main contractor and the statutory body concerned. If the work is carried out by way of a domestic sub-contract, then the contractor takes the risk of delay in respect of non-statutory work. However, where such work is covered by a provisional sum item, then although it forms part of the contract in that the contract sum includes the value of provisional items, it was nevertheless held in the case of *Henry Boot Construction Ltd* v. *Central Lancashire New Town Development Corporation* (1980) that such work did not form part of the contract works for which the contractor was responsible under JCT 63 as the employer entered into separate contracts with the statutory undertakers. In such cases the contractor will be given an extension of time under clause 2.4.8 and will be entitled to reimbursement of loss and expense under clause 4.12.3 if the regular progress of the works has been materially affected.

[23] '... *the deferment* ...'.

This must relate to the delay in giving the initial possession of the site. The wording is not appropriate to cover interruptions in possession of the site or parts of it once the contractor has obtained possession. Such interruptions should be covered wherever possible by an instruction as to postponement of work pursuant to clause 3.15.

[24] '... *possession of the site* ...'.

These words are used also in clause 2.1. In the context of clause 2.1 they appear to refer to the whole of the site and cannot easily be read to mean part only of the site. There seems no reason to give the words any different meaning in this sub-clause. Accordingly the deferment cannot relate to part only of the site.

[25] '... *written cancellation* ...'.

This will signal the operation of clause 2.8 (repayment of liquidated damages).

[26] '. . . as may be necessary.'
If the extension of time which has led to the written cancellation brings the extended time for completion in line with the date of practical completion, there will be no further certificate. If the reason for the cancellation also extends the time for completion to a date ahead of the date of issue of the written cancellation, then again no further certificate will be necessary at that time, though it may be later on, should the date of practical completion be later than the extended time for completion. This requirement for a written cancellation is a logical step which, in JCT 80, is left to implication – see the case of *A. Bell & Son (Paddington) Ltd* v. *C.B.F. Residential Care and Housing Association* (1989).

[27] 'Subject to the issue of a certificate under clause 2.6 . . .'.
This is a condition precedent to the right to deduct liquidated damages (see *Amalgamated Building Contractors Ltd* v. *Waltham Holy Cross UDC* (1952); *Token Construction Co. Ltd* v. *Charlton Estates Ltd* (1973)). The right to deduct may be lost if the architect wilfully fails to perform his functions in relation to the proper consideration of clause 2.4 events and the making of extensions of time as this would affect the validity of the certificate. However, the fact that the architect is guilty only of tardiness in his consideration of possible extensions of time will not of itself cause an existing clause 2.6 certificate to be invalid: see the case of *Temloc Ltd* v. *Errill Properties Ltd* (1987).

[28] '. . . as the Employer may require in writing . . .'.
This appears to be another condition precedent to the employer's right to claim liquidated damages under this clause. If, following the employer's requirement in writing that liquidated damages are to be deducted or claimed the architect makes further extensions of time, the employer's written requirement will fall along with the architect's existing clause 2.6 certificate. Before the employer can then deduct or claim any liquidated damages at all he must make a further written requirement to tie in with the architect's replacement clause 2.6 certificate. See the case of *A. Bell & Son (Paddington)* v. *CBF Residential Care and Housing Association* (1989).

[29] '. . . not later than the date of the final certificate . . .'.
The contractor is entitled to know before the other financial aspects of the contract are concluded whether or not it is the employer's

intention to deduct or recover any liquidated damages to which he may be entitled. Employers may seek to guard against the loss of their liquidated damages through oversight, clerical error or sheer administrative incompetence, by stating in the tender documents that they intend to deduct liquidated damages. Such a step is certainly not within the spirit or intention of this clause and is almost certainly ineffective.

[30] '... at the rate stated in the Appendix ...'
In the case of *Temloc Ltd* v. *Errill Properties Ltd* (1987) under a JCT 80 contract the appendix entry was '£nil'. This was held to be fully binding upon the employer and was treated as exhaustive of the employer's remedies against the contractor for delay in completion.

[31] '... due or to become due to the Contractor under this Contract ...'.
The employer may also be able to deduct the liquidated damages from any other money due to the contractor from the employer e.g. under another closely connected contract. This will be in accordance with the employer's general legal right to set-off and counterclaim. It is a right under the general law which applies in certain situations.

Clauses 2.9 and 2.10

[h]   Practical Completion

2.9   When in the opinion of the Architect/the Contract Administrator Practical Completion of the Works is achieved he shall forthwith issue a certificate to that effect. Practical Completion of the Works shall be deemed for all the purposes of this Contract[1] to have taken place on the day named in such certificate.[2]

Defects liability

2.10   Any defects, shrinkages or other faults which appear[3] and are notified by the Architect/the Contract Administrator to the Contractor not later than 14 days[4] after the expiry of the defects liability period named in the Appendix from the date of Practical Completion, and which are due to materials or workmanship[5] not in accordance with the Contract or frost occurring before Practical Completion[6] shall be made good by the Contractor at no cost to the Employer unless the Architect/the Contract Administrator with the consent of the Employer[7] shall otherwise instruct; and if the Architect/the Contract Administrator does so otherwise instruct then an appropriate deduction[8] in respect of any such defects, shrinkages or other faults not made good shall be made from the Contract Sum.

The Architect/The Contract Administrator shall, when in his opinion the Contractor's obligations under this clause 2.10 have been discharged, issue a certificate to that effect.[9]

---

[h]   If provision is required for Partial Possession before Practical Completion see Practice Note on IFC84:IN/1.

COMMENTARY ON CLAUSES 2.9 AND 2.10

For a summary of the general law in relation to practical completion see page 59.

Clause 2.9 does not expressly deal with those outstanding items of work or minor defects which exist but which the architect regards as insufficient to hold up the issue of a certificate of practical completion. In practice, the contractor will be asked to deal with such matters during the defects liability period. As it could be argued by a contractor that clause 2.10 does not cover items of outstanding or defective work which were known to exist at practical completion, it is advisable for the architect in such cases to refer to such items by way of list or other suitable record and to obtain the contractor's confirmation as to how they will be dealt with before actually issuing the certificate of practical completion (but see *[3]* below under Notes to clauses 2.9 and 2.10).

This clause in no way affects the employer's right to recover under the general law the consequential losses flowing from defective work. Furthermore, even after the expiry of the defects liability period, the contractor still has a legal liability in respect of defective work, subject to the effects of the issue of a final certificate (see clause 4.7).

Clause 2.10 deals with the contractor's obligations during the defects liability period. It requires the contractor to make good any defect, shrinkages or other faults which appear and are notified by the architect to the contractor not later than 14 days after the expiry of the defects liability period named in the appendix. If no period is stated in the appendix then it is to be treated as being six months from the day named in the certificate of practical completion.

For a general discussion on defects liability clauses see page 59.

NOTES TO CLAUSES 2.9 AND 2.10

[1]   '. . . for all the purposes of the contract . . .'.

The achieving of practical completion has some very important contractual consequences as follows:

(i)   Within fourteen days the architect must issue a certificate including 97.5 per cent of the value of work as against a previous certified payment of 95 per cent which thus effectively releases one moiety of retention. See clause 4.3.

(ii)   The defects liability period commences on the day named in the certificate in accordance with clause 2.10.

(iii)   The procedures leading up to the issue of a final certificate commence. See clauses 4.5 and 4.6.

(iv)   The liquidated damages payable for delay in completion in accordance with clause 2.7 will stop on the day named in the certificate of practical completion.

(v)   The contractor's risk in respect of loss or damage to the works will switch to the employer on this date. Up until practical completion the contractor will generally be responsible for the risk of damage or loss to the works by virtue of his general obligation to complete in clause 1.1. However, in respect of certain risks, the contractor will not be liable if clause 6.3B or clause 6.3C has been chosen whereby the risk of loss or damage to the works for the events described will fall upon the employer. The contractor's obligation to insure, where clause 6.3A is selected, will usually cease at this date.

(vi)   This date marks the time beyond which variation instructions may not be validly given (except possibly in relation to items of outstanding work).

(vii)   This date is usually the last date on which time will run under the Limitation Acts for actions for breach of contract (except in relation to outstanding items completed or defects remedied during the defects liability period, in which case the date of their completion or remedy will be the appropriate date).

[2] '. . . the day named in such certificate.'
It is the day named in the certificate which is relevant and not the date of the certificate itself. There is no need therefore for the architect to issue a certificate bearing a retrospective date.

[3] '. . . which appear . . .'.
This clause does not as it might have done (see, e.g. JCT 80 clause 17.2) go on to refer to defects etc which appear during the defects liability period i.e. after practical completion. It may just be open to the employer therefore to contend that defects etc. which are known to exist at the time of practical completion must be remedied by the contractor under this clause. The position is not however clear.

[4] '. . . not later than 14 days . . .'
The notification can be at any time from the date of practical completion to 14 days after the expiry of the defects liability period. It can be notification from time to time during this period. There is no express provision for a schedule of defects although in practice one will often be prepared at the end of the defects liability period. The architect clearly has an important duty to his client to ensure that the works are thoroughly examined and tested as appropriate before expiry of the defects liability period (literally up to 14 days after that date) and that these are notified to the contractor.

[5] '. . . materials or workmanship . . .'
It should be noted that there is no reference to defects in design. These are not of course the contractor's responsibility. The remedying of such defects falls outside this clause and poses problems generally e.g. a variation instruction for remedial works owing to defective design cannot be given after the date of practical completion.

[6] '. . . occurring before Practical Completion . . .'
In relation to damage caused by frost, this must relate to injury or potential injury suffered during the contract period but which reveals

itself during the defects liability period. Frost damage caused after practical completion will be at the employer's risk.

[7] '. . . with the consent of the employer . . .'
After practical completion the architect has no inherent power under this clause to accept work not in accordance with the contract or to omit work required by the contract documents. If the employer is to accept something different from that for which he contracted, then he must consent to this. No doubt this provision will prove very useful where the defect is either irremediable or would involve unreasonable cost to put right e.g. if it would involve the undoing and redoing of a considerable amount of good work.

[8] '. . . an appropriate deduction . . .'
It was thought that in some of the JCT contracts e.g. JCT 80, that the architect in exercising his power to 'otherwise instruct' could produce a result whereby the defect could be made good at the employer's expense. It is unlikely that this was ever intended and this clause now makes it clear beyond doubt that it is only an appropriate deduction from the contract sum which is permitted in cases where the contractor is relieved from his obligation to make good the defects etc.

An appropriate deduction should generally, it is submitted, be based upon what would have been the cost to the contractor of remedying the defects etc., excluding for this purpose any further saving to the contractor as a result of his not having to incur related costs e.g. removal of good work and its replacement which is necessary in order to remedy defective work, or meeting the employer's dislocation costs. There should be no ransom element in the valuation of an appropriate deduction. The valuation of the appropriate deduction is taken into account by way of an adjustment to the contract sum thus ensuring that it is the quantity surveyor rather than the employer who determines what it should be – see last paragraph of clause 4.5.

[9] '. . . issue a certificate to that effect.'
For the relationship between this certificate and the release of the last moiety of retention and the issue of the final certificate see clause 4.6.

Optional clause 2.11

Partial Possession by the Employer

2.11   If at any time or times before the date of issue by the Architect/the Contract Administrator of the certificate of Practical Completion the Employer wishes to take possession of any part or parts of the Works and the consent of the Contractor (which consent shall not be unreasonably withheld) has been obtained then, notwithstanding anything expressed or implied elsewhere in this Contract, the Employer may take possession thereof. The Architect/the Contract Administrator shall thereupon issue to the Contractor on behalf of the Employer a written statement identifying the part or parts of the Works taken into possession and giving the date when the Employer took possession (in clauses 2.11, 6.1.4 and 6.3C.1 referred to as 'the relevant part' and 'the relevant date' respectively); and

– for the purpose of clause 2.10 *(Defects liability)* and 4.3 *(Interim payment)* Practical Completion of the relevant part shall be deemed to have occurred and the defects liability period in respect of the relevant part shall be deemed to have commenced on the relevant date;

– when in the opinion of the Architect/the Contract Administrator any defects, shrinkages or other faults in the relevant part which he may have required to be made good under clause 2.10 shall have been made good he shall issue a certificate to that effect;

– as from the relevant date the obligation of the Contractor under clause 6.3A or of the Employer under clause 6.3B.1 or clause 6.3C.2 whichever is applicable, to insure shall terminate in respect of the relevant part but not further or otherwise; and where clause 6.3C applies the obligation of the Employer to insure under clause 6.3C.1 shall from the relevant date include the relevant part;

– in lieu of any sum to be paid or allowed by the Contractor under clause 2.7 *(Liquidated damages)* in respect of any period during which the Works may remain incomplete occurring after the relevant date there shall be paid or allowed such sum as bears the same ratio to the sum which would be paid or allowed apart from the provisions of clause 2.11 as the Contract Sum less the amount contained therein in respect of the said relevant part bears to the Contract Sum.

The consequential amendments are:

Amendment

6.1.3   Line 1 **delete** 'The' **insert** 'Subject to clause 6.1.4 the'.

Add as clause 6.1.4:

'6.1.4 If clause 2.11 has been operated then, in respect of the relevant part, and as from the relevant date such relevant part shall not be regarded as 'the Works' or 'work executed' for the purpose of clause 6.1.3.'

6.3.3 Line 19, after 'existing structures' **insert** '(which shall include from the relevant date any relevant part to which clause 2.11 refers)'.

6.3A.1 Line 7, after 'shall' **insert** '(subject to clause 2.11)'.

6.3B.1 Line 6, after 'and shall' **insert** '(subject to clause 2.11)'.

6.3C.1 Line 3, after 'structures' **insert** '(which shall include from the relevant date any relevant part to which clause 2.11 refers)'

6.3C.2 Line 6, after 'and shall' **insert** '(subject to clause 2.11)'.

6.3D.3 Line 3, after '6.3D.2' **insert** '(or any revised sum produced by the application of clause 2.11)'.

COMMENTARY ON OPTIONAL CLAUSE 2.11 AND CONSEQUENTIAL AMENDMENTS

The appendix to Practice Note IN/1 sets out an extra clause, namely 2.11, for use where the employer with the contractor's consent wishes to take possession of part or parts of the works before completion of the whole. If it is to be used, clause 2 of IFC 84 must clearly be amended by the addition to it of clause 2.11.

Where it is intended before the contract is entered into that completion of parts of the works is required before other parts, this cannot be satisfactorily achieved by using clause 2.11. The IFC 84 Sectional Completion Supplement should be used in such a case (this is dealt with under the next heading below page 103).

Clause 2.11 enables the employer, with the contractor's consent, to take possession of part only of the works which part, once taken over, must be made the subject of a written statement issued by the architect on behalf of the employer identifying the part taken over ('the relevant part') and specifying the date when it was taken over ('the relevant date').

The contractor's consent must be obtained and must not be unreasonably withheld. In most instances the contractor will be prepared to consent as this has certain advantages as will be seen below when looking at the contractual effects of partial possession. However, occasionally there may be reasonable grounds upon which the contractor can withhold consent, e.g. where consenting to partial possession would result in difficulties of access or working which would inevitably cause the contractor to overrun on the remainder of the contract into a period of time when his resources were to be fully committed elsewhere.

The three key elements are, firstly, the identity of the relevant part; secondly, the valuation of the relevant part, and thirdly, the relevant date.

In deciding whether to seek partial possession the employer and the architect must carefully consider the practicability of it. While it may pose few problems in some situations, e.g. individual housing or industrial units, it may raise very considerable difficulties in others. Of particular importance are common services such as light, heat and power where to take over part of a building may require taking over the common services for the whole. Examples where there could be potential difficulties in relation to common parts are central staircases or roof structures in relation to blocks of flats. Even in relation to individual units there could be common roads, accesses etc. to be considered. If only part of a common service is to be taken over there may be difficulties both in defining and valuing it.

Once taken into possession the relevant part is deemed to be practically completed. Great care is therefore required to identify the relevant part precisely.

The architect should consider the possibility in appropriate circumstances of utilising clause 2.1 where the requirement is for storage or very limited use or occupation as, if clause 2.1 is used, this will not be deemed to amount to practical completion – see page 62.

The precise identity of the relevant part can be vital as the defects liability period under clause 2.10 operates in respect of the relevant part from the relevant date. For example, the whole of the heating system may have to be taken over to obtain beneficial use of part. All of the relevant part will be subject to the defects liability period. This could prejudice the employer if it turns out that there is a defect in that part of the heating system not utilised.

The architect must issue a certificate of making good defects for the relevant part when in his opinion any defects, shrinkages or other faults have been made good pursuant to clause 2.10.

The valuation of the relevant part and the fixing of the relevant date are important for a number of reasons including:

(i)   One half of the retention is released for the relevant part as it is deemed to be practically completed;

(ii) The insurance of the works obligations under clause 6.3A, 6.3B or 6.3C ceases. Where clause 6.3C operates the employer's obligation to insure existing structures will extend to the relevant part as from the relevant date. Strangely enough, where clause 6.3B applies and the employer takes part into possession there is no requirement for him to insure it against specified perils;

(iii) The liquidated damages figure stated in the appendix is reduced by the proportion which the value of the relevant part bears to the unadjusted contract sum. The operation of this reduction or apportionment formula makes no allowance for the different effects in terms of financial loss which any delay in completion of outstanding parts may have. For instance, a retail unit may be taken over in part with only, say, £20,000 worth of landscaping outstanding. Here any delay in the outstanding part of the works may cause little or no loss in terms of the unit opening for trade and maintaining it. On the other hand, such a unit may be taken into possession for the employer to arrange to have it fitted out ready for trade leaving the public access roads still to be finished off by the original contractor at a cost of, say, £15,000. Here any delay in finishing off the balance of work beyond the employer's fitting out period, could prevent trading and the estimated loss could be just as much as if no part had been taken over at all and yet the liquidated damages rate will have been significantly reduced. The operation of this formula is therefore inherently unlikely to produce a liquidated damages figure which is based upon a genuine pre-estimate. As a result it has been suggested by some that a contract with this type of clause in it may cause the liquidated damages provision to be a penalty.

If clause 2.11 is adopted, there are also consequential amendments to various clauses which must also be adopted in order to ensure that clause 2.11 operates satisfactorily in practice.

SECTIONAL COMPLETION SUPPLEMENT FOR IFC 84

The Tribunal has issued for use with IFC 84 a Sectional Completion Supplement (July 1985) – see Appendix 5 to this book, page 429. It provides a series of amendments to the IFC 84 Conditions to enable completion of the works by sections.

Where it is known in advance of the contract being entered into that either possession of parts of the site is to be given at different times or that completion and use by the employer of some parts of the works is required before others, the sectional completion supplement should be used. To rely on the optional clause 2.11 partial possession provision is not appropriate in such circumstances. Further, to try and achieve a similar result by including somewhere in the contract documentation a schedule of handovers and handbacks e.g. for a housing refurbishment scheme, is very likely to fail in its purpose. At best it is unlikely to deal adequately with such matters as practical completion of the part handed back; release of retention; insurance; apportioned liquidated damages and separate extension of time provisions for each part. At worst it will not even get off the ground and could well invalidate the liquidated damages provisions of the contract. Further, by virtue of clause 1.3 of IFC 84 seeking to achieve sectional completion by using contract documents other than the conditions is likely to fail. Clause 1.3 provides that:

> 'Nothing contained in the Specification/Schedules of Work/Contract Bills shall override or modify the application or interpretation of that which is contained in the Articles, Conditions, Supplemental Conditions or Appendix.'

Examples of unsuccessful attempts are to be found in such cases as *Gleeson* v. *Hillingdon Borough Council* (1970) and *Bramall and Ogden Ltd* v. *Sheffield City Council* (1983).

Where this supplement is to be used the works must be divided into sections. This must be clearly done in the contract documents. The sections should be numbered and the numbers inserted into the amended appendix entries which will also state the date for possession, completion, rate of liquidated damages, defects liability period and value for each section.

Each section has its own start and completion date, its own certificate of practical completion and its own defects liability period. Each section also has its own liquidated damages and extension of time provision applicable to it.

Care must be taken to ensure that where a section is to be completed and taken into possession for use by the employer, any necessary enabling services are considered, as it may be necessary to take over part or the whole of such services as part of the first section to be completed. Much will depend upon whether the common elements e.g. heating or power can be divided into a number of sections.

There is only one final certificate for the whole of the works even though sectional completion applies. Accordingly its issue will usually be dependent upon the issue of a certificate of making good defects for the last section in which defects are made good.

Where possession of one section is dependent upon progress on any other section, the appendix entries for related sections need to be considered with great care in order to keep the liquidated damages clause intact. The solution with dependent sections is to give them fixed completion dates by reference to the calendar but to fix the date for possession of the section by reference to a critical date for an earlier section, such as the achieving of practical completion. For example, a date could be inserted for possession of section 2 which is 'ten days after practical completion of section 1'.

Finally, it sometimes happens in practice that while sectional completion is required, it is not possible at the time of contracting to identify the section easily, e.g. in a housing 'decant' scheme where the particular batch of houses to be handed over is dependent upon when tenants are prepared to vacate to alternative temporary accommodation. There may be no easy answer to this problem. It may be possible to identify a section within the contract documents by reference to, say, any four adjoining houses. Even so difficulties can easily arise, for instance in identifying and valuing the individual sections which the amended appendix to IFC 84 requires to be done at the outset.

Chapter 5

Control of the works (Part I)

CONTENT

This chapter looks at clauses 3.1 to 3.3 of section 3 of IFC 84 which deal with matters relating to assignment and sub-contracting, including consideration of named persons as sub-contractors.

SUMMARY OF GENERAL LAW

(A) Assignment

Assignment in relation to a contract means the transfer of contractual rights or benefits by a party to the contract to some other person who was not originally a party to it, i.e. a stranger to the contract.

A contract is made up in part of benefits and in part of burdens. What amounts to a benefit to one party will very often be a burden to the other e.g. under a building contract the employer obtains the benefit of a building and the contractor assumes the burden of it or the liability for building it. The law treats the transfer of benefits and burdens differently.

The benefits
Subject to what a contract may expressly provide, a contracting party may assign the benefits due to him under the contract e.g. the contractor under a building contract may assign the benefit of the retention money or the employer may assign the partly constructed building together with the right to have it completed. If a party to a contract assigns any benefit under it, he should give notice of that fact to the other party. Once notified the other party must, if the benefit being assigned represents a burden to him, discharge that burden in favour of the new beneficiary i.e. the assignee.

Certain benefits of a truly personal nature cannot be assigned e.g. the right to litigate where the cause of action is purely personal.

The burdens

The law does not permit the unilateral assignment of contractual liabilities. The burdens under a contract can only be transferred with the consent of the other contracting party. Once that consent is forthcoming the original party can be released from the burdens. The person to whom the burdens are transferred becomes liable in the place of the person who transferred them. This in law is called a novation. It must be distinguished from vicarious performance of contractual liabilities, i.e. someone else actually performing the obligation whilst the contracting party remains fully liable under the contract. Vicarious performance is dealt with below under the next heading.

Many building contracts expressly prohibit even the assignment of benefits without the consent of the other contracting party.

(B)   Sub-contracting

Sub-contracting of contractual obligations i.e. vicarious performance of the actual work, is permitted under the general law unless the obligations are personal in nature (see next paragraph). In many instances, so long as the original party remains responsible for any failure to properly and fully perform the burden or obligation, it matters not who actually carries it out. For example, a main contractor may of his own choice arrange for a sub-contractor to carry out certain of the work. Subject to what the express terms of the contract may say, the main contractor will be permitted to do this while remaining fully liable to the employer for any failure of the sub-contractor to adequately fulfil his contractual obligations where this leads to a similar failure of the main contractor *vis-à-vis* the employer. The employer will be able to claim from the main contractor who must in turn claim from the sub-contractor. The relationship between the main contractor and the sub-contractor is of no concern to the employer. This arrangement is known in the construction industry as domestic sub-contracting to distinguish it from nominated sub-contracting. If it is the employer who selects a particular sub-contractor to the main contractor then different consideratons apply. This topic is dealt with under the next heading.

If the employer can demonstrate that he is placing reliance on the particular skill or expertise of a given contractor then even a

building contract may be regarded in such a case as being of a
personal nature thus preventing the sub-contracting of any of the
work.

(C)   Selection by employer of a sub-contractor to the main contractor

The general principle of English law that a main contractor will be
responsible to the employer for the defaults and breaches of
contract of a sub-contractor in respect of workmanship and
materials can be significantly qualified by the parties' contractual
arrangements.

(i)   Full nomination
It may be that by the terms of the main contract the employer
reserves to himself the right, through his architect or other agent
named under the contract, to nominate, name or otherwise select a
sub-contractor with whom the main contractor is obliged to enter
into contract for part of the contract works. The main contract
often achieves this by the use of a provisional or prime cost sum
expended on the instruction of the architect. A typical building
contract will probably go on to say that this provisional or prime cost
sum is adjustable in line with the actual sub-contract final account
for the work done, barring only such increases in the sub-contract
sum as are due to the default of the main contractor. Thus, the
actual cost of such work will be met by the employer, the main
contractor not being required or even permitted to price the work
for himself.

MAIN CONTRACTOR'S RESPONSIBILITY FOR
NOMINATED SUB-CONTRACTOR'S WORK

The particular contract being used must of course be considered.
Where there is full nomination by the use of a provisional or prime
cost sum, then in most standard forms of construction contract there
is likely to be some restriction placed on the liability of the main
contractor for the nominated sub-contractor's defaults, even though
this could leave the employer without a remedy in respect of losses
suffered due to unfinished work.

   In most building contracts, unless the contractor is also the
designer, he will not be responsible for failures of design or fitness
for purpose of the work: see *Norta Wallpapers (Ireland) Ltd* v. *John
Sisk & Sons (Dublin) Ltd* (1976).

It may be otherwise if the contractor expressly accepts a design liability in his contract with the employer as did the contractor in the case of *Independent Broadcasting Authority* v. *EMI Electronics Ltd and BICC Construction Ltd* (1980) where his quotation was stated to be 'for the design, supply, and delivery . . .'

The contract will possibly contain some express terms about the quality of workmanship and materials. Instead or in addition, there may be implied terms covering all or part of this ground.

Where the nominated sub-contractor completes the sub-contract work required of him but that work is defective, the main contractor will be responsible for it to the employer unless it is a design defect. However, if the nominated sub-contractor fails to complete the sub-contract works, the employer will often have to re-nominate and the cost of completing the outstanding work will fall on the employer. The question of whether the contractor is responsible for a nominated sub-contractor's defective workmanship and materials prior to any re-nomination can be a difficult question. Clearly the terms of the contract have to be carefully considered. While the terms of the particular contract may prohibit the main contractor from himself carrying out work covered by a prime cost item, this may not absolve him from the responsibility of meeting the cost of remedying defects in that work. On the other hand, it may. For example, in the case of *Fairclough Building Ltd* v. *Rhuddlan Borough Council* (1985) the contractor under a JCT 63 contract was held not liable for part of the cost of remedial works required when a nominated sub-contractor failed to complete his sub-contract work, leaving behind defective work.

While the employer may have to meet the cost of having the unfinished work of the nominated sub-contractor completed, the contractor will also no doubt suffer losses which he must bear if he cannot recover them from the nominated sub-contractor.

If the employer has no direct contract between himself and the nominated sub-contractor, he will have no contractual rights at all to recover this extra cost of completing the work. The main contractor will not be liable (he generally cannot be made to finish off the work at his own cost if it is a prime cost item) to the employer for the nominated sub-contractor's breach in failing to complete and the employer will have no contract with the nominated sub-contractor.

The chances of the employer recovering directly from the nominated sub-contractor in negligence in relation to defective work is negligible unless the negligently created defect causes separate physical damage to property of the employer. The House of Lords case of *Junior Books Ltd* v. *The Veitchi Co. Ltd* (1982) to the contrary, allowing recovery of purely economic loss in such circumstances, seems unlikely to be followed, at any rate for the foreseeable future.

In recent years on a number of occasions the courts have been called on to consider the legal position following defaults by nominated suppliers or nominated sub-contractors. Three particularly important decisions are to be found in the cases of *Young and Marten Ltd* v. *McManus Childs* (1969); *Gloucester County Council* v. *Richardson* (1969); and *North-West Metropolitan Regional Hospital Board* v. *T.A. Bickerton & Son Ltd* (1970).

In the case of *Young and Marten Ltd* v. *McManus Childs* a main contractor experienced in roofing and tiling work requested an estimate from a sub-contractor to supply and fix a high grade tile known as 'Somerset 13'. In doing so the main contractor relied entirely on his own skill and judgment in the choice of tile. These tiles were in fact made by only one manufacturer. The sub-contractor obtained and fixed the tiles through a sub-sub-contractor who in turn purchased the tiles from the sole manufacturer. The particular batch of tiles purchased was defective although this could not be detected on a reasonable examination. When the contractor was sued by the building owners, he joined in as a third party to the action his sub-contractor. The sub-contractor could not bring in the sub-sub-contractor because the statutory limitation period had by then expired, so the issue before the courts was who was responsible as between the main contractor and his sub-contractor for the defective roof tiles. The House of Lords held that, while the main contractor may have relied on his own judgment, and accordingly any warranty of fitness for purpose was excluded, this did not exclude the warranty of merchantable quality so that the main contractor was entitled to recover from the sub-contractor.

In the case of *Gloucester County Council* v. *Richardson* (1968), which was based on the then RIBA 1957 form of contract, the court held that a main contractor was not liable to the employer for defective concrete columns supplied by a nominated supplier, firstly because the nominated supplier had restricted his liability for defective goods by his contract with the main contractor, and therefore it would have been unfair to impose a liability as between the main contractor and employer when the main contractor could not follow the usual chain of liability through to the person actually at fault; and secondly because there was no provision in the main contract for the main contractor to make reasonable objection to the nomination by the employer of a supplier, unlike the position in relation to nominated sub-contractors.

It is clear that the opposite conclusions reached in these two cases was the result of their particular facts, and in particular in the second case, there is little doubt that if there had been no restriction on the supplier's liability, and if the main contractor had been given a right to raise reasonable objection to entering into a contract with a nominated supplier, then the decision would have been similar to that in the first case. Generally speaking therefore, it can be said that, so far as the obligation to carry out work with care and skill and to supply goods which are of merchantable quality is concerned, the main contractor will be liable for defective workmanship or materials of the nominated sub-contractor or supplier, unless, that is, the nominated sub-contractor or supplier fails to complete, in which case this liability may well be diluted – see *Fairclough Building Ltd* v. *Rhuddlan Borough Council* (1985). Under JCT 80, so far as nominated suppliers are concerned, the architect cannot without the consent of the contractor nominate a supplier whose contract in any way restricts, limits or excludes the liability of the nominated supplier to the contractor in respect of materials or goods supplied, unless the architect has specifically approved in writing these restrictions etc., in which case the liability of the contractor to the employer in respect of the materials or goods is likewise restricted.

The position in relation to responsibility for defective workmanship or materials of named sub-contractors under IFC 84 is dealt with later in this chapter (see particularly the Commentary to clause 3.3.4, page 144).

In the case of *North-West Metropolitan Hospital Board* v. *T.A. Bickerton & Sons Ltd* (1970) the contracts before the court were JCT 63 and the standard 'green' form of sub-contract for nominated sub-contractors. The nominated heating sub-contractor went into liquidation before commencing work. The main contractor contended that the employer was bound to nominate a second sub-contractor and to pay the main contractor the amount of the second sub-contractor's account. The employer on the other hand contended that there was no duty to re-nominate or to pay more than would have been payable had the original sub-contractor not defaulted. It was held that the wording of the contract meant that the contractor was forbidden to carry out works for which a prime cost item was provided in the contract and accordingly, the contractor's argument was accepted.

It is clear that the particular contractual arrangements made between the parties will be of the essence and no general statement can cover the multifarious situations which arise. The contract may seek to deal expressly with the sub-contractor's failure to complete as does JCT 80 in clause 35, or it may leave the situation open, in which case it is tentatively submitted that, if the sub-contractor fails to complete for whatever reason, then in most instances where a prime cost item is used, a re-nomination will be required and the employer will generally have to meet the subsequent nominated sub-contractor's account.

However, the more recent and very important House of Lords case of *Percy Bilton Ltd* v. *Greater London Council* (1982) (dealt with in detail on page 132) demonstrates that the point has not yet been reached whereby the employer warrants to the main contractor that a nominated sub-contractor will complete his sub-contract work without interruption.

What is very clear from the present state of the law in relation to defaulting nominated sub-contractors is that the employer needs to create a direct contract between himself and the nominated sub-contractor in order that he can claim directly against him in respect of losses brought about by a failure of the nominated sub-contractor to perform his contractual obligations.

POSSIBLE REMEDIES OF EMPLOYER AGAINST NOMINATED SUB-CONTRACTOR

(1)   Direct contract

It is now very regular practice for the employer to enter into a direct contract with the nominated sub-contractor. By this means the employer can have a remedy directly against the nominated sub-contractor if he fails to complete the contract works, thus necessitating a re-nomination which causes loss to the employer. There are other very good reasons for having this direct contractual relationship, the most important being where the nominated sub-contractor is engaged to carry out specialist design work or where reliance is placed on him in the selection of particular goods or materials. As the main contractor will have no responsibility in respect of design or fitness for purpose of the sub-contract work, the employer needs this direct contractual link in order to have a remedy in the event of there being a failure in design or a failure of the sub-contract work to be reasonably fit for its purpose. This direct contract can also conveniently cover many other aspects of the relationship between the employer and the nominated sub-contractor e.g. payment for pre-nomination work; provision of a right of recovery of lost liquidated damages suffered by the employer where a nominated sub-contractor causes delay giving rise to an extension of the main contract completion date; early final payment of nominated sub-contractor; and direct payment of the nominated sub-contractor.

(2) Duty of care

It should not be forgotten that, although there may be no contractual relationship between a nominated sub-contractor and employer, the nominated sub-contractor may nevertheless owe a duty of care to the employer under the general law of negligence. If in breach of this duty of care, the sub-contractor will be liable to the employer in respect of any forseeable losses flowing from this breach.

The starting point of the modern law of negligence was the case of *Donoghue* v. *Stevenson* (1932) which established the 'neighbour principle'. This basically means that you must take reasonable care to avoid acts or omissions which you can reasonably foresee would be likely to injure your neighbour. Your neighbour is someone so closely and directly affected by your act that you ought reasonably to have them in contemplation as being so affected when directing your mind to the acts or omissions in question.

If a nominated sub-contractor negligently supplies defective work, goods or materials which cause physical damage to the separate property of the employer, this will give the employer a right to sue for damages, which could, though only incidentally, include the cost of remedying the defective work. However, the employer would have no such remedy where the defect only caused physical damage to the very work which the sub-contractor carried out. If the defect causes any physical damage to the remainder of the contract works it is a matter of considerable debate as to whether the rest of the contract works is to be treated as separate property of the employer so as to impose a negligence liability on the sub-contractor – see *D & F Estates Ltd* v. *Church Commissioners* (1988). It is suggested that where negligently created defects in the nominated sub-contractor's work cause physical damage to other parts of the contract works, the construction of which was not the contractual responsibility of the sub-contractor, this can fairly be treated as other property of the employer.

Where it is negligent advice rather than a negligent act or omission of which the nominated sub-contractor is guilty, the employer will be able to recover purely financial loss even though no physical damage has been caused – see *IBA* v. *EMI Electronics and BICC Corporation Ltd* (1980). The House of Lords case of *Junior Books Ltd* v. *The Veitchi Co. Ltd* (1982) appeared to have extended the ambit of the law of negligence to provide for recovery of financial loss where there has been a negligent act or omission even in the absence of physical damage to property. However, since then a whole line of cases has made it clear that the *Junior Books* case will not be followed and is not an accurate or reliable statement of the law – see for example *D & F Estates Ltd* v. *Church Commissioners* (1988).

Further, if there exists a direct express contract between employer and nominated sub-contractor, this will govern the scope of any duty of care owed by the nominated sub-contractor and the tort of negligence cannot be used to fill any gaps in the contractual remedies available – see *Greater Nottingham Co-Operative Society Ltd* v. *Cementation Piling and Foundations Ltd* (1988).

(3)   Responsibility under statute

It should not be forgotten that in certain circumstances a supplier of work, goods or materials may well owe a duty to others for whom the benefit of that work etc. is carried out to ensure that it is reasonably safe. For example, such duty may be owed by a nominated sub-contractor to an employer under the Defective Premises Act 1972, the Consumer Protection Act 1987 or under the Public Health Act 1961 and the Health and Safety at Work etc. Act 1974 in respect of building regulations. The extent and nature of the duty which may be owed to the employer under such legislation is outside the scope of this book. However, more is said concerning the Defective Premises Act and building regulations when considering section 5 of IFC 84 (statutory obligations etc.) in chapter 8.

(4)   Collateral warranty or contract

Occasionally, although there appears to be no direct contractual relationship between employer and nominated sub-contractor, the courts may decide that there is a relationship between the two parties which establishes what is called a collateral contract. A leading case on this is *Shanklin Pier Co. Ltd* v. *Detel Products* (1951).

Facts:

A specialist painting sub-contractor went out of his way to advertise the suitability of his paint to protect the pier at Shanklin from rust. He persuaded the employer that his paint would be admirably suited for such an application and on the strength of this warranty the employer nominated him to the main contractor. The paint did not do its job. The sub-contractor argued that there was no contract between him and the employer and therefore no remedy.

Held:

There was separate contract containing all the usual elements of a binding contract, namely offer, acceptance and consideration, directly between employer and sub-contractor. The consideration was that the employer agreed to nominate the sub-contractor in exchange for the nominated sub-contractor's assurances that his product was suitable. There was a collateral warranty or collateral contract.

Similarly, a collateral warranty may exist between the employer and a nominated or uniquely specified supplier – see *Greater London Council* v. *Ryarsh Brick Co. Ltd and Others* (1985).

(ii)   Less than full nomination

There are a number of ways in which the employer can be involved in the selection of a sub-contractor to the main contractor without the full machinery of nomination being involved. For instance, the employer may provide a list of sub-contractors from whom the main contractor can choose one. The work will be fully described in the contract documents and will be priced by the contractor. In such circumstances the resulting sub-contract may be very close to a domestic sub-contract. For an example of this see JCT 80 clause 19.3.

Alternatively, the employer may, without using a provisional or prime cost sum, identify a named sub-contractor while providing the main contractor with sufficient information to enable him to price the work himself, even though it will be carried out by the named sub-contractor. The consequences of this will depend on the wording of the contract, although it may be said that if the main contractor has full or fairly full information before tendering as to the identity, sub-contract conditions, sub-contract price, programme, attendances etc., so far as these can be finalised before a sub-contract is entered into, it is the more likely that the main contractor will carry some responsibility for the sub-contractor's failure to perform.

While the conditions of the sub-contract cannot in themselves affect any plain express terms in the main contract as to the main contractor's responsibility for the work of a nominated or named sub-contractor, it is clear that such conditions may be relevant when considering the liabilities of the various parties. The case of *Gloucester County Council* v. *Richardson*, discussed earlier (page 110), demonstrates this.

CONSIDERATION OF THE RELEVANT CLAUSES OF IFC 84

Clause 3.1

Assignment

3.1   Neither the Employer nor the Contractor shall, without the written consent of the other, assign this Contract.

COMMENTARY ON CLAUSE 3.1

For a summary of the general law see page 106. This clause is in a typical form for a building contract. Even without such a clause, neither the employer nor contractor could transfer the burden or liabilities arising under this contract without the consent of the other party. However, benefits can always be assigned unless the contract prohibits it without consent, as does this one.

The consent must be in writing and so should any assignment or novation itself.

Clause 3.2

Sub-contracting

The Contractor shall not sub-contract any part of the Works other than in accordance with clause 3.3 without the written consent of the Architect/the Contract Administrator whose consent shall not unreasonably be withheld.

It shall be a condition in any sub-contract to which clause 3.2 refers that

3.2.1   the employment of the sub-contractor under the sub-contract shall determine immediately upon the determination (for any reason) of the Contractor's employment under this Contract; and

3.2.2   the sub-contract shall provide that

(a) Subject to clause 1.10 of these Conditions (in clauses 3.2.2(b) to (d) called 'the Main Contract Conditions'), unfixed materials and goods delivered to, placed on or adjacent to the Works by the Sub-Contractor and intended therefor shall not be removed except for use on the Works unless the Contractor has consented in writing to such removal, which consent shall not be unreasonably withheld.

(b) Where, in accordance with clause 4.2.1(b) (*Interim payments*) of the Main Contract conditions, the value of any such materials or goods shall have been included in any interim payment certificate under which the amount properly due to the Contractor shall have been discharged[1] by the Employer in favour of the Contractor, such materials or goods shall be and become the property of the Employer and the Sub-Contractor shall not deny that such materials or goods are and have become the property of the Employer.

(c) Provided that if the Main Contractor shall pay the Sub-Contractor for any such materials or goods before the value therefor has, in accordance with clause 4.2.1(b) (*Interim payments*) of the Main Contract Conditions, been included in any interim payment certificate under which the amount properly due to the Contractor has been discharged by the Employer in favour of the Contractor, such materials or goods shall upon such payment by the Main Contractor be and become the property of the Main Contractor.

(d) The operation of sub-clauses (a) to (c) hereof shall be without prejudice to any property in any materials or goods passing to the Employer as provided in clause 1.11 (*Off-site materials or goods*) of the Main Contract Conditions.

COMMENTARY ON CLAUSE 3.2

For a summary of the general law see page 106. Clause 3.2 permits the contractor to enter into domestic sub-contracts only with the consent of the employer. However, such consent must not be unreasonably withheld. The test of unreasonableness is likely to be objective rather than subjective. However, if an employer chooses a particular contractor because of his skill and judgment in a certain field it may be that a withholding of consent to sub-contracting would be a reasonable stance to take. Such a situation is likely to be rare. Provided that the contractor will remain as fully liable to the employer for the sub-contractor's defaults as if he had carried out the work himself, the reason for withholding consent would have to be compelling to prevent it being unreasonable. It is submitted that the correct position is that the contractor is fully liable to the employer in such circumstances despite apparent aberrations from this fundamental principle implicit in some recent decisions, e.g. *John Jarvis Ltd* v. *Rockdale Housing Association* (1987); *Greater Nottingham Co-Operative Society Ltd* v. *Cementation Piling and Foundations Ltd* (1988); *Scott Lithgow* v. *Secretary of State for Defence* (1989).

It is submitted that the reason for withholding consent must rest on the ability of the sub-contractor to perform the work competently, efficiently and on time and it would be a very liberal interpretation of the clause for an employer to validly withhold consent, as some public sector employers apparently do, to a sub-contract with anyone who does not appear on an approved list of contractors.

Any consent which is given is conditional on the sub-contract containing, as a minimum, certain conditions:

(1)   Automatic determination of sub-contract

The sub-contract must contain a condition that the employment of the sub-contractor shall determine immediately on a determination, for any reason, of the employment of the contractor under the main contract.

If the sub-contractor's employment is determined in this fashion, the question arises as to whether or not this determination is a breach of the sub-contract by the contractor? In itself, it will not be. However, if the determination of the employment of the main contractor under the main contract came about as a result of the contractor's breach of that main contract, it is at least arguable that the contractor is in breach of an implied term in the sub-contract that he will not give the employer grounds for determining his employment under the main contract. If such is the case a main contractor will be liable in damages for breach of contract.

Additionally, the sub-contract itself may well make provision for what is to happen following such determination. For instance, the recommended form of sub-contract for use in conjunction with IFC 84, i.e. IN/SC clauses 28.3 and 29, provides that in such an event the sub-contractor is to receive the total value for completed work and an ascertained value for partly finished work; the cost of the materials or goods properly ordered and either paid for or involving the sub-contractor in a legal obligation to pay and which have not been incorporated into the sub-contract works; the cost of removing plant, temporary buildings, tools, goods or materials etc. from the site; and also any loss and expense incurred in respect of which the sub-contractor became entitled to reimbursement prior to the determination of his employment; and also any direct loss or damage (except where the main contractor's employment is determined under clause 7.8 of IFC 84) caused by the determination.

If the determination of the sub-contractor's employment under IN/SC is due to a proper determination by the main contractor of his own employment under IFC 84, then this expenditure will be recoverable from the employer under IFC 84 – see clause 7.7(b).

It will be seen therefore that the main contractor will, at any rate initially, and even though himself innocent, suffer financial consequences similar to those which would flow from a serious breach of contract by him. The reason for this is that it results eventually in the financial losses being met by the party at fault i.e. the employer in this particular example. The innocent sub-contractor claims from the innocent main contractor who in turn claims from the employer who is at fault. The sub-contract conditions for use between the named sub-contractor and the main contractor in relation to contracts under IFC 84 are on the whole similar in this respect to the recommended domestic sub-contract conditions IN/SC (see clauses 28.3 and 29).

If there were no conditions in the sub-contract for an automatic determination of the sub-contractor's employment under the sub-contract, the effect on the sub-contract of a determination of the main contractor's employment under IFC 84 would fall to be considered under the common law and this could produce uncertainty. For instance, if, through no fault of the main contractor, he lost possession of the site, due to the employer's wrongly retaking it, it might be argued that the sub-contract was thereby frustrated. It is better therefore to have an immediate determination and the implications of it dealt with expressly in the sub-contract conditions.

(2)   Materials and goods
By clause 3.2.2, the sub-contract must contain certain terms relating to materials and goods.

Firstly, there must be a term in the sub-contract that any unfixed materials and goods delivered to, placed on or adjacent to the works by the sub-contractor shall not be removed without the consent in writing of the contractor, which consent shall not be unreasonably withheld. This provision is made expressly subject to clause 1.10 which means that, so far as the contractor is concerned, he must obtain the consent of the architect before giving consent to the sub-contractor.

Secondly, there must be a term in the sub-contract to the effect that where the value of any materials or goods has been included in any interim payment certificate under which the amount properly due to the contractor has been discharged by the employer, such materials or goods shall become the property of the employer and the sub-contractor shall not deny that this is so. This provision is included to deal, so far as the contract terms can do so, with the type of problem experienced by the employer in the case of *Dawber Williamson Roofing Ltd* v. *Humberside County Council* (1979).

Facts:

The main contract was based on JCT 63. There was a domestic sub-contractor and his contract was based on the standard form of domestic sub-contract (the 'blue' form). By clause 14 of the main contract, it was stated that any unfixed materials or goods delivered to, and placed on or into the works should not be removed without consent, and that when the value of those goods had been included in a certificate under which the contractor had received payment, such materials or goods should become the employer's property. By clause 1 of the domestic sub-contract the sub-contractor was deemed to have notice of all the provisions of the main contract (other than prices). (It should be noted that there was no express provision in the sub-contract as to if and when the property in the sub-contractor's materials or goods was to pass to the main contractor.)

The sub-contractor delivered 16 tons of roof slates to the site and submitted invoices to the main contractor. Under the main contract an interim certificate was issued including the value of the slates. The employer paid the appropriate sum to the main contractor. According therefore to the main contract, as the amount had been certified and paid, the ownership in the slates would vest in the employer. The main contractor did not pay the domestic sub-contractor and went into liquidation. The sub-contractor claimed that he was still the owner of the slates and therefore entitled to possession of them.

Held:

The slates were still owned by the sub-contractor. They were never at any time owned by the main contractor so he could not pass the title in them to the employer.

This case has been criticised as not being in line with sound commercial practice in that an employer had to pay for goods without ownership in them thereby being transferred to him, even though in the overall scheme of main contract and sub-contract it may have been thought that this was the intention. There is thus a gap in JCT 63 and also in JCT 80. However, so far as JCT 80 is concerned, Amendment 1.1984 now deals with the situation in a similar way to IFC 84.

Thirdly, the sub-contract must contain a term whereby, if the contractor has paid the sub-contractor for any such materials or goods *before* their value has been included in a payment certificate and the amount discharged by the employer, then such materials or goods shall, on payment by the contractor, become his property.

By virtue of these provisions appearing in the sub-contract, the sub-contractor cannot in such circumstances challenge the title either of the employer or the contractor, as the case may be, to the materials or goods in question. While this will provide some protection for the employer where he has discharged the amount due to the contractor in respect of the materials and goods, albeit the contractor has not paid the sub-contractor, this provision will not assist the employer where the sub-contractor is not himself the owner of the materials or goods e.g. where there is a retention of title clause in his contract with the supplier of the materials or goods. (For a summary on the general law relating to retention of title see the Commentary to clause 1.10 on page 45.)

The third provision referred to above relating to payment by the contractor to the sub-contractor before the value of materials or goods has been included in a payment certificate under the main contract and payment made, fills a gap which appears to have existed until now in the JCT forms and related sub-contract documents e.g. DOM/1, which did not expressly provide that materials or goods should become the property of the main contractor even where he had paid for them. The provision which is now required to be contained in the sub-contract should prevent the sub-contractor from claiming that ownership remains with him even though he has been paid e.g. under a retention of title arrangement covering all supply contracts generally between that sub-contractor and main contractor.

Of course, once the materials or goods are incorporated in the works, they will become the property of the landowner in any event.

Fourthly, the operation of the three provisions referred to above is expressly stated to be without prejudice to any property in any materials or goods passing to the employer as provided in clause 1.11 of IFC 84.

The standard domestic sub-contract form for use with IFC 84, namely IN/SC, contains all of the provisions required to be included by clause 3.2.2.

By virtue of clause 7.1(d) (see page 344) any failure by the contractor to comply with the provisions of clause 3.2 by reason of a default on his part is a ground for determination of the contractor's employment by the employer, subject however to the proviso that the notice of determination shall not be given unreasonably or vexatiously.

NOTES TO CLAUSE 3.2

[1]   '. . . shall have been discharged . . .'.
See note [3] to clause 1.10 – see page 47).

NAMED PERSONS AS SUB-CONTRACTORS

Introduction
Clauses 3.3.1 to 3.3.8 inclusive provide for a system of selection by the employer of a sub-contractor to carry out part of the contract work. This involves the selection by the employer of a named person to carry out work as a sub-contractor to the main contractor and it therefore represents a system of nomination though very different in parts from the concept of nomination embodied in JCT 80. The sub-contractors concerned are known as 'named' sub-contractors and not 'nominated' sub-contractors.

The system is structured so as to provide two methods of naming:

(1)   The person who it is intended shall be the sub-contractor is named in the main contract tender documents and the work to be done is described in detail there. The main contractor puts in his own tender price for the work, after being informed of the sub-contract price and most of the relevant conditions relating to the sub-contract works e.g. conditions of contract, attendances and possibly progamme details of the proposed sub-contractor. The contractor is therefore in a position to price the sub-contract works as he thinks fit but with due regard to the information which has been made available to him. The detailed sub-contract information is obtained by the employer through the architect obtaining from the sub-contractor whom it is proposed to name, the completed Parts I and II of a sub-contract tender document known as Form of Tender

and Agreement NAM/T. Once the named sub-contract is duly entered into, the main contractor accepts a greater degree of responsibility for delays and defaults by the named sub-contractor than does the contractor in respect of a nominated sub-contractor under JCT 80. There is no liability on the contractor for design failures where the named sub-contractor has design obligations and the contractual provisions for naming expressly exclude design responsibility (see clause 3.3.7). The architect is thus not concerned with questions of payment or delay on the part of a named sub-contractor. These will be matters between the named sub-contractor and the main contractor.

However, following on a determination of the employment of the named sub-contractor by the contractor, under clause 27.1 or clause 27.2 of NAM/SC, the extent to which the contractor is responsible for completing the outstanding work and the financial implications flowing therefrom will depend upon which of the three options listed in clause 3.3.3 is chosen by the architect. (Clause 27.1 or 27.2 of NAM/SC basically provides for the contractor to determine the employment of the named sub-contractor due to the latter's default or bankruptcy etc. The provisions of clause 27.1 and 27.2 are considered in more detail later when considering the operation of clause 3.3.4 of IFC 84.) The consequences of each instruction are set out in clause 3.3.4.

If the named sub-contractor's employment is determined either as a result of the contractor's default or by the contractor accepting a repudiation of the sub-contract following a repudiatory breach by the named sub-contractor, the architect will be required to choose one of the three options listed in clause 3.3.3, but the costs associated with this will be borne by the contractor. In such circumstances the financial consequences in clause 3.3.4 apply only to the extent that they result in a reduction in the contract sum e.g. where there has been an omission of work under clause 3.3.3(c) or where the cost of completing the outstanding balance of work is less than it would have been had there been no default by the first named sub-contractor. Furthermore, the provision for an extension of time and for loss and expense will not apply – see clause 3.3.6(a). This is discussed in more detail in the commentary to clause 3.3.6 on page 153:

(2) A named sub-contractor can be introduced by means of an instruction as to the expenditure of a provisional sum after the main

contract has been entered into. Here, the contractor has a right of reasonable objection. In the event of the contractor determining the employment of the named sub-contractor under clause 27.1 or 27.2 of NAM/SC, the subsequent instruction of the architect under clause 3.3.3 is to be regarded as a further instruction issued under the provisional sum. The provisional sum will therefore generally be adjusted as appropriate to take into account any variation in the cost of completing the work. This second system resembles more closely therefore the traditional form of nomination under JCT 80. However, should the sub-contractor's employment be determined in any other manner than under clause 27.1 or clause 27.2 of NAM/SC, the instruction under clause 3.3.3 will not be regarded as a further instruction under the provisional sum unless it results in a reduction in the contract sum – see clause 3.3.6(a).

Of the two methods, naming in the main contract tender document will generally be more advantageous to the employer but the use of this method presupposes that it is possible for the architect to name the proposed sub-contractor at that stage and this will not always be the case.

Provided the named sub-contractor's employment is determined under clause 27.1 or 27.2 of NAM/SC, the main contractor will be entitled, depending on which of the three types of instruction is issued under clause 3.3.3, to payment from the employer of the cost of completing the work and/or an extension of time and/or direct loss and expense. The employer therefore incurs extra costs and suffers losses. Clause 3.3.6(b) attempts to provide an indirect method whereby the employer recovers such increased cost and losses from the named sub-contractor through the main contractor. This is discussed in detail later in the Commentary on this particular sub-clause.

On a determination of the employment of the named sub-contractor by the contractor, express provision is made in clause 3.3.3(a) for the architect to name another person to finish off the outstanding work. However, it will be seen, when the detailed drafting is considered later, that the possibility of a second named sub-contractor's employment being determined does not appear to have been considered so far as work described in the contract documents for execution by a named person is concerned. Certainly, the opening paragraph of clause 3.3.4 restricts the operation of the financial consequences flowing from a re-naming to the situation where the instruction to re-name arises 'in respect of a person

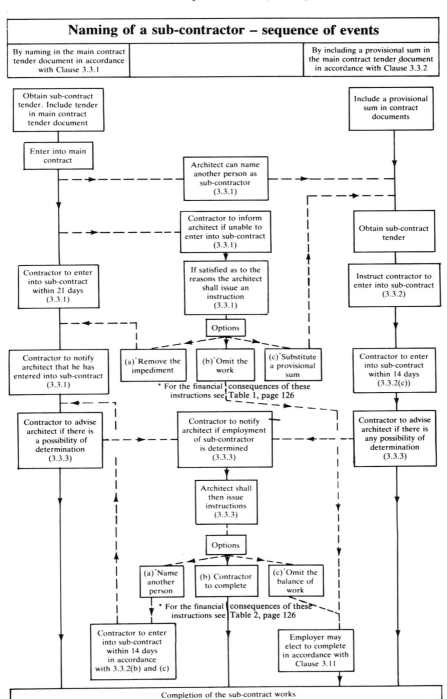

Naming of a sub-contractor – sequence of events

| By naming in the main contract tender document in accordance with Clause 3.3.1 | | By including a provisional sum in the main contract tender document in accordance with Clause 3.3.2 |

Obtain sub-contract tender. Include tender in main contract tender document

Include a provisional sum in contract documents

Enter into main contract

Architect can name another person as sub-contractor (3.3.1)

Contractor to inform architect if unable to enter into sub-contract (3.3.1)

Obtain sub-contract tender

Contractor to enter into sub-contract within 21 days (3.3.1)

If satisfied as to the reasons the architect shall issue an instruction (3.3.1)

Instruct contractor to enter into sub-contract (3.3.2)

Options

(a) Remove the impediment | (b) Omit the work | (c) Substitute a provisional sum

Contractor to notify architect that he has entered into sub-contract (3.3.1)

* For the financial consequences of these instructions see Table 1, page 126

Contractor to enter into sub-contract within 14 days (3.3.2(c))

Contractor to advise architect if there is a possibility of determination (3.3.3)

Contractor to notify architect if employment of sub-contractor is determined (3.3.3)

Contractor to advise architect if there is any possibility of determination (3.3.3)

Architect shall then issue instructions (3.3.3)

Options

(a) Name another person | (b) Contractor to complete | (c) Omit the balance of work

* For the financial consequences of these instructions see Table 2, page 126

Contractor to enter into sub-contract within 14 days in accordance with 3.3.2(b) and (c)

Employer may elect to complete in accordance with Clause 3.11

Completion of the sub-contract works

Table 1	**Clause 3.3.1** Consequences of the architect's instruction following the failure of the contractor to enter into a sub-contract with the person named in the contract documents		
Clause 3.3.1 architect's instructions – Options	(a) Remove the impediment by changing the particulars	(b) Omit the work (see note below)	(c) Omit the work and substitute a provisional sum

Consequences of
the instruction
	Clause 3.3.1
	— Valued as a variation – clause 3.7
	— Extension of the time – clause 2.3
	— Disturbance of progress – clause 4.11
			Any subsequent instruction as to the expenditure of a provisional sum treated as follows:
			— Valued as a variation
			    – Clause 3.7
			— Extension of time
			    – Clause 2.3
			— Disturbance of progress
			    – Clause 4.11

Table 2	**Clause 3.3.3** Consequence of the architect's instruction following the contractor's determination of the employment of the named sub-contractor in accordance with the provisions of Clause 27.1 or 27.2 of NAM/SC		
Clause 3.3.3 architect's instructions – Options	(a) Name another person to execute the outstanding balance of work	(b) Contractor to complete (can sub-contract with consent Clause 3.2)	(c) Omit the outstanding balance of work (see note below)
Consequences of the instruction where the sub-contractor is named in the contract documents	**Clause 3.3.4(a)** — Contract sum adjusted by\ net difference in price of first and second named sub-contractor for the remaining work (discounting any cost included for rectification of defects in the work of the first named sub-contractor) — Extension of time – Clause 2.3 N.B. Clause 4.11 (disturbance of progress) does not apply	**Clause 3.3.4(b)** — Valued as a variation – Clause 3.7 — Extension of time – Clause 2.3 — Disturbance of progress – Clause 4.11	
Consequences where the sub-contractor is named in an instruction as to the expenditure of a provisional sum	**Clause 3.3.5** — Valued as a variation – Clause 3.7 — Extension of time – Clause 2.3 — Disturbance of progress – Clause 4.11		
Recovery of monies from first named sub-contractor	**Clause 3.3.6(b)** Contractor to take reasonable action to recover from the named sub-contractor the additional amounts payable to the contractor by the employer together with employer's lost liquidated damages		
Where the employment of the named person is determined otherwise than in accordance with Clause 27.1 or Clause 27.2 of NAM/SC	**Clause 3.3.6(a)** The provisions of Clause 3.3.4(a) or 3.3.4(b) or 3.3.5 as applicable (see above) shall apply in any adjustment arising out of the instruction but only to the extent that they result in a reduction of the contract sum N.B. Clause 2.3 (extension of time) and Clause 4.11 (disturbance of progress) do not apply		

Note: Where the architect's instruction omits work then the employer can complete under the provisions of Clause 3.11

named in the [contract documents] . . .'. The second named person is not a person so named.

The system of naming adopted in IFC 84 is represented in the diagram and tables on pages 125 and 126.

Clause 3.3.1

Named persons as sub-contractors

3.3.1   Where it is stated in the Specification/Schedules of Work/Contract Bills that work described therein for pricing by the Contractor is to be executed by a named person who is to be employed by the Contractor as a sub-contractor the Contractor shall not later than 21 days after entering into this Contract enter into a sub-contract with the named person using Section III of the Form of Tender and Agreement NAM/T referred to in the 1st Recital.

If the Contractor is unable so to enter into a sub-contract in accordance with the particulars given in the Contract Documents he shall immediately inform the Architect/ the Contract Administrator and specify which of the particulars have prevented the execution of such sub-contract. Provided the Architect/the Contract Administrator is reasonably satisfied that the particulars so specified have prevented such execution the Architect/the Contract Administrator shall issue an instruction which may

(a) change the particulars so as to remove the impediment to such execution; or

(b) omit the work; or

(c) omit the work from the Contract Documents and substitute a provisional sum.

An instruction under clause 3.3.1(a) or 3.3.1(b) shall be regarded as an instruction under clause 3.6 requiring a Variation which shall be valued under clause 3.7 and the provisions of clauses 2.3 *(Extension of time)* and 4.11 *(Disturbance of progress)* as relevant shall apply. Where the instruction is under clause 3.3.1(b) the Employer may, subject to the terms of clause 3.11, have the work omitted executed by a person to whom clause 3.11 refers. An instruction under clause 3.3.1(c) shall be dealt with in accordance with clause 3.3.2.

The Contractor shall notify the Architect/the Contract Administrator of the date when he has entered into the sub-contract with the named person. The Architect/the

Contract Administrator may, but, subject to clause 3.3.3,
not after the date so notified, issue an instruction that the
work to which this clause 3.3.1 refers is to be carried out
by a person other than the person named in the Specifi-
cation/Schedules of Work/Contract Bills. Such instruction shall omit the work from the Contract Documents
and substitute a provisional sum which shall be dealt
with in accordance with clause 3.3.2.

COMMENTARY ON CLAUSE 3.3.1

This clause deals with one of only two available methods which this
contract provides for the employer to select a sub-contractor to
carry out part of the work, the other method being described in
clause 3.3.2.

The employer must provide a detailed description of the work in
the specification/schedules of work/contract bills. This description
of the work, together with other necessary information, is communicated to the proposed sub-contractor by use of a standard form
of tender known as Tender and Agreement NAM/T. Section I of
NAM/T is the invitation to tender and it will contain such matters
as a list of those potential main contractors invited to tender and the
main contract information, if available e.g. the appendix entries
from the proposed main contract (and departures from this for the
purposes of the sub-contract where these are known). It will contain
the expected commencement dates and periods of sub-contract
work and other relevant information. The sub-contractor whom it is
proposed to name will, in section II, provide information as to his
requirements.

Once the selection has been made by the employer, details of the
sub-contract price and all the other relevant information are
included in the main contract tender documents to form part of the
information on which the contractor tenders. All the documents on
which the sub-contract tender is based should be provided to the
main contractor.

These will include, as appropriate, drawings, bills of quantities and
a specification or schedule of works (see the last paragraph of the
first recital to IFC 84). If a specification is included on its own and is
not priced, the sub-contractor's tender being based on drawings and
a specification only, the sub-contractor should supply a sub-contract
sum analysis or a schedule of rates in support of his tender.
Wherever possible of course, the sub-contract documentation
should mirror that used for the main contract.

The contractor prices the sub-contract work himself and it is this price which forms part of the contract sum of the main contract. If the main contractor and named sub-contractor enter into a sub-contract and the named sub-contractor completes the sub-contract works, the named sub-contractor's price will be of little concern to the employer. However, should the employment of the named sub-contractor be determined, the sub-contract price could, as will be seen later, become of considerable significance to the employer. Once the main contract has been entered into, the contractor has just 21 days in which to enter into a sub-contract with the named person on the basis of sections I and II of NAM/T and this is achieved by completing the Agreement in Part III by which the standard sub-contract for named persons, known as NAM/SC, is incorporated. The contractor must notify the architect of the date when the named sub-contract has been entered into.

This rather short period of 21 days is likely to put time at a premium. It may well be therefore that on occasions the proposed main contractor will make contact with the person who has been named in the main contract tender documents during the main contractor's own tender period in order to clarify any uncertain areas and also, if the main contract is being let some considerable time after the completion of sections I and II of NAM/T, to confirm that the details contained therein are still applicable. Such a practice may well be regarded unfavourably by the sub-contracting side of the industry.

It is advisable for the architect, just before accepting the tender of a main contractor, to check with the person whom it is proposed to name as a sub-contractor, that there is no departure from the basis on which he has agreed to carry out the work, or that if there are, that this can be quickly resolved. This should reduce to a minimum the chances of a disagreement arising between the main contractor and proposed named sub-contractor during the 21 day period, particularly in those cases where there has been no contact between main contractor and named sub-contractor before the commencement of that period.

Since the contractor tenders on the basis of the particulars given in the contract documents in relation to the sub-contract works, it is to be hoped that those particulars will remain unchanged and form the basis of the sub-contract. However, should any difficulty occur in the proposed named sub-contractor and contractor agreeing the particulars which is not resolved between them during the 21 day period, the contractor must immediately inform the architect of this fact, specifying which of the particulars in the contract documents has caused the impasse.

If the proposed named sub-contractor withdraws his offer completely during the 21-day period for reasons unconnected with any problem concerning the particulars e.g. a liquidation of the sub-contractor, the architect is able to issue an instruction requiring that the work be carried out by some other person. This will be done by omitting the work from the contract documents and substituting a provisional sum to be dealt with in accordance with clause 3.3.2 – see last paragraph of clause 3.3.1.

Contractor's inability to enter into a sub-contract in accordance with the particulars given in the contract documents

The second and third paragraphs of clause 3.3.1 deal with the situation where the contractor and sub-contractor cannot agree to enter into a sub-contract based upon the particulars given in the contract documents. The drafting of these paragraphs probably proceeds on the assumption that the contractor and proposed sub-contractor cannot agree with one another as to the particulars. What if they are in agreement that the particulars should change, e.g. programme times for the sub-contract works being varied from that in the particulars given in NAM/T? There is no impasse in such a situation. The problem is that such a change is strictly speaking a variation and must therefore be the subject of an instruction from the architect. No doubt this can be done by the architect issuing an instruction under clause 3.3.1(a) but this does appear to be using a sledgehammer to crack a nut. Provided the terms of the main contract remain unaffected, particularly in terms of the contractor's price for the named sub-contract work, it would have been sensible to build into clause 3.3 a facility for the contractor and proposed sub-contractor to agree to change the particulars subject to approval by the architect, not to be unreasonably withheld, and for this approval not to amount to a variation requiring a formal instruction.

Provided the architect is reasonably satisfied that the particulars specified by the contractor as preventing him from entering into a sub-contract have in fact caused the impasse, he may issue an instruction which may:

(a)   change the particulars so as to remove the impediment;

(b) omit the work;
(c) omit the work from the contract documents and substitute a
     provisional sum;

These three possibilities will be considered in turn:

(a) Change the particulars so as to remove the impediment
Such a change will very often have financial implications. So far as
the contractor is concerned, the instruction changing the particulars
is to be regarded as a variation and will therefore be valued
accordingly and clauses 2.3 (extension of time) and 4.11 (loss and
expense) will apply where appropriate. The extent to which they
will apply is not easy to discern.

To take an example, suppose there is a failure to agree between
main contractor and proposed named sub-contractor on the periods
required for the sub-contract works in relation to a proposed sub-
contract for a small amount of piling work to be carried out in the
very early stages of the contract. The employer may have been late
in going out to tender and what was originally envisaged by the
proposed named sub-contractor as a summer contract is now a
winter contract so that the proposed named sub-contractor requires
further time in which to carry out the sub-contract works and further
money in respect of it. If the architect now changes the particulars in
order to accommodate the proposed named sub-contractor's longer
period during which to carry out the works and increased sub-
contract price, what is the contractor's position? Do the changed
particulars have to be agreed by both the proposed named sub-
contractor and the contractor in order 'to remove the impediment'
or must the contractor accept the changed particulars even though
they will have a direct bearing on the overall completion date and
the figure at which he priced the sub-contract work? Presumably, if
he must accept these changed particulars, whether he likes them or
not, his recompense will be found in the valuation of the instruction,
the extension of time and the loss and expense provisions.

(i) Valuation of the instruction
It is highly likely that the valuation will be made under clause 3.7.4.
This will often be based on the new sub-contract price with perhaps
a percentage added for the contractor's pricing of the work. By
clause 3.7.7 no loss and expense can be included in this valuation
although it can be recovered, if appropriate, under clause 4.11.

(ii)   Extension of time

It is only a delay in the date for completion due to compliance with an instruction which changes the period for the sub-contract works which will be allowed, should this change mean that the contract works will not be completed by that date. As clause 2.4.5 is expressly restricted to *compliance* with instructions under clause 3.3, delays prior to the instruction cannot, of course, be attributed to compliance with it.

(iii)   Loss and expense

The loss and expense must arise as a result of the regular progress of the works being materially affected due to the instruction. If the changed particulars affect the contractor's programme then it is quite probable that his regular progress will be disturbed and he will accordingly be entitled to reimbursement of any loss and expense incurred. Again, the period up to the issue of the instruction changing the particulars so as to remove the impediment is not covered – see clause 4.12.7.

Delays and disturbances to regular progress occurring between the contractor notifying the architect that a sub-contract cannot be entered into and the architect issuing an instruction changing the particulars etc.

It is submitted that, provided the architect acts within a reasonable time in issuing the instruction changing the particulars etc., there can be no claim for an extension of time or for loss and expense. To the extent that any delay is unreasonable there is a possibility that the contractor can claim an extension of time under clause 2.4.7 (late instructions etc.) and loss and expense under clause 4.12.1 (late instructions etc.), provided the appropriate application for instructions referred to therein is made by the contractor. (The notification given by the contractor in accordance with the second paragraph of clause 3.3.1 may in itself be a sufficient application for the necessary instructions required by clauses 2.4.7 and 4.12.1. To be certain the contractor should expressly request instructions.) Support for this view, it is suggested, is to be found in the important case of *Percy Bilton Ltd* v. *Greater London Council* (1982), which was a House of Lords' case dealing with a form of contract based on JCT 63.

Facts:

The contract was on the GLC's own conditions of contract but based on JCT 63. It involved the question of extensions of time and liquidated damages following non-performance by a nominated sub-contractor. The nominated sub-contractor for mechanical services went into liquidation. The main contractor, relying on the principle laid down in the case of *North-West Metropolitan Regional Hospital Board* v. *T.A. Bickerton & Sons Ltd* (1970), called on the employer through his architect to make a fresh nomination. However, there was a considerable delay in making the new nomination due mainly to the fact that the firm first proposed for the renomination withdrew without ever starting work. Another sub-contractor was eventually nominated some four months after the repudiation by the first nominated sub-contractor. There were thus two parts to the delay: the first part which was regarded as a reasonable time and the second part which was regarded as an unreasonable time between repudiation and renomination.

Held:
The employer had a major responsibility for the unreasonable delay and under clause 23(f) (extension of time for late instructions under JCT 63) the contractor should have had a reasonable extension of time in respect of this part of the delay. In relation to the first part of the delay i.e. that which was a reasonable period after the repudiation by the first nominated sub-contractor, there was nothing in JCT 63 which imposed what would, in effect, be a warranty on the employer that the nominated sub-contractor would invariably carry on work continuously. This would place an unreasonable burden on the employer. It was the duty of the employer, acting through his architect, to give instructions for a renomination within a reasonable time. The contractor was therefore not entitled to an extension of time for the first part of the delay, and if this meant that the contractor was late in completing then he would suffer liquidated damages.

It is submitted that there is nothing in IFC 84 to suggest that any principles other than those stated in this case would apply.

If owing to the absence of the required application for instructions by the contractor he is not entitled to claim under those clauses, it is still possible that the main contractor can sue for damages for breach of contract by the employer because of the architect's unreasonable delay in issuing instructions.

(b) Omit the work
An instruction omitting the work is to be regarded as a variation. It

will be valued as appropriate under clause 3.7.2 and will result in an adjustment downwards of the contract sum. Again the extension of time and loss and expense provisions apply but it is unlikely that any delay or disruption will be caused by a straight-forward omission of work.

Where an instruction requiring such an omission is given, the omitted work may be carried out by another contractor engaged directly by the employer. The contractor's consent (which shall not be unreasonably withheld) will be required to this course of action by virtue of clause 3.11.

(c) Omit the work from the contract documents and substitute a provisional sum

It may not be possible for the proposed named sub-contractor and contractor to reach an agreement which results in a concluded sub-contract. The architect cannot simply instruct the contractor to enter into a sub-contract with another named sub-contractor e.g. the next lowest sub-contracting tenderer, as perhaps the architect may do in nominating a sub-contractor under JCT 80. The contractor will not have priced the work on this changed basis. The option open to the architect is to omit the sub-contract work and then to substitute a provisional sum giving instructions naming a sub-contractor in accordance with clause 3.3.2 (dealt with below).

This instruction may follow protracted negotiations which may have caused delay and disruption. It is submitted that, provided the architect acts with reasonable expedition, there can be no claim by the contractor for an extension of time or loss and expense (see *Percy Bilton Ltd* v. *Greater London Council* (1982)). But there could be other problems. For example, the delays running prior to the instruction could produce a result whereby whoever is named under the provisional sum cannot complete the sub-contract work within suffi-cient time to enable the main contractor to complete by the contract completion date. If this is so, and if no extension of time can be given under either clause 2.4.5 (clauses 3.8 or 3.3) because in truth the effective delay was already suffered before the instruction was issued so that *compliance* with it did not as such *cause* any delay, the contractor can most probably object under clause 3.3.2(c) within 14 days of the date of the issue of the instruction.

Independently of any failure to agree between the main con-tractor and the proposed named sub-contractor, at any time before a sub-contract with a named person has been entered into, the architect may issue an instruction omitting the work from the contract documents and substituting a provisional sum to which clause 3.3.2 will apply.

NOTES TO CLAUSE 3.3.1

[1] '. . . for pricing by the Contractor . . .'.
It is clear that for the purposes of this contract the employer will generally only be concerned with the contractor's price and not that of the named sub-contractor. Provided the named sub-contractor enters into a sub-contract and subsequently completes the work, the employer has no interest in the price submitted by the named sub-contractor. It is only in the event of the sub-contract not being entered into or of the employment of the named sub-contractor being determined that the sub-contractor's price becomes of interest to the employer. The extent of the employer's interest in the named sub-contractor's price is dealt with in the Commentary to clause 3.3.3.

[2] '. . . not later than 21 days after entering into this Contract . . .'
This could be a tight timetable and it is envisaged that, prior to the main contract being let, the contractor, sub-contractor and architect may have liaised to ensure that the basic agreement holds good in relation to essential terms and particulars in the main contract tender documents.

[3] '. . . in accordance with the particulars given in the Contract Documents . . .'
The particulars are those in the tender of the named person contained in sections I and II of the Form of Tender and Agreement NAM/T (see first recital to IFC 84). The particulars required are extensive, covering not only main contract information but also the sub-contractor's price, dates of anticipated commencement on site and of the periods required to carry out the work, periods of notice required, periods required for the submission of sub-contractor's drawings, insurances, fluctuations etc.

[4] '. . . which shall be valued under clause 3.7 . . .'
If an instruction is given under clause 3.3.1(a) then no doubt the valuation will often be carried out under clause 3.7.4.

[5] '. . . not after the date so notified . . .'
In practice, whether or not the architect has been so notified, if the

sub-contract has been entered into, it will be too late for the architect to instruct that the work is to be carried out by some other person. The architect has no power of course to instruct the contractor to breach the sub-contract conditions.

Clause 3.3.2

> 3.3.2   (a) In an instruction as to the expenditure of a provisional sum under clause 3.8 the Architect/the Contract Administrator may require work to be executed by a named person who is to be employed by the Contractor as a sub-contractor.
>
> (b) Any such an instruction shall incorporate a description of the work and all particulars of the tender of the named person for that work in a Form of Tender and Agreement NAM/T with Sections I and II completed together with the Numbered documents referred to therein.
>
> (c) Unless the Contractor shall have made reasonable objection[1] to entering into a sub-contract with the named person within 14 days of the date of issue of the instruction he shall enter into a sub-contract with him using Section III of the Form of Tender and Agreement NAM/T for the execution of the said work.

COMMENTARY ON CLAUSE 3.3.2

By the use of this clause, the architect may, by issuing an instruction as to the expenditure of a provisional sum after the contract has been entered into, select a named person to carry out sub-contract work after the contract has been entered into. The instruction must provide the detailed information contained in sections I and II of NAM/T. However, if the work concerned is crucial to the contractor's programme, it is the better practice to name the sub-contractor when the main contract tenders are being invited and allow the contractor to price for the work in his own tender.

If compliance with the instruction involves the contractor in delay or disturbance to regular progress e.g. where the named sub-contractor requires a period for completing the sub-contract works which will necessarily cause delay and disruption to the contractor, the contractor will be entitled to an extension of time under clauses 2.3 and 2.4.5 and loss and expense under clauses 4.11 and 4.12.7.

Once the sub-contract has been entered into, any delay or disturbance suffered by the contractor due to the activity or inactivity of the named sub-contractors will not attract an extension of time or loss and expense as it will not be due directly to compliance with the instruction under the provisional sum.

Many of the risks associated with a failure by the named sub-contractor to complete the sub-contract works fall squarely on the employer (see clause 3.3.5).

There is no provision in IFC 84 either for the contractor to make his own allowance for attendance and profit on the expenditure of provisional sums in favour of named sub-contractors or for payment thereof. In practice some suitable system is often devised to cater for this. If nothing is provided it is difficult to see how the contractor can recover anything for attendance and profit.

In practice some suitable system is often devised to cater for this. One method is to include an item for the contractor to price, but bearing in mind the domestic nature of the subsequent sub-contract, it may be unwise to describe the item as being an allowance for attendance and profit. However, the inclusion of a priceable item has the merit of providing a firm basis for subsequent adjustment and of making the amount inserted by the contractor subject to competition.

NOTES TO CLAUSE 3.3.2

[1] 'Unless the Contractor shall have made reasonable objection . . . '.

Under clause 3.3.1 if the main contractor is dissatisfied with the selection arrangements or conditions, he can always elect not to submit a tender for the main contract work. This opportunity does not arise under clause 3.3.2. Accordingly, the contractor has a right of reasonable objection which, if he intends to exercise it, must be exercised within 14 days of date of the issue of the instruction. This period does not give the contractor long to assess the particulars given in NAM/T. If any of the particulars are not acceptable, or if any outstanding items are not settled within the 14 day period, the contractor should report this fact to the architect within the period.

If the right to reasonable objection is not exercised, it might be felt by some employers that thereafter the contractor should take the financial and other risks of a failure to perform by the sub-contractor. However, this is not the case in IFC 84, for although the contractor does take the risk of delays and disruption caused by the named sub-contractor up to the point of the determination of the named sub-contractor's employment, thereafter he is entitled to seek an extension of time and loss and expense, if any, suffered as a result of the instruction given by the architect under clause 3.3.5.

Further, any increased cost of completing the sub-contract works will itself fall upon the employer.

It is submitted that, in considering what is a reasonable ground for objection, factors other than the sub-contract particulars may be considered e.g. the contractor's previous experience with the proposed named sub-contractor by which he can demonstrate that the sub-contractor concerned has in the past had difficulty in achieving a reasonable standard of work or in achieving reasonable progress.

Clause 3.3.3

> 3.3.3 The Contractor shall not determine the employment of the named person otherwise than under clause 27.1 or 27.2 of the Sub-Contract Conditions NAM/SC, nor let that employment be determined by accepting repudiation[1] of the sub-contract.
>
> The Contractor shall advise the Architect/the Contract Administrator as soon as is reasonably practicable of any events which are likely[2] to lead to any determination of the employment of the named person howsoever arising.[3]
>
> Whether or not such advice has been given if, before completion[4] of the sub-contract work, the employment of the named person is howsoever determined, the Contractor shall notify the Architect/the Contract Administrator in writing stating the circumstances. The Architect/the Contract Administrator shall issue instructions as may be necessary in which he shall:
>
> (a) name another person[5] to execute the work or the outstanding balance of work in accordance with clause 3.3.2(b) and subject to clause 3.3.2(c), or
>
> (b) instruct the Contractor to make his own arrangements for the execution of the work or the outstanding balance of the work, in which case the Contractor may sub-contract the work in accordance with clause 3.2, or
>
> (c) omit the work or the outstanding balance of work.

COMMENTARY ON CLAUSE 3.3.3

As it is the employer who has selected the named person to be a sub-contractor to the contractor in the event that the sub-contract works are not completed by the named sub-contractor, this clause requires the employer through the architect to give appropriate instructions as to the unfinished work. Certain financial consequences flow depending on which option contained within this clause is chosen and the method by which the named sub-contractor's employment is determined. These financial consequences operate as a safeguard to the contractor in the event of the employment of the named sub-contractor being determined by the contractor under clause 27.1 or clause 27.2 of NAM/SC. Looked at another way, it is the cost to the employer in exercising the privilege of selection of a sub-contractor.

The first paragraph of clause 3.3.3 provides that the contractor shall not determine the employment of the named sub-contractor otherwise than under clause 27.1 or clause 27.2 of NAM/SC. Nor must he let that employment be determined by accepting a repudiation by the named sub-contractor of the sub-contract itself. The reasons for this are tied in with clause 3.3.6(b) regarding the obligation on the contractor to pursue the sub-contractor for the extra costs incurred by the employer following the architect's issuing an instruction under clause 3.3.3 in relation to the unfinished work. It will be seen later that there are certain financial consequences flowing from an instruction under this clause, and depending on the type of instruction issued, the employer will have to pay for the increased costs of having the work finished and is likely to be deprived of liquidated damages. Furthermore he may have to pay the contractor loss and expense.

The method of recovery is dealt with in the commentary to clause 3.3.6(b). However it is worth noting at this point that the main contractor is required to seek recovery of the employer's losses etc. from the named sub-contractor under clause 27.3.3 of NAM/SC which imposes a contractual obligation on the named sub-contractor to pay these sums to the main contractor who then holds any sums recovered to the account of the employer. However, as the claim by the main contractor will be one under, and in accordance with, the contract rather than for breach of it, it is essential that when the named sub-contractor's employment is determined, the sub-contract conditions themselves remain intact in order that clause 27.3.3 can be relied on by the main contractor. If the sub-contract itself is brought to an end by the contractor accepting a repudiation of it by the named sub-contractor, the effect will be that with one or two exceptions e.g. the arbitration clause, the sub-contract will fall to the ground and with it the contractor's right to recover the employer's losses. It is for this reason that the contractor must not determine the employment of the named sub-contractor otherwise than under clause 27.1 or 27.2 of NAM/SC. It will be seen in the commentary to clause 3.3.4 and 3.3.5 that, should the named sub-contractor's employment be determined in any way other than under clause 27.1 or 27.2 of NAM/SC, the financial consequences to the contractor can be very severe indeed.

Futhermore, the restriction as to the manner in which the contractor can bring the named sub-contractor's employment to an end can pose the contractor with serious problems.

Firstly, it may be that the factual situation does not easily lend itself to a system of notices which is required in order for the contractor to rely upon the provisions of clause 27.1 or 27.2 of NAM/SC. The relationship between named sub-contractor and contractor may have deteriorated to a level whereby the named sub-contractor has left the site.

The main contractor may well be keen to prevent any delays occurring which could affect the overall completion date. However in such a situation he must still serve the appropriate notices and wait the appropriate time before he can determine the employment of the named sub-contractor.

Secondly, it may be that for good reason the contractor may wish to keep his options open in relation to the determination of the named sub-contractor's employment. He may wish to purport to determine not only in accordance with the provisions of clause 27.1 or 27.2 but also, if it turns out that for some reason the sub-contractor's employment was not validly determined, on the basis of accepting a repudiation of the contract on the part of the named sub-contractor.

Thirdly, the contractor could inadvertently fail to follow the correct procedures for serving the appropriate notices under clause 27.1 or clause 27.2. This may not become apparent until some considerable time after the notices have been served and the named sub-contractor has left the site. If the determination was invalid because of a defective notice then the contractor will have failed to determine the named sub-contractor's employment in accordance with clause 27.1 or 27.2 and potentially grave financial consequences follow for the contractor.

Fourthly, there may be situations where the contractor simply cannot determine the named sub-contractor's employment under clause 27.1 or clause 27.2 e.g. the death of a named sub-contractor who is an individual or the frustration in law of the sub-contract. These situations are likely to be extremely rare and pre-sumably in such a case it would not amount to a breach by the contractor of clause 3.3.3. It is uncertain in such a case who will have to meet the increased costs etc. of having the work completed. By a literal reading of clause 3.3.6(a) the costs would appear to fall on the contractor. This might be regarded by some as harsh.

There is no doubt that the restriction imposed on the contractor as to the method by which the named sub-contractor's employment may be determined represents a radical step which could well be a cause of considerable concern to contractors.

The second paragraph of clause 3.3.3 requires the contractor to advise the architect as soon as reasonably practicable of any events which are likely to lead to a determination of the employment of the named person, howsoever that determination may arise. This will enable the architect and employer to consider or take appropriate steps, perhaps to contain or improve a deteriorating situation. Further, the third paragraph of clause 3.3.3 requires the contractor to notify the architect in writing of the determination of the employment of the named sub-contractor. The written notice must state the circumstances in which the determination has come about.

Once notified of the determination of the named sub-contractor's employment, the architect must issue any necessary instructions which must include one of the three options set out in this clause. He must:

(a) name another person to execute the work or the outstanding balance of work in accordance with clause 3.3.2(b) and subject to clause 3.3.2(c); or

(b) instruct the contractor to make his own arrangements for the execution of the work, in which case the contractor may sub-contract the work in accordance with clause 3.2; or

(c) omit the work requiring to be completed.

Although these alternatives are separated by what appears to be a disjunctive 'or', there seems no good reason why more than one should not be used for different elements of the outstanding work e.g. the naming of a specialist under (a) to test and commission some specialist plant with the contractor himself being instructed under (b) to finish off other associated work relating thereto. However, it must be said that the words 'the work' used in (a), (b) and (c) of this clause do not facilitate this interpretation.

The financial consequences flowing from an instruction under this clause are dealt with under clause 3.3.4 and 3.3.6 where the naming of a sub-contractor arises under clause 3.3.1, and in clause 3.3.5 and 3.3.6 where a provisional sum has been used under clause 3.3.2.

NOTES TO CLAUSE 3.3

[1]   '. . . nor let that employment be determined by accepting repudiation . . .'.
It is clear that the sub-contractor's employment can come to an end either by the operation of clause 27.1 or 27.2 or by the contractor (albeit wrongly under this clause) accepting a repudiation of contract on the part of the named sub-contractor. Without these words it might have been arguable whether the acceptance of a repudiation by the contractor which brought the sub-contract itself to an end involved a 'determination of the employment' of the named sub-contractor as that phrase is used throughout this contract.

[2]   '. . . which are likely . . .'
This advice could be useful to the architect in relation to his general supervision and control of the works. There will no doubt be a temptation to mediate between the contractor and named sub-contractor which could in certain circumstances be advantageous, provided it is done with extreme care so as not to either directly influence the contractor's decisions or prejudice the employer's position. Whatever else he does, the architect should familiarise himself with the problem. A failure by the contractor to so advise the architect could lead to the contractor being the author of his own misfortune. For instance, advance warning to the architect of a possible determination of the named sub-contractor's employment may enable the architect to make some preliminary investigations or enquiries in advance of the possible issue of an instruction under clause 3.3.3(a), (b) or (c) and, should a determination in fact occur, such advance investigations or enquiries could reduce the time reasonably taken by the architect between the determination of the named sub-contractor's employment and the issue of the instruction in relation to the unfinished work, thereby reducing what would otherwise be the delay for which the contractor, and not the employer, is responsible (see *Percy Bilton Ltd* v. *Greater London Council* (1982)).

[3]   '. . . howsoever arising'.
i.e. whether under clause 27.1 or 27.2 of NAM/SC or in any other way.

[4]   '. . . before completion . . .'.
This means 'apparent completion'. See the case of *City of*

Westminster v. *Jarvis* (1970) in which the House of Lords held that completion meant completed ready to hand over, even if in reality, but unknown to the parties, there was defective work. If the named sub-contractor's work is apparently completed i.e. it has achieved practical completion within the meaning of clause 15 of NAM/SC, then it is suggested that it is completed for the purposes of this clause even if this is before practical completion of the works as a whole. If this is so and the named sub-contractor's employment is determined after such apparent completion but before practical completion of the whole of the works, then the provisions of this clause do not apply so that e.g. any failure by the named sub-contractor to remove or remedy work found to be defective during this period will (unless it is a question of defective design for which the contractor is not responsible) be solely the contractor's responsibility and cannot be dealt with by an instruction under this clause.

[5] . . . name another person . . .'.
This appears to preclude the possibility of renaming the same person. This might be desirable e.g. where a receiver is appointed to manage the business of a named sub-contractor and the contractor decides to determine the named sub-contractor's employment on this ground (see clause 27.2 of NAM/SC), the named sub-contractor may be in a position, with the agreement of the receiver, to complete the sub-contract works in accordance with the sub-contract as to time and price. In such a case the employer might very much like to rename the same sub-contractor. However, in such circumstances, the employer does have the option of issuing an instruction under clause 3.3.3(c) to omit the outstanding balance of work which can then form the basis of a direct contract between the employer and the sub-contractor (see last paragraph of clause 3.3.4).

Clause 3.3.4

3.3.4   The following provisions of this clause 3.3.4 shall apply where an instruction referred to in clause 3.3.3(a)–(c) arises in respect of a person named[1] in the Specification/Schedules of Work/Contract Bills under clause 3.3.1 whose employment has been determined under clause 27.1 or 27.2 of the Sub-Contract Conditions NAM/SC.

(a) such an instruction under clause 3.3.3(a) shall be

regarded as an event to which clause 2.3 (*Extension of time*) applies, but not as a matter to which clause 4.11 (*Disturbance of progress*) applies, and the Contract Sum shall be adjusted by the amount of the increase or the reduction in the price of the second named sub-contractor for the work not carried out by the first named sub-contractor when compared with the price of the first named sub-contractor for that work;[2] provided that in the foregoing adjustment there shall be excluded from the price of the second named sub-contractor any amount included therein for the repair of defects[3] in the work of the first named sub-contractor;

(b) such an instruction under clause 3.3.3(b) or (c) shall be regarded as one requiring a Variation which shall be valued under clause 3.7 and as an event to which clause 2.3 (*Extension of time*) applies and as a matter to which clause 4.11 (*Disturbance of progress*) applies.

Where the instruction is under clause 3.3.3(c) the Employer may, subject to the terms of clause 3.11, have the omitted work executed by a person to whom clause 3.11 refers.

COMMENTARY ON CLAUSE 3.3.4

This clause deals with the financial consequences which arise when the employment of a sub-contractor named under clause 3.3.1 has been determined by the contractor under clause 27.1 or clause 27.2 of NAM/SC, and the architect issues instructions with regard to the unfinished sub-contract work under clause 3.3.3.

Clause 27.1 and 27.2 of NAM/SC

The operation of clause 3.3.4, dealing as it does with payment by the employer to the contractor in respect of the costs of, and associated with the completion of the outstanding work etc. is of vital importance to the contractor. However the provisions of clause 3.3.4 operate only where the named sub-contractor's employment has been determined by the contractor under clause 27.1 or clause 27.2 of NAM/SC. It is of the utmost importance therefore that the contractor takes great care to ensure that any determination is in accordance with those clauses.

Clause 27.1 provides that the contractor may determine the employment of the sub-contractor if the latter defaults in any of the following ways:

(i)   without reasonable cause suspends the carrying out of the works;

(ii)   without reasonable cause fails to proceed with the works reasonably in accordance with the progress of the main contract works;

(iii)   refuses or neglects after notice in writing to remove defective work or improper materials thereby materially affecting the works or wrongfully failing to rectify defects etc.;

(iv)   failing to obtain consent to sub-letting.

Clause 27.1 provides for the issue of a notice specifying the default. It must be sent by registered post or recorded delivery. If the default then continues for 10 days after receipt of the notice or at any time thereafter is repeated, the contractor may within 10 days of that continuance or repetition, again by notice sent by registered post or recorded delivery, forthwith determine the employment of the sub-contractor. Such notice shall not however be given unreasonably or vexatiously. The question of the service of notices is discussed in some detail in chapter 10 dealing with the determination provisions of IFC 84.

Clause 27.2 of NAM/SC entitles the contractor by written notice to determine forthwith the employment of the sub-contractor (without the need for a prior warning notice) in the event of the sub-contractor becoming bankrupt or making a composition or arrangement with creditors, or having a proposal in respect of his company for a voluntary arrangement for a composition of debts or arrangement in accordance with the Insolvency Act 1986, or having an application made under the Insolvency Act 1986 in respect of his company to the court for the appointment of an administrator, or having a winding-up order made, or a resolution for voluntary winding-up being passed, or a provisional liquidator appointed or having an administrative receiver as defined in the Insolvency Act 1986 appointed, possession being taken of any debentures secured by floating charge etc.

The financial consequences set out in clause 3.3.4 depend on which of the three options contained in clause 3.3.3 is chosen by the architect. The position is as follows:

Case (A): The instruction names another person to execute the work or the outstanding balance of work
The instruction does not amount to a variation. Express provision is

made for the adjustment of the contract sum in relation to the cost of completing the outstanding work and also for an extension of time in relation to any delays due to compliance with the instruction. It is also expressly stated that clause 4.11 (dealing with reimbursement of loss and expense) shall not apply, although these words are probably inserted *ex abundanti cautela* (out of an abundance of caution) as nothing in the wording of clause 4.11 or 4.12 would suggest that loss and expense could be claimed arising out of an instruction under clause 3.3.3(a). If therefore the architect's instruction causes the regular progress of the works to be materially affected, any loss and expense in which the contractor is thereby involved cannot be recovered from the employer. No doubt the contractor will look to the first named sub-contractor for recompense. Of course, if the architect takes an unreasonable time in issuing the instruction then clause 4.12.1 (late instructions etc.) may apply.

Adjustment of contract sum
The contract sum is to be adjusted by adding to, or subtracting from it, the amount by which the new sub-contract price for the outstanding work exceeds or is below the price, as the case may be, of the original named sub-contract for that work. It should be noted that, in making the adjustment, it is the price of the first named sub-contractor which must be used in the valuation and not the price for that work included by the contractor in his own tender. Generally, it will involve an addition to the contract sum. It is provided in clause 3.3.4(a) that any amount included in the second named sub-contractor's price for the repair of defects in the first-named sub-contractor's work is to be excluded from the adjustment as the main contractor is responsible for the defective workmanship executed and defective materials supplied by the named sub-contractor. These words put the matter beyond doubt although it is submitted that the wording in clause 3.3.4(a) namely '. . . the price of the second named sub-contractor for the work not carried out by the first named sub-contractor . . .' would in any case be sufficient to enable the architect or quantity surveyor in calculating the additional cost, if any, due to the contractor to discount any part of the second named sub-contractor's price which relates to the cost of remedying defects in the first named sub-contractor's work. This position is in marked contrast to that on the renomination of a sub-contractor under JCT 63 and JCT 80 – see *Fairclough Building Ltd* v. *Rhuddlan Borough Council* (1985).

It will be essential for the architect to obtain from the second named sub-contractor a separate price for the repair of defects so that the adjustment to the contract sum can be made. Even then, the architect or quantity surveyor on his behalf has, it is suggested, considerable problems of evaluation to cope with. A valuation must be made of that part of the first named sub-contractor's price which relates to unfinished work. This is no simple task.

It can be seen that the employer is therefore concerned with the sub-contract price of the named sub-contractor in this instance rather than the contractor's own price for the work.

The Form of Tender and Agreement NAM/T does not in parts readily lend itself to use on renaming, though it is required by clause 3.3.3(a) to be used. For example, there is no express provision requiring the price of the second named sub-contractor to be split between outstanding work and repairing defects, though no doubt this can be required in the 'Numbered Documents' (see NAM/T section I).

The situation where the sub-contractor's employment is determined otherwise than under clause 27.1 or 27.2 of NAM/SC is dealt with in clause 3.3.6(a). Briefly, where the determination is otherwise than under these two clauses, the contract sum cannot be adjusted upwards but only downwards e.g. where a second named sub-contractor's cost for completing the outstanding balance of work results in a saving. Furthermore, the extension of time provisions will not apply.

The financial benefits to the contractor flowing from clause 3.3.4 in the determination of the named sub-contractor's employment are likely to be of immense importance and for them to be lost perhaps because of a technical defect in the serving of an appropriate notice under clause 27.1 or 27.2 of NAM/SC or because the factual situation is such that the contractor is obliged to accept the named sub-contractor's repudiation of the sub-contract seems indefensible.

Extension of time

As clause 2.3 (extension of time) applies, the architect must make an extension of time for completion of the works if such completion has been or is likely to be delayed owing to the contractor's compliance with the instruction (see clause 2.4.5). Any such delay must therefore occur after the instruction is issued. The operation of clause 2.4.5 does not relate to any delay occurring before the instruction e.g. delay on the part of the first named sub-contractor. Once the architect is notified of the determination of the employment of the first named sub-contractor it will of necessity take some time to reach a position whereby an instruction to rename can be issued. The time taken could cause delay and disruption to the contractor's programme. It is submitted that provided the architect acts with reasonable expedition the risks associated with this delay i.e. the imposition of liquidated damages, must be borne by the contractor. However, if entering into a second named sub-contract would inevitably result in the contractor being unable to complete by his contract completion date where the real cause of the delays occurred prior to the instruction (so that no extension of time event under clause 2.4 is appropriate), the contractor could raise a reasonable objection pursuant to clause 3.3.2(c). The implications of the House of Lords' decision in *Percy Bilton Ltd* v. *Greater London Council* (1982) (dealt with earlier on page 134) are again important here.

Difficulties in assessing extensions of time are bound to arise e.g. where compliance by the contractor with the instruction involves the second named sub-contractor completing unfinished work sufficiently late or out of sequence to cause the contractor to complete the total works late.

The delay may be due to one or more of the following factors:

(i)   delay on the part of the first named sub-contractor;
(ii)   the time taken by the architect between determination of the employment of the first named sub-contractor and the instruction naming a second sub-contractor;
(iii) delay on the part of the second named sub-contractor.

Whichever one of these delays is appropriate, it is not compliance with the instruction which caused a delay in completion of the works as a whole. However, if the second named sub-contractor's programme for the unfinished work, to be carried out under similar conditions, shows a longer period for completion of the unfinished work than did that of the first named sub-contractor, this would, if it delayed completion of the work as a whole, entitle the contractor to an appropriate extension of time – similarly, if the sequence of working of the second named sub-contractor differed from that of the first in such a manner that the main contractor was delayed in completing the works.

It is submitted that the delay caused to completion of the works brought about by the time taken by the second named sub-contractor repairing defects in the first named sub-contractor's work will not entitle the contractor to an extension of time.

Where the determination of the employment of the named sub-contractor is not in accordance with the provisions of clause 27.1 or 27.2 of NAM/SC, no extension of time can be given (see clause 3.3.6(a)). For a comment on the potential injustice of this see the last paragraph under the previous heading, page 147.

Case (B): The contractor is instructed to make his own arrangements for the execution of the work or the outstanding work is omitted
This instruction is to be regarded as a variation and will accordingly be valued under clause 3.7. Clause 2.3 (extension of time) and clause 4.11 (loss and expense) apply.

Valuation of variations
No doubt the contractor's rates and prices will form the basis in any valuation. Any changes in the conditions under which any work is to be carried out must be considered. However, there may well be some significant limitations to this. For instance, if prior to the determination of the employment of the named sub-contractor he was guilty of delay which resulted in the outstanding work having to be carried out in a winter period rather than in the summer period for which it was programmed and priced, can that factor be taken into account in the valuation or is it a matter for which the contractor under the contract is wholly responsible? It is submitted that under the contract the main contractor is responsible for the delays of his named sub-contractor and he cannot benefit from his own default so that such factors cannot be taken into account. Where the architect's instruction omits the outstanding balance of work it will be valued under clause 3.7.2.

Where the employment of the named sub-contractor is determined otherwise than under clause 27.1 or clause 27.2 of NAM/SC the variation can only be valued to the extent that it reduces the contract sum – see clause 3.3.6(a). The potential injustice of this provision has already been commented on earlier on page 147.

Extension of time
Clause 3.3.4(b) expressly provides that clause 2.3 (extensions of time) applies to an instruction issued under clause 3.3.3 (b) or (c). The implications of the decision of the House of Lords in *Percy Bilton Ltd* v. *Greater London Council* (1982) dealt with earlier on page 134

are again relevant here. It may be difficult in a given situation to determine whether or not a late finish by the contractor was as a result of complying with an instruction to finish off the sub-contractor's work or was in truth due to the time taken by the architect in issuing the instruction in the first place. The comments under the heading of 'Extensions of time' under case (A) above are also relevant here.

Where the determination of the sub-contractor's employment was not brought about by the contractor under the provisions of clause 27.1 or clause 27.2 of NAM/SC the extension of time provisions do not apply (see clause 3.3.6(a)). This particular aspect of the clause has already been discussed earlier on page 144.

Loss and expense
Clause 3.3.4(b) expressly provides that clause 4.11 (disturbance of progress) shall apply to an instruction issued under clause 3.3.3(b) or (c).

Only such loss and expense as is due to the issue of the instruction is reimbursible. Difficulties in establishing the effective cause of the disturbance are bound to arise and the example instanced under the heading 'Extension of time' under case (A) above should be considered here.

Once again, the implications of the House of Lords' decision in *Percy Bilton Ltd* v. *Greater London Council* (1982) are relevant.

The loss and expense provision does not apply where the employment of the named sub-contractor is determined otherwise than under clause 27.1 or 27.2 of NAM/SC.

Where an instruction under clause 3.3.3(c) is issued, the employer may employ another direct contractor to finish off the work or he may even finish it off himself provided the contractor's consent is obtained in accordance with clause 3.11.

NOTES TO CLAUSE 3.3.4

[1] '*. . . arises in respect of a person named . . .*'
Surely the instruction should relate to the work of the person named and not to the person. The effect of this inappropriate language could prevent this clause applying when it is desired to issue an instruction involving a further naming where it is a second named person whose employment has been determined. This is only compounded by the reference in clause 3.3.4(a) to the second named sub-contractor.

[2] '. . . *the price of the first named sub-contractor for that work* . . .'

The architect, or where appointed the quantity surveyor on his behalf, must endeavour to value the outstanding work. Where there are bills of quantities as a sub-contract document there is unlikely to be much difficulty. If there are no bills then the task is likely to be more difficult. In practice the quantity surveyor is likely to try and agree this value with the contractor if at all possible.

[3] '. . . *for the repair of defects* . . .'
Does repair include complete replacement? The normal meaning of the word repair would tend to suggest something less than complete replacement. Further, elsewhere in the contract e.g. clause 2.10 (defects liability) reference is made to making good defects, which would cover replacement if this was necessary to make good the defect. The choice of word is perhaps unfortunate. It is the defective work rather than the defect itself which is repaired. It is with some diffidence therefore that it is suggested that, in order to make sense of the risk sharing between employer and contractor in the event of the employment of a named sub-contractor being determined, the words used should be construed to extend to replacement as well as mere repair of defective work. There appears to be no ulterior reason for construing the words in a narrow sense. It is further suggested that the words used can extend even to carrying out of work not done by the first named sub-contractor where the defective work is defective because of an omission.

Clause 3.3.5

> 3.3.5 Where an instruction referred to in clause 3.3.3(a)–(c) arises in respect of a person named in an instruction as to the expenditure of a provisional sum under clause 3.3.2 whose employment has been determined under clause 27.1 or 27.2 of the Sub-Contract Conditions NAM/SC, such instruction shall be regarded as a further instruction issued under the provisional sum.

COMMENTARY ON CLAUSE 3.3.5

Where the employment of a named person following on an instruction as to the expenditure of a provisional sum under clause 3.3.2 is determined, the instruction under clause 3.3.3 will be regarded as a further instruction issued under the provisional sum. The financial consequences set out in clause 3.3.4 do not apply as they are applicable only to the situation where the determination is of the employment of a sub-contractor whose work was described as being required to be executed by a named person in the specification/schedules of work/contract bills. However, the further instruction as to the expenditure of the provisional sum issued under clause 3.8 will be valued under clause 3.7 and can give rise to an extension of time and reimbursement of loss and expense – clause 2.4.5 and clause 4.12.7.

Where the sub-contractor's employment is determined otherwise than under clause 27.1 or 27.2 of NAM/SC the valuation will only apply to the extent that the contract sum is adjusted downwards. The extension of time and loss and expense provisions will not apply. This may prove in some situations to be unfair to the contractor – see earlier page 147.

The wording of clause 3.3.2, 3.3.3 and this clause does not appear to rule out the possibility of a determination of the employment of a second or further named person followed by a further instruction. Also, as it remains an instruction as to the expenditure of a provisional sum under clause 3.3.2, the contractor retains a right of reasonable objection as provided therein.

Clause 3.3.6

3.3.6   (a) Where the employment of the named person is determined otherwise than under clause 27.1 or 27.2 of the Sub-Contract Conditions NAM/SC in respect of the instructions under clause 3.3.3(a) or (b) or (c) the provisions of clause 3.3.4(a) or (b) or 3.3.5 as appropriate shall apply but only to the extent that they result in a reduction in the Contract Sum and the instruction shall not be regarded as an event to which clause 2.3 (*Extension of time*) applies or a matter to which clause 4.11 (*Disturbance of progress*) applies.

(b) The following provisions of this clause 3.3.6(b) shall apply where the employment of the named person is determined under clause 27.1 or 27.2 of the Sub-Contract Conditions NAM/SC:
– the Contractor shall take such reasonable action as is necessary to recover from the named sub-contractor under clause 27.3.3 of the Sub-Contract Conditions NAM/SC any additional amounts payable to the Contractor by the Employer as a result of the application of clause 3.3.4(a) or (b) or 3.3.5 together with an amount equal to any liquidated damages that would have been payable or allowable by the Contractor to the Employer under clause 2.7 but for the application of clause 3.3.4(a) or (b) or 3.3.5;
– the Contractor shall account to the Employer[1] for any amounts so recovered;

- in taking such action the Contractor shall not be required to commence arbitration or other proceedings unless the Employer shall have agreed to indemnify the Contractor against any legal costs reasonably incurred in relation thereto;
- if the Contractor has failed to comply with this provision[2] he shall repay to the Employer any additional amounts paid as a result of the application of clause 3.3.4(a) or (b) or 3.3.5 and shall pay or allow an amount equal to the liquidated damages referred to herein.

COMMENTARY ON CLAUSE 3.3.6

By virtue of clause 3.3.6(a) if the named sub-contractor's employment is determined otherwise than under clause 27.1 or clause 27.2 of NAM/SC, the obvious financial benefits to the contractor from the operation of clause 3.3.4 and clause 3.3.5 are not to apply. Clauses 3.3.4 and 3.3.5 will only operate to the extent that their application would in any event result in a reduction in the contract sum e.g. an instruction under clause 3.3.3(c) omitting the outstanding balance of work. The opening words of clause 3.3.3 forbid the contractor from determining a named sub-contractor's employment otherwise than under clause 27.1 or clause 27.2. These two clauses and also the possible difficulties which could be experienced by a contractor faced with a defaulting sub-contractor have been discussed earlier on pages 123 and 147.

No doubt the purpose of the first paragraph of clause 3.3.3 and clause 3.3.6(a) is to try to ensure that the named sub-contractor's employment is determined under clause 27.1 or 27.2 of NAM/SC in order that the main contractor can pursue the sub-contractor under clause 27.3.3 of NAM/SC to recover on behalf of the employer the extra cost incurred by the employer by the operation of the financial provisions in clauses 3.3.4 and 3.3.5. These of course only come into operation when the named sub-contractor is in default. A determination of the named sub-contractor's employment by any other method is likely to prevent recovery under NAM/SC by the contractor of the employer's extra costs. See below. IFC 84 in such a situation therefore deprives the contractor of the financial benefits contained in clause 3.3.4, leaving the contractor to seek recovery from the named sub-contractor of the extra costs incurred by the contractor in complying with the architect's instructions issued under clause 3.3.3(a), (b) or (c).

Clause 3.3.6(b) imposes on the contractor an obligation to

recover from the named sub-contractor under clause 27.3.3 of NAM/SC any additional amounts payable to the contractor by the employer as a result of the application of clauses 3.3.4(a) or (b) or clause 3.3.5, together with an amount equal to any liquidated damages which the employer would have deducted or recovered but for the operation of the extension of time provisions.

Clause 27.3.3 of NAM/SC cross refers to clause 3.3.6(b) and imposes an undertaking on the named sub-contractor not to contend that the contractor has suffered no loss or that the sub-contractor's obligation to pay the contractor should in any way be reduced or extinguished by reason of the operation of clause 3.3.4(a) or (b) or clause 3.3.5 of IFC 84.

It can be seen therefore that in truth the contractor receives the financial benefits provided in clause 3.3.4 but nevertheless has an obligation to obtain a sum equivalent to these benefits from the named sub-contractor which must then be handed over to the employer.

Bearing in mind that the main contractor does not suffer any loss in this regard, it might conceivably be argued that clause 27.3.3 of NAM/SC represents something akin to a penalty provision in that it requires the payment of a sum of money from the named sub-contractor to the contractor which bears no relationship whatsoever to any actual loss suffered or incurred by the contractor. See *Campbell Discount Co. Ltd* v. *Bridge* (1962), particularly the judgments of Lord Denning and Lord Devlin.

However, if the overall scheme is considered, in particular the absence of a direct contractual relationship between employer and named sub-contractor under which the employer can obtain redress for the increased expenditure incurred as a result of the defaults of the named sub-contractor, it is clear that the provisions are reasonable and it would involve a strict application of the doctrine of privity of contract to deprive these provisions of their intended effect. It is hoped that a court or an arbitrator may not adopt too strict an approach based on privity of contract alone, as the overall commercial as well as contractual scheme of things should be considered.

Some, albeit limited, support for the argument that this type of provision is not a penalty is to be found in the Australian case of *Corporation of Adelaide* v. *Jennings* (1985) CLJ 1984–85 Vol. 1 No. 3 at page 205 where Wilson J in the Australian High Court said:

'A sub-contractor who refuses or fails to complete his sub-contract does not escape liability for the reasonable cost of doing the sub-contract work under a plea that a third party (the proprietor) will become liable under the head contract to pay an amount equal to that cost. The natural consequence of his failure to complete is the incurring of the reasonable cost of completing the sub-contract less any unpaid balance of the sub-contract price. Those damages are recoverable under the first branch of the rule in *Hadley* v. *Baxendale* (1954) and it is immaterial that the builder has no claim under the second branch. The builder's right under the head contract to recoup from the proprietor any increase in the cost of completing the sub-contract does not relieve the sub-contractor from liability to the builder; the measure of that liability depends, of course, on the sub-contract. But if a proprietor pays the builder a sum representing an increase in the contract price occasioned by the sub-contractor's default, the proprietor is subrogated to such rights as the builder may have to recover that sum from the sub-contractor.'

Admittedly this judgment was given in relation to a case based on an Australian standard form of building contract known as the RAIA Form of Contract. Even so the principle is sound.

These potential difficulties could have been avoided if a different method of recovery of the employer's increased costs had been employed e.g. either a direct contract between employer and named sub-contractor dealing with the question of the named sub-contractor's default, or for the contractor to suffer the increased costs as a result of the failure of the named sub-contractor to complete the sub-contract works, supported perhaps by an indemnity provision on the part of the employer in favour of the contractor should the contractor be unable, after taking reasonable steps, to obtain recompense from the defaulting named sub-contractor. Of course, the latter alternative has from the contractor's point of view, the distinct disadvantage, when compared with the actual scheme of recovery in IFC 84 and NAM/SC, that the purse strings would be held by the employer and not by the contractor.

Clause 3.3.6(b) requires the contractor to account to the employer for any amounts recovered from the named sub-contractor. The contractor is not required to commence arbitration or other proceedings unless the employer provides an indemnity against legal costs reasonably incurred. If the contractor fails to comply with clause 3.3.6(b) he is required to repay the employer any additional amounts paid by the employer as a result of the application of clause 3.3.4(a) or (b) or clause 3.3.5 together with an amount equal to the lost liquidated damages because of the application of those provisions.

These provisions are complicated and cumbersome and may well prove rather difficult to operate in practice.

NOTES TO CLAUSE 3.3.6

[1] '. . . account to the Employer . . .'.
These words are probably sufficient to impose a trust or fiduciary status on the amount so recovered. In such a case, provided the amounts so recovered are in a separate and identifiable fund in the contractor's hands, the employer will have a right to the sum recovered in priority to the claim of any trustee in bankruptcy, liquidator, or administrative receiver of the contractor.

[2] '. . . with this provision . . .'
This presumably refers to the whole of clause 3.3.6(b). Having regard to the significance of failing to comply, the drafting could do with being made clearer, e.g. by replacing these words with '. . . *with the provisions of this clause 3.3.6(b) . . .'.*

Clause 3.3.7

> 3.3.7 Whether or not a person who has been named as a sub-contractor under any of clauses 3.3.1 to 3.3.5 is responsible to the Employer for exercising reasonable care and skill in:
> — the design of the sub-contract works insofar as the sub-contract works have been or will be designed by the named person;
> — the selection of the kinds of materials and goods for the sub-contract works insofar as such materials and goods have been or will be selected by the named person; or
> — the satisfaction of any performance specification or requirement relating to the sub-contract works;
>
> the Contractor shall not be responsible to the Employer under this Contract for anything to which the above terms relate, nor, through the Contractor, shall the person so named or any other sub-contractor be so responsible;[1] provided that this shall not be construed so as to affect the obligations of the Contractor or any sub-contractor in regard to the supply of goods and materials and workmanship.
>
> The provisions of this clause 3.3.7 shall apply notwith-

standing that the Sub-Contract Sum stated in Article 2 of Section III of the Tender and Agreement NAM/T referred to in clause 3.3.1 or 3.3.2 included for the supply of any design, selection or satisfaction as referred to herein, and that such Sub-Contract Sum is included for within the Contract Sum or the Contract Sum as finally adjusted.

COMMENTARY ON CLAUSE 3.3.7

IFC 84 is a work and materials contract. It is not a design contract and the contractor is not responsible under this contract for design. Therefore, if a named sub-contractor has a design element to perform this clause attempts to make it clear that the contractor is not responsible for failures in relation to such design. The contractor's liability is only in relation to the quality of goods and materials supplied and the standards of workmanship.

The three insets all relate to aspects of design or fitness for their purpose of the sub-contract works. It is perhaps a criticism of this clause that it may cover all matters of design however trivial or insignificant. Even if the named sub-contractor has a certain level of discretion in relation to quite minor matters which could be classified as design e.g. the type of nails to be used or the type of joint in woodwork, this clause may effectively exclude any liability of the contractor in respect of failures in relation to such matters. Such examples are arguably matters of design or fitness for purpose.

It is submitted that the contractor, even under IFC 84, will ordinarily be responsible for the exercise of his discretion in relation to such minor details. Why therefore should he not be responsible where such minor details are left to the discretion of a named sub-contractor? It may be possible for the employer to successfully contend that such minor matters are part of the contractor's obligations in regard to the supply of goods and materials and workmanship and which will therefore be caught by the proviso at the end of the first paragraph of this clause. If this is not the case, the employer must seek a direct design warranty from the named sub-contractor in every case, even where there does not appear to be any element of traditional design in the named sub-contract. A form of design agreement has been produced for use between employer and named sub-contractor by two of the constituent bodies on the Joint Contracts Tribunal, namely the Royal Institute of British Architects and the Committee of Associations of Specialist Engineering Contractors. It is known as ESA/1 and is discussed very briefly later on page 160.

The last paragraph of clause 3.3.7 attempts to make it clear that the contractor has no liability in connection with the design matters referred to in the first paragraph of the clause, even where the contract sum contains, as it is likely to, the design fee of the named sub-contractor within the main contractor's pricing of the work.

NOTES TO CLAUSE 3.3.7

[1] '. . . *nor, through the Contractor, shall the person so named or any other sub-contractor be so responsible . . .'.*
These words prevent the traditional legal chain of liability being established i.e. the employer looking to the contractor, who in turn would look to the sub-contractor, who may look to a sub-sub-contractor etc. through a contractual link down to the point where the ultimate responsibility rests with the person by whom it should properly be borne (though the clause would probably have this effect in any event).

Clause 3.3.8

    3.3.8   Clauses 3.3.1–3.3.7 shall not apply to the execution of part of the Works by a local authority or a statutory undertaker executing such work solely in pursuance of its statutory rights or obligations.

COMMENTARY ON CLAUSE 3.3.8

This is an important clause which states that local authorities or statutory undertakers who execute part of the works cannot become named sub-contractors where their work is carried out in pursuance of statutory obligations i.e. work which they must do; or statutory rights i.e. work which they are not bound to do but which they have the right to do. The word 'rights' is difficult to construe in this context. Does it mean that if a certain kind of work is to be carried out, the local authority or statutory undertaker can insist that they and no one else may do it? Or does it mean work which they have a power to do but only by agreement with the employer? Perhaps the word 'powers' would have been more appropriate.

If work relating to part of the contract works falls to be carried out by a local authority or statutory undertaker in pursuance of its statutory obligations or rights, it may be carried out by a direct arrangement with the employer. If included as a provisional sum for the purpose of placing the responsibility for delays or failures of the local authority or statutory undertaker on the contractor, it will probably not work. See the case of *Henry Boot Construction Ltd* v. *Central Lancashire New Town Development Corporation* (1980).

Contractor's failure to comply with the provisions of clause 3.3

A failure on the part of the contractor to comply with any of the provisions of clause 3.3 due to his default is a ground for the determination of his employment by the employer, subject however to the proviso that the notice of determination must not be given unreasonably or vexatiously – see clause 7.1(d).

Standard documents for use in connection with the naming of a sub-contractor

Where clause 3.3.1 (named persons to carry out sub-contract work described in the specification/schedules of work/contract bills) is to be used, the contractor in tendering for the main contract must price the named sub-contractor's work as described in the tender documents and, unlike the situation regarding traditional nomination under JCT 80, the contract sum will not subsequently be adjusted to take account of changes in the sub-contract sum. For this reason it is important that contractors are fully aware of the basis on which the proposed named sub-contractor has tendered so that a proper assessment of the proposed named sub-contractor's price can be made by the contractor in building up his own price for the work.

For example, he needs to know as much as possible about the period or periods required for the completion of the sub-contract work, special attendances, main contract appendix details, as well as the proposed sub-contract price. The contractor takes a significant proportion of the financial risks of poor performance by the proposed named sub-contractor. It is only right therefore that as much information as possible is provided. It is clearly better that this be provided in a standard form and it is to be found in Parts I and II of the Form of Tender and Agreement NAM/T. It is also obviously desirable that there be a standard set of sub-contract conditions for named sub-contractors and this is provided in NAM/SC.

Form of Tender and Agreement NAM/T[1]

This is the form of tender on which tenders will be invited by the architect from persons who may subsequently be named to carry out sub-contract works, whether the works are described in the specification/schedules of work/contract bills or under an instruction as to the expenditure of a provisional sum. The form consists of 15 pages and comprises three sections. Section I is the invitation to tender for completion by the architect and contains main contract information and sub-contract information, there being 17 items in all. Section II comprises the tender by the sub-contractor in which there are six items for completion. Finally, section III comprises the articles of agreement which is the actual agreement for execution incorporating the sub-contract documents including the Tender and Agreement NAM/T, the Sub-Contract Conditions NAM/SC and the numbered documents which are listed on the first page of section I of NAM/T.

The Sub-Contract Conditions NAM/SC

These are the standard conditions of sub-contract between the named sub-contractor and the contractor. They are considerably detailed and are of course designed to be compatible with IFC 84. The conditions are based to a large extent on the standard domestic form of contract associated with JCT 80 i.e. DOM/1. There are some differences to take account of the fact that a named sub-contractor is not purely domestic.

ESA/1[2]

This is an agreement prepared outside the Joint Contracts Tribunal by two of the constituent bodies, namely the Royal Institute of British Architects and the Committee of Associations of Specialist Engineering Contractors. It deals with design responsibilities undertaken by named sub-contractors in relation to the sub-contract works and accordingly, as the main contractor will generally have no responsibility for design under IFC 84, this agreement is between the named sub-contractor and the employer only. The design obligation is expressed in similar, though not identical, terms to that

1. Printed in full in Appendix 2
2. Printed in full in Appendix 3

to be found in the JCT Standard Form of Employer/Nominated Sub-Contractor Agreement (NSC/2) for use between employer and nominated sub-contractor under JCT 80 (see clause 2.1 of NSC/2).

While IFC 84 is designed for works of a simple nature not involving complicated electrical or mechanical services, nevertheless in practice named sub-contractors will undoubtedly be used whose work will carry within it a design element, often of a specialist nature. Clause 3.3.7 of IFC 84 expressly relieves the contractor from responsibility to the employer for any defects in the design of a named sub-contractor's work. It is absolutely imperative therefore that in every contract in which a named sub-contractor is used, the employer or those advising him consider carefully the question of whether or not a named sub-contractor is providing any design service, and if so the separate agreement ESA/1 or some other suitable agreement should be used.

If such an agreement is not used the employer will find himself without a contractual remedy in respect of defective design by a named sub-contractor. While the employer may be able to establish a cause of action against the named sub-contractor for breach of a duty of care or by virtue of some collateral warranty, this will certainly not automatically be the case, and the absence of a suitable direct agreement between the employer and the named sub-contractor in such a situation will leave a serious gap in the employer's panoply of remedies in respect of defective work.

Chapter 6

Control of the works (Part II)

CONTENT

This chapter looks at clauses 3.4 to 3.15 of section 3 of IFC 84 which deals with the following matters:

Contractor's person-in-charge
Architect's instructions
Variations (including their valuation)
Instructions to expend provisional sums
Levels and setting out
Clerk of works
Work not forming part of the contract
Instructions as to inspection – tests
Instructions as to the removal of work etc.
Instructions as to postponement.

SUMMARY OF GENERAL LAW

(A)   Contract administrator's instructions

Building contracts which provide for an architect will generally empower him to issue instructions. While as between the employer and the contractor the validity of the architect's instruction will depend on the terms of the building contract, as between the architect and the employer the authority of the architect to issue certain instructions e.g. as to variations in the contract works, will depend on his conditions of engagement with the employer. These may impose restrictions on his power to issue certain instructions particularly where this would add to the contract sum.

The building contract will often require the architect's instructions to be in writing. Such instructions can and very often do cover a wide variety of matters e.g. variations to the contract work or to the conditions under which they are carried out, the removal of defective work, testing, expenditure of provisional sums, postponement, resolution of inconsistencies in the requirements laid down in the contract documents and re-nomination of sub-contractors. Compliance with an instruction of the architect may be a good ground for the contractor to seek an extension of time for completion where it causes delay, and monetary compensation where it causes financial loss.

The contract will often provide a mechanism for the contractor to challenge the validity of an instruction and also an express power for the employer to have any work required by an instruction carried out by someone other than the contractor if he fails to comply with it, the additional cost being borne by the contractor.

(B) Variations

When the word 'variation' is used in relation to building contracts, it generally refers to a variation in the contract works or occasionally to the conditions (not contract conditions) under which the contract work is to be carried out e.g. working hours, working space, access to site and suchlike. It does not relate to a variation in the terms of the contract itself. It cannot be stressed too often that the architect has no inherent power to vary the terms or conditions of the contract. If the architect agrees with the contractor that the terms of the contract will in some way be varied, the contractor will be taking a risk as the employer will not be bound by the agreement. The contractor should therefore check the actual authority of the architect to agree to such a variation.

The building contract between employer and contractor will bestow on the architect the power to order variations in relation to the contract works themselves. To the extent therefore that the instruction requiring a variation under the main contract is in accordance with its terms, the contractor need not check with the employer that the architect is empowered by the employer to order such a variation. The employer, in appointing the architect to act under the contract, is bestowing on him the power and authority which the main contract gives to the architect. The contract can of course expressly contain limits as to the contract administrator's authority in this connection e.g. see clause 2 and Part II of the

FIDIC contract (4th edition). Often a restriction is placed on the architect's freedom to issue instructions ordering variations where the contract sum is exceeded by more than a given percentage.

It is imperative that the architect, when issuing instructions requiring a variation to the contract works, does so strictly in accordance with the terms of the contract giving him that authority. A failure to do so can cause severe difficulties with the possibility of the contractor being disentitled to payment for such unauthorised, though ordered, variations, and the architect running a risk of personal liability in such circumstances for losses suffered by the contractor.

In the absence of an express power for variations in the contract works, there is no implied right of the employer to have variations carried out by the contractor. The drawings, specification etc. would in fact be frozen at the date that the contract was entered into, and if the employer required any departure from the requirements of these documents, this could only be achieved by a separately negotiated agreement between himself and the contractor.

Work indispensably necessary and standard methods of measurement
As a general principle, an obligation to do specified work includes an obligation to do all necessary ancillary work. Even where a detailed bill of quantities is prepared in accordance with a standard method of measurement, it may not descend to every detail of construction. There may be some items or some work processes which are not fully described in the standard method of measurement, and it is at least arguable that this work will not amount to a variation and will have to be carried out within the overall contract sum without adjustment therefor.

CONSIDERATION OF THE RELEVANT CLAUSES OF IFC 84

Clause 3.4

Contractor's person-in-charge

3.4   The Contractor shall at all reasonable times keep upon the Works a competent person-in-charge and any written instructions given to him by the Architect/the Contract Administrator shall be deemed to have been issued to the Contractor.

COMMENTARY ON CLAUSE 3.4

Clearly, the contractor should appoint a responsible person in this position. Under JCT 63 the person was called a foreman-in-charge but this was changed in JCT 80 and also in IFC 84, no doubt to avoid any suggestion of sexual discrimination. It spoils the position somewhat to find therefore a reference in the clause to the word 'him'. Any instruction of the architect given to the person-in-charge is deemed to have been given to the contractor. This is especially important wherever time limits are related to an instruction e.g. clauses 3.3.2 (entering into a named sub-contract), 3.13.1 (instructions on opening up and testing) and 3.5 (compliance with the architect's instructions generally).

Clauses 3.5.1 and 3.5.2

Architect's/Contract Administrator's instructions

3.5.1   All instructions of the Architect/the Contract Administrator shall be in writing. The Contractor shall forthwith comply with such instructions issued to him which the Architect/the Contract Administrator is empowered by the Conditions to issue; save that where such instruction is one requiring a Variation within the meaning of clause 3.6.2 the Contractor need not comply to the extent that he makes reasonable objection in writing to the Architect/the Contract Administrator to such compliance.

If within 7 days after receipt of a written notice from the Architect/the Contract Administrator requiring compliance with an instruction the Contractor does not comply therewith then the Employer may employ and pay other persons to execute any work whatsoever which may be necessary to give effect to such instruction and all costs incurred thereby may be deducted by him from any monies due or to become due to the Contractor under this Contract or shall be recoverable from the Contractor by the Employer as a debt.

3.5.2   Upon receipt[1] of what purports to be an instruction issued to him by the Architect/the Contract Administrator the Contractor may request the Architect/the Contract Administrator to specify in writing the provision of the Conditions which empowers the issue of the said instruction. The Architect/the Contract Administrator shall forthwith comply with any such request, and if the Contractor shall thereafter comply with the said instruction (neither party before such compliance having given to the other a written request to concur in the appointment of an arbitrator under Clause 9 in order that it may be decided[2] whether the provision specified by the Architect/the Contract Administrator empowers the issue of the said instruction), then the issue of the same shall be deemed for all the purposes of this Contract to have been empowered by the provision of the Conditions specified by the Architect/the Contract Administrator in answer to the Contractor's request.

COMMENTARY ON CLAUSE 3.5

For a summary of the general law see page 162.

All the architect's instructions must be in writing. This contract makes no provision whatsoever for oral instructions even if the contractor confirms them in writing and they are not dissented from by the architect. However, no doubt compliance by the contractor with an oral instruction varying the works, followed by a written ratification by the architect, will produce an effective variation from the date of the oral instruction (see clause 3.6 first paragraph). Nevertheless, the contractor should be very cautious indeed about acting on a verbal instruction.

Having received the written instruction (and this includes receipt by the person-in-charge referred to in clause 3.4), the contractor is bound to comply *forthwith*, except that if the instruction is one requiring a variation under clause 3.6.2 (imposition of or changes from the original contract documents in relation to obligations or restrictions in regard to access, working space, working hours or sequence of work), the contractor need not comply to the extent that he makes reasonable objection in writing to such compliance. What would be a reasonable objection in these circumstances is difficult to define, particularly as an instruction under clause 3.6.2 can give rise to reimbursement of direct loss and expense under clause 4.12.7 and an extension of time under clause 2.4.5, as well as involving a fair valuation of the variation itself. Presumably therefore a reasonable objection must generally relate to a matter for which money is insufficient recompense.

If within seven days of receipt by the contractor of a written notice from the architect requiring him to comply with an instruction the contractor does not comply therewith, the architect can employ others to do the work and the costs involved can be deducted from money due to the contractor or can be recovered from him as a debt. This is an extreme remedy which can cause serious practical problems on site and hinder the smooth running of the contract. It should be remembered that a failure by the contractor to implement the instruction *forthwith* is in itself a breach of contract. The second paragraph of this clause relating to the seven day time limit does not have the effect of giving the contractor a further seven days in which to comply with the instruction. If he fails to comply forthwith then he will be liable to the employer for breach of contract even for the period before as well as within the seven day period and will become responsible for any foreseeable losses thereby suffered by the employer.

Instructions can only be issued so far as they are empowered under the conditions of contract. Clause 3.5.2 enables the contractor to test the validity of the instruction by requesting the architect to specify in writing the provision of the conditions which empowers the issue of the instruction in question. The architect must thereupon comply with the request. If the contractor thereafter complies with the instruction (neither party having before compliance given notice of arbitration to test the instruction), it shall be deemed to have been empowered by the provision of the conditions specified by the architect in his response to the contractor's request.

If dissatisfied, the contractor would generally be well advised to comply with the instruction but may preserve his right to challenge the architect's answer by giving notice of arbitration before compliance. Of course if the instruction is demonstrably not a valid one, the contractor would not be in breach of contract in refusing to comply with it. However, before refusing, the contractor should be very clear as to his ground for refusal.

If the architect delays in giving an answer to the contractor's request, it is difficult to see what remedy the contractor has. The contractor should comply with the instructions forthwith and the fact that he has invoked the provisions of clause 3.5.2 does not affect this, so that any delay on the architect's part in responding to the contractor's clause 3.5.2 request, should neither delay nor disrupt work. Accordingly there is no extension of time event or loss and expense matter to cover such a delay in response. In extreme cases it may amount to a breach of contract on the part of the employer.

NOTES TO CLAUSE 3.5

[1] 'Upon receipt . . .'.
The contractor's request must be made on receipt of a purported instruction i.e. straightaway.

[2]   '*... in order that it may be decided ...*'.
The reference to arbitration may be opened as can any other
reference to arbitration under this contract (per article 5 and
clause 9) immediately, and does not have to await practical completion or alleged practical completion of the works.

Clauses 3.6 and 3.7

Variations

3.6   The Architect/The Contract Administrator may subject to
           clause 3.5.1 issue instructions[1] requiring a Variation
           and sanction[2] in writing any Variation made by the Con-
           tractor otherwise than pursuant to such an instruction.
           No such instruction or sanction shall vitiate the Contract.

           The term Variation as used in the Conditions means:

3.6.1   the alteration or modification of the design or quality or
           quantity of the Works as shown upon the Contract Draw-
           ings and described[3] by or referred to in the Specifica-
           tion/Schedules of Work/Contract Bills including
           – the addition, omission or substitution of any work,
           – the alteration of the kind or standard of any materials
             or goods to be used in the Works,
           – the removal from the site of any work executed or
             materials or goods brought thereon by the Contractor
             for the purposes of the Works other than work mate-
             rials or goods which are not in accordance with this
             Contract;

3.6.2   the imposition by the Employer of any obligations or res-
           trictions or the addition to or alteration or omission of any
           such obligations or restrictions so imposed or imposed by
           the Employer in the Specification/Schedules of Work/
           Contract Bills in regard to:
           – access to the site or use of any specific parts of the
             site;
           – limitations of working space;
           – limitations of working hours;
           – the execution or completion of the work in any specific
             order.

Valuation of Variations and provisional sum work –
Approximate Quantity, measurement and valuation

3.7   The amount to be added to or deducted from the Contract
           Sum in respect of instructions requiring a Variation and
           of instructions on the expenditure of a provisional sum
           may be agreed between the Employer and the Contractor

prior to[4] the Contractor complying with any such instruction but if not so agreed there shall be added to or deducted from the Contract Sum an amount determined by a valuation made by the Quantity Surveyor in accordance with the following rules in clauses 3.7.1 to 3.7.9.

All work executed by the Contractor for which an Approximate Quantity is included in the Contract Documents shall be measured and valued by the Quantity Surveyor in accordance with the rules in clauses 3.7.1 to 3.7.9.

3.7.1   'priced document' as referred to in clauses 3.7.2 to 3.7.8 means, where the 2nd Recital, alternative A applies, the Specification or the Schedules of Work as priced by the Contractor or the Contract Bills; and, where the 2nd Recital, alternative B applies, the Contract Sum Analysis or the Schedule of Rates;

3.7.2   omissions shall be valued in accordance with the relevant prices in the priced document;

3.7.3.* for work of a similar[5] character to that set out in the priced document the valuation shall be consistent with the relevant values therein[6] making due allowance for any change in the conditions[7] under which the work is carried out and/or any significant[8] change in quantity of the work so set out;

3.7.4   a fair valuation shall be made
         – where there is no work of a similar character set out in the priced document, or
         – to the extent that the valuation does not relate to the execution of additional, or substituted work or the execution of work for which an Approximate Quantity is included in the Contract Documents or the omission of work, or
         – to the extent that the valuation of any work or liabilities directly associated with the instruction cannot reasonably be effected by a valuation by the application of clause 3.7.3;

3.7.5   where the appropriate basis of a fair valuation is Daywork, the valuation shall comprise
         – the prime cost of such work (calculated in accordance with the 'Definition of Prime Cost of Daywork carried out under a Building Contract' issued by the Royal Institution of Chartered Surveyors and the Building Employers Confederation, which was current at the Base Date) together with percentage additions to each section of the prime cost at the rates set out by the

*Refer to footnote on page 176.

Contractor in the priced document, or [i]
- where the work is within the province of any specialist trade and the said Institution and the appropriate body representing the employers in that trade have agreed and issued a definition of prime cost of daywork, the prime cost of such work calculated in accordance with that definition which was current at the Base Date together with percentage additions on the prime cost at the rates set out by the Contractor in the priced documents;

3.7.6   the valuation shall include, where appropriate, any addition to or reduction of any relevant items of a preliminary nature, provided that where the Contract Documents include bills of quantities no such addition or reduction shall be made in respect of compliance with an instruction as to the expenditure of a provisional sum for defined work* included in such bills;

3.7.7   no allowance shall be made in the valuation for any effect upon the regular progress of the Works or for any other direct loss and/or expense[9] for which the Contractor would be reimbursed by payment under any other provision in the Conditions;

3.7.8   if compliance with any such instructions substantially changes the conditions under which any other work is executed, then such other work shall be treated as if it had been the subject of an instruction of the Architect/ the Contract Administrator requiring a Variation under clause 3.6 to which clause 3.7 shall apply;
Clause 3.7.8 shall apply to the execution of work for which an Approximate Quantity is included in the Contract Documents to such extent as the quantity is more or less than the quantity ascribed to that work in the Contract Documents and, where the Contract Documents include bills of quantities, to compliance with an instruction as to the expenditure of a provisional sum for defined work* only to the extent that the instruction for that work differs from the description given for such work in such bills;

3.7.9   where the priced document is the Contract Sum Analysis or the Schedule of Rates and relevant rates and prices

---

[i] There are three Definitions to which the second sub-paragraph of clause 3.7.5 refers, namely those agreed between the Royal Institution and the Electrical Contractors Association, the Royal Institution and the Electrical Contractors Association of Scotland and the Royal Institution and the Heating and Ventilating Contractors Association.

are not set out therein so that the whole or part of clauses 3.7.2 to 3.7.8 cannot apply, a fair valuation shall be made.

---

* See footnote to clause 8.3 (Definitions).

COMMENTARY ON CLAUSES 3.6 AND 3.7

For a summary of the general law see earlier page 163.

Clause 3.6 empowers the architect to issue instructions requiring a variation. In addition, it permits the architect to sanction in writing any variation made by the contractor but not authorised by an instruction. Variations therefore can be the result of a prior instruction or a sanction after the event.

The meaning of the term variation is given in clauses 3.6.1 and 3.6.2. Briefly, it means firstly (under clause 3.6.1):

(1) alteration or modification of design;
(2) alteration or modification of quality;
(3) alteration or modification of quantity;

and expressly includes:

(a) additions, omissions or substitutions of any work;
(b) alterations to kinds or standards of materials or goods;
(c) the removal of completed work, or of materials or goods from the site other than work, materials or goods which are not in accordance with the contract.

Secondly (under clause 3.6.2) it means:

the imposition, if not previously imposed by the employer in the specification/schedule of work/contract bills and the addition to, or alteration or omission of, such obligations or restrictions or of those imposed by the employer in the specification/schedule of works/contract bills in regard to:
(a) access to the site;
(b) limitations of working space;
(c) limitations of working hours;
(d) the execution or completion of work in any specific order.

Note that the architect may issue an instruction changing any of the 'particulars' in relation to named persons as sub-contractors which may be contained in the contract documents. This will be 'regarded' as a variation, presumably even where the nature of the change in the particulars is outside the meaning given to variation in this clause e.g. a change in the named sub-contractor's price for the work between the main contract being let and the sub-contract being entered into (see clause 3.3.1). The position is similar in relation to instructions given under clause 3.3.3(b).

Although there is no express restriction on the extent to which an instruction can vary the works, there must, it is submitted, be some point at which a so-called variation would change the contract so fundamentally that it would not be regarded as a variation within the meaning of the contract: *Sir Lindsay Parkinson and Co.* v. *Commissioners of Works* (1949).

In relation to the matters set out in clause 3.6.2, the contractor can make reasonable objection to any instruction imposing or varying such matters – see the commentary to clause 3.5.1 on page 166.

Although it is common practice to issue a certificate of practical completion when some relatively unimportant items still remain to be completed, it is generally thought that the architect has no power to order variations after the issue of this certificate. Certainly the overall structure of the contract provides ample logic for this view.

Firstly, a variation instruction after practical completion of the works, for example, could result in the works being rendered thereby not sufficiently complete to warrant a certificate of practical completion i.e. a variation in such circumstances could appear to undo an existing practical completion certificate. Secondly, the extension of time and loss and expense provisions are geared either to delays in completion i.e. delays to practical completion, or disturbance to regular progress respectively, whereas it is difficult to see how the progress of the works can be affected once they are complete. It clearly makes good sense that the works should not be capable of being varied by a variation after they have been completed.

However, the consequence of this is that where the architect issues a certificate of practical completion even though there are some outstanding items of work, he probably cannot issue variation instructions relating thereto. As the outstanding work, if any, will often relate to final finishes which are a frequent area of variations, it could in strict contractual terms, though perhaps not often in practice, give rise to problems. The use of the optional clause 2.11

permitting partial possession and providing for deemed practical completion of part may be a solution in appropriate circumstances (see page 100).

Valuation of variations and provisional sum work

Clause 3.7 deals with the valuation of the architect's instructions requiring a variation issued under clause 3.6 and the valuation of the architect's instructions issued under clause 3.8 covering the expenditure of provisional sums included in the contract documents. The clause lays down a series of steps which have to be followed when valuing variations. They are as follows:

(i)   a valuation may be agreed between the employer and the contractor prior to the contractor complying with the instruction;

(ii)   a valuation may be made by the quantity surveyor using as a basis the relevant prices in the 'priced document';

(iii)   where there are no relevant prices then a valuation by means of a fair rate or value for the work;

(iv)   by daywork.

Added emphasis has been placed on the possibility of agreement between employer and contractor as to the price for the work, presumably to stress the simple nature of the work for which IFC 84 is intended, as compared with the valuation provisions of JCT 80 (clause 13.5). Although that possibility is covered in clause 13.4.1 of JCT 80, it is not given the same prominence as in this clause where it appears as the first option open to the parties to the contract. It is important to note that it is the employer, and not the architect or the quantity surveyor on his behalf, who must be a party to the agreement, though the architect or quantity surveyor may well act as agent of the employer in reaching an agreement if so authorised for this purpose. If this is the case it is suggested that the contractor should obtain confirmation of this authority for his own protection.

If an attempted agreement between employer and contractor in relation to the price to be paid for a variation does not materialise, the contractor must nevertheless 'forthwith comply' with the architect's instruction (clause 3.5.1) unless it is an instruction requiring a variation of the kind described in clause 3.6.2. In the absence of an agreement, the instruction will be valued in accordance with certain rules contained in clauses 3.7.1 to 3.7.9, and these may be summarised as follows:

(1) Clause 3.7.2: Omissions shall be valued in accordance with the relevant prices in the priced document.

(2) Clause 3.7.3: Where the work is of a similar character to that set out in the priced document, the valuation shall be consistent with the values therein but allowance must be made for:

   (a)   changes in the conditions under which the work is carried out;
   (b)   any significant change in quantity.

(3) Clause 3.7.4: A fair valuation shall be made:

   (a)   where there is no work of a similar character in the priced document;
   (b)   to the extent that the instruction calling for valuation does not involve the addition, substitution or omission of work e.g. variations under clause 3.6.2;
   (c)   to the extent that the valuation of the work or liabilities directly associated with the instruction cannot reasonably be achieved by a valuation under 2. above.

(4) Clause 3.7.5: Where the appropriate basis of a fair valuation is daywork, the valuation shall comprise:

   (a)   the prime cost of the work calculated in accordance with the 'Definition of Prime Cost of Daywork carried out under a Building Contract' issued by the RICS and BEC current at the base date stated in the appendix to IFC 84 (see clause 8.3), plus any percentage addition to each section of the prime cost at the rates set out by the contractor in the priced document, or
   (b)   where there is specialist work and the RICS and the appropriate body representing the employers in that trade have issued an agreed definition of prime cost daywork then calculated in accordance with that definition plus any percentage as in (a) above.

There is no express provision for the submission of daywork vouchers by the contractor, unlike JCT 80. The submission of vouchers will, however, almost certainly be necessary and the lack of a time limit for their submission could create difficulties unless it is covered elsewhere in the contract documents.

(5) Clause 3.7.6: The valuation shall include, where appropriate, any addition to or reduction of any relevant items of a preliminary nature.

If it can be established that the instruction has had an effect on the allowances included by the contractor in his preliminaries, then an element of the valuation should include for relevant items from those preliminaries. This could prove difficult if a breakdown of the preliminaries has not been established prior to entering into the contract.

(6) Clause 3.7.7: There shall be no allowance made in the valuation for any effect on the regular progress of the works or for any other loss and expense reimbursible under any other provision in the conditions.

(7) Clause 3.7.8: If compliance with the variation instruction substantially changes the conditions under which any other work is executed, such other work shall be treated as if it had been the subject of a variation instruction.

It is not difficult to envisage an instruction from the architect which affects the whole operation of the contract, e.g. an instruction under clause 3.6.2 affecting the 'execution or completion of the work in any specific order'. The architect would therefore be well advised to consider the effect that his instructions requiring a variation may have on the valuation of other work before issuing the instruction, particularly when the instruction may result in a complete revaluation of all the work remaining to be done if the conditions under which it is executed change substantially. The employer's protection is in the word 'substantially'. The fact that the conditions for the carrying out of the other work change is not enough. The change must be substantial which, it is submitted, in this context means more than just significant. It is only a little way short of requiring completely different conditions.

(8) Clause 3.7.9: If the priced document is a contract sum analysis

or schedule of rates not containing any relevant rates or prices so that either the whole or part of the valuation rules summarised in (1) to (7) above cannot apply, then a fair valuation shall be made.

Somewhat oddly, the contractor is given no express right to be present whenever it is necessary to measure work for valuation purposes – c.f. clause 13.6 of JCT 80.

*Approximate Quantities**

Where an item of work has been made the subject of an approximate quantity in the contract documents, that work, when executed, must be measured and valued by the quantity surveyor. The requirement for it to be measured is an exception to the general lump sum nature of IFC 84.

The valuation is to be made in accordance with the rules in clauses 3.7.1 to 3.7.9 referred to above.

It must be appreciated that approximate quantity work is valued even where it has not been the subject of a variation instruction. It is not, however, easy to see how, in what is admittedly a strictly legalistic approach to the precise wording of the contract, any of the rules set out in clauses 3.7.1 to 3.7.9 apply.

Clause 3.7.2 cannot apply as the amount cannot be an omission of work where the requirement is that it be measured. There will have never been a lump sum for a firm quantity of work.

Clause 3.7.3 cannot apply as the work to be valued is the very work set out in the priced document and cannot be 'work of a similar character to that set out in the priced document'.

Clause 3.7.4 provides for a fair valuation to be made in certain circumstances none of which fit approximate quantity work.

Clause 3.7.5 cannot apply where the approximate quantity work is priced in the priced document.

Clause 3.7.6 cannot apply as the contractor in the priced document will have already made allowance or be deemed to have made allowance for preliminary items.

Clause 3.7.8 which deals with substantial changes to the conditions under which any other work is executed is expressly stated to apply to approximate quantity work to the extent that the approximate quantity is greater or less than the approximate quantity stated in the contract documents. It does not therefore deal with the valuation of the actual approximate quantity work contained in the contract documents.

* The difficulties pointed out in this section have now been dealt with in the 1990 reprint of IFC 84 which, however, was not available when this book went to press.

Clause 3.7.9 could apply but only where the approximate quantity work is contained in the contract sum analysis or a schedule of rates and provided there are no relevant rates and prices set out therein, in which case a fair valuation is to be made.

Where, therefore, approximate quantity work is priced and the work as measured can be regarded as a reasonably accurate forecast of the quantity of work required it does not appear to be capable of valuation under the rules set out in clauses 3.7.1 to 3.7.9.

If the quantity as measured reveals that the approximate quantity stated in the contract documents was not a reasonably accurate forecast of the work required, then by virtue of clause 1.4 it will be an error in quantity and the architect must issue an instruction to correct the error. That instruction will change '. . . the quantity of work deemed to be included in the contract sum . . .' and accordingly '. . . the correction shall be valued under clause 3.7'. (See last paragraph of clause 1.4.) It would have been preferable if clause 1.4 also provided expressly that such an instruction should be treated as a variation (c.f. JCT 80 clause 2.2.2.2). In this case clauses 3.7.2, 3.7.3, 3.7.6 and 3.7.7 can apply. Presumably, the first sentence of clause 3.7.8 will apply to the instruction. The second sentence could then still apply to the extent that the measured quantity was greater or less than the corrected quantity.

The intention must have been for the reasonably accurate approximate quantities as measured to be simply valued at the rates and prices provided in the contract documents, though clause 3.7, which expressly encompasses approximate quantity work, nowhere actually provides for this. Under JCT 80 the valuation of approximate quantity work under clause 13.5 is the subject of much clearer drafting, with clause 13.5.1.4 providing that 'where the Approximate Quantity is a reasonably accurate forecast of the quantity of work required the rate or price for the Approximate Quantity shall determine the Valuation'.

In practice, no doubt quantity surveyors will have little difficulty in simply applying the rate contained in the contract documents, but its contractual basis seems to be missing.

Provisional Sums for Defined Work under SMM 7

Where the valuation relates to an instruction for the expenditure of a provisional sum for defined work included in the bills of quantities, the valuation rules in clauses 3.7.1 to 3.7.9 make it clear that:

– by clause 3.7.6 the valuation shall not include an addition to or reduction of any relevant preliminary items as these are deemed to have been allowed for in the contractor's programming, planning and pricing of preliminaries – see General Rule 10.4 of SMM 7. Clearly, if the instruction differs from the description of the work in the defined provisional sum then this may well amount to a variation to which clause 3.7.6 and indeed the other sub-clauses in clause 3.7 dealing with the valuation of variations will apply as appropriate;

– by clause 3.7.8 compliance with the instruction shall not be treated as changing the conditions under which any other work is executed so as to lead to that other work being treated as if it had been the subject of an instruction requiring a variation, except to the extent that the instruction for the defined provisional sum work differs from the description given for such work in the contract bills. Quite apart from this express exception, it is difficult to see how compliance with an instruction to expend a provisional sum for defined work which is strictly in line with the description given for such work could change the conditions under which other work is executed, any more than carrying out any fully described and measured item could. Clearly, if the instruction changes the description this would in any event be a variation under clause 3.6 so that the valuation rules in clauses 3.7.1 to 3.7.9 (including therefore clause 3.7.8) would apply anyway.

Reference should also be made to the Commentary on clauses 1.4 and 1.5 (see page 39).

NOTES TO CLAUSES 3.6 AND 3.7

[1] '*. . . may subject to clause 3.5.1 issue instructions . . .*'.
These must be in writing and be authorised by the conditions of contract, as required in clause 3.5.1.

[2] '*. . . may . . . sanction . . .*'.
This vests in the architect a considerable discretion. It is a limited licence to accept the contractor's unauthorised departures from the contract works. Clearly however, as with the power to vary generally, the authority of the architect to give such sanctions is limited in two ways. Firstly the sanction must be issued in accordance with this clause. Secondly, as between the architect and his client, the employer, his authority will depend on his terms of engagement

which will need to contain some form of express authority from his employer/client. This is not a matter with which the contractor need concern himself, unless perhaps he is put on notice that the architect does not have such authority. It also of course enables the architect to regularise a previous orally requested variation with which the contractor has complied. If the architect declines to exercise the discretion inherent in the word 'may', it is difficult to see any remedy for the contractor under the contract. The arbitration provisions, though giving the arbitrator very wide powers – see clause 9.3 – may not extend that far.

[3] '. . . shown upon the Contract Drawing and described . . .'.
The word used is 'and' not 'or'. The alteration or modification to design or change in quality or quantity must therefore strictly relate to the works shown in both contract drawings and in one of the other contract documents referred to. This produces apparently odd results e.g. the quality of an item may appear in only one document e.g. the specification or the bills of quantities. It may not appear on the contract drawings at all. Literally such an item of work cannot thereafter be varied under the terms of this contract. It is necessary to strain considerably the language used in order that a 'variation' can be interpreted to include the sort of changes it is clearly intended to cover.

[4] '. . . prior to . . .'.
There would be nothing to prevent the employer and contractor agreeing the amount after compliance with the instruction. Strictly speaking, such an agreement would be outside the terms of the contract. In such a case the agreement will override any valuation by the quantity surveyor in accordance with the rules in clause 3.7.1 to 3.7.9.

[5] '. . . similar . . .'.
For this provision to operate fairly between the employer and the contractor, this word must in practice be treated as meaning of the same character. This can be demonstrated by looking at any particular item of work. The item of work will be covered by a description. If the instruction requiring a variation alters in any way that description of the work, it must thereby become different and may well in fairness be deserving of a different rate. To interpret the words in any other way will be to prevent the quantity surveyor from

applying a fair valuation where the varied description of the item, although arguably remaining of a similar character to the original, justifies a different rate.

[6] '. . . consistent with the relevant values therein . . .'.
If the contractor has made a mistake in his rate he will nevertheless generally be bound by it. This could be to his advantage or disadvantage: *Dudley Corporation* v. *Parsons and Morrin* (1967). This view is reinforced by the provisions of clause 4.1, which provides that any error or omission, whether of arithmetic or not, in the computation of the contract sum shall be deemed to have been accepted by the parties to the contract.

[7] '. . . change in the conditions . . .'.
This can include a change in site conditions e.g. more restricted access, or in the elements i.e. carrying out work at winter rates rather than the summer rates contained in the priced document. The location of the work may also be relevant here.

[8] '. . . significant . . .'.
This is a matter of degree. Clearly a substantial change in quantity can affect the unit rate in the priced document.

[9] '. . . or for any other direct loss and/or expense . . .'.
The word 'other', and indeed the whole phrase, is somewhat confusing. To begin with, it equates the words 'any effect upon the regular progress of the works' with 'direct loss and/or expense', whereas they are of course different things. Direct loss and/or expense may or may not result from the regular progress of the works being affected. Further, if there is a disturbance of regular progress due to a variation instruction which involves the contractor in direct loss and/or expense, this is reimbursable under clause 4.12.7 and it is difficult to see how any other direct loss and/or expense (note the same phrase is used) can be suffered, let alone reimbursed by payment under any other provision in the conditions. Perhaps the words used are a somewhat abstruse reference to clauses 4.2 to 4.5 which provide the actual mechanism for payment in respect of loss and expense.

Clause 3.8

Instructions to expend provisional sums

3.8 The Architect/The Contract Administrator shall[1] issue instructions as to the expenditure of any provisional sums.

COMMENTARY ON CLAUSE 3.8

This is the express power for the architect to issue instructions with regard to the expenditure of provisional sums whether by way of naming a sub-contractor under clause 3.3.2 or otherwise.

Since the introduction into IFC 84 of SMM 7 where bills of quantities are used (Amendment 4 of July 1988) provisional sums may be defined or undefined (see clause 8.3 – Definitions and the footnote to that clause where SMM 7 General Rules 10.1 to 10.6 are set out – page 375 of this book). Defined provisional sums have been commented upon earlier in relation to clause 1.4 (page 39); when considering clauses 2.4.5 and 2.4.7 (see pages 79 and 80) and when considering the valuation of provisional sums (see page 177).

A provisional sum for undefined work is the sum provided for work where the information required in accordance with General Rule 10.3 (defined work) cannot be given. The possibility of a quantity surveyor intending to provide for an item as a provisional sum for defined work and unwittingly creating a provisional sum for undefined work and the possible significance of this has been considered earlier when commenting upon clause 1.4 (see page 39).

NOTES TO CLAUSE 3.8

[1] '... shall ...'.
This word is peremptory, and it is submitted that the architect is bound to issue instructions. It may be, and often is, decided not to expend a provisional sum item at all, in which case an instruction must be issued omitting it.

Clause 3.9

Levels and setting out

3.9 The Architect/The Contract Administrator shall determine any levels which may be required for the execution

of the Works, and shall provide the Contractor by way of
accurately dimensioned drawings with such information
as shall enable the Contractor to set out the Works.[1] The
Contractor shall be responsible for, and shall, at no cost
to the Employer, amend, any errors arising from his own
inaccurate setting out. With the consent of the
Employer[2] the Architect/the Contract Administrator may
instruct that such errors shall not be amended and an
appropriate deduction[3] for such errors not required to be
amended shall be made from the Contract Sum.

COMMENTARY ON CLAUSE 3.9

While it is for the architect to determine any levels which may be
required and to provide the contractor, by accurately dimensioned
drawings, with sufficient information for him to be able to set out
the works, it is for the contractor to actually set them out accurately.
If he fails to do so, any errors must be corrected, unless the employer
consents to adopting the error, at no cost to the employer. In
reaching any such compromise the employer should weigh carefully
the possible implications of accepting an error in setting out e.g. if it
contravenes any planning consent or condition or if it could give rise
to a third party claim such as for trespass or infringement of a right
of light. If the error is to be adopted by the employer then an
appropriate deduction is to be made from the contract sum. Such an
adjustment will often be agreed upon, if not, it will not necessarily be
easy to assess, and can, it is submitted, be the subject of a reference
to arbitration. See also Note [3] which follows.

From the wording of the clause it would appear that the contractor has no right to insist on amending his error instead of suffering a
reduction in the contract sum.

NOTES TO CLAUSE 3.9

[1]   '. . . to set out the Works.'
In JCT 80 (clause 7) these words were followed by the words 'at
ground level'. The absence of these words is no doubt meant to make
it clear that the contractor's responsibility for accurately setting out
also covers the situation where the contract works are built below
ground level or above it e.g. an underground or rooftop car park.

[2]   'With the consent of the Employer . . .'.
If the contractor fails to set out the works accurately he will be in
breach of contract. In such a case the architect would not normally

have any apparent or ostensible authority to relieve the contractor from his breach of contract. These words make it clear that if the contractor is to be relieved from his contractual obligation to correct his inaccuracy, then the instruction of the architect can only operate with the consent of the employer who is, unlike the architect, a party to the contract. As the consequences of errors in setting out can vary from the insignificant to the disastrous, it is submitted that the express requirement for the employer's consent is reasonable. Even so, it may be pointed out that perhaps it is not appropriate to introduce something which should be dealt with in the terms of engagement between the employer and his architect into a contract between the employer and the contractor, which is where the extent of the architect's authority is to be found so far as the contractor is concerned.

[3] '. . . an appropriate deduction . . .'.
If the architect instructs the contractor not to amend an error in his setting out, he must make an adjustment downwards of the contract sum. He cannot add to the contract sum. Under some earlier wording of JCT 80 (clause 7 before Amendment 4 of July 1987) it was just about conceivable that the architect had power, on ordering the contractor to amend his error, to adjust the contract sum upwards. Clearly the words used in JCT 80 could have been interpreted as giving the architect apparent or ostensible authority, particularly in an extreme case e.g. demolish and rebuild, to pay the contractor for his own error. Employers would have been rightly dismayed at such a possibility.

What is an appropriate deduction? The cost to the contractor of correcting a serious setting out error could be very large, e.g. the whole building being one metre too far to the west. To correct this could require total demolition and rebuilding. If the architect with the consent of the employer instructs the contractor not to correct the error can this enormous saving to the contractor be taken into account? It is submitted not as this introduces a ransom element into the valuation. It will nevertheless be very difficult on occasions for the architect or quantity surveyor to discern the correct principles to apply in every case in carrying out the valuation. The valuation itself takes the form of an adjustment to the contract sum (see clauses 4.2.2 and 4.5). It is therefore a matter for the architect or quantity surveyor and not the employer to determine what is an appropriate deduction.

Clause 3.10

Clerk of Works

3.10   The Employer shall be entitled to appoint a clerk of works
        whose duty shall be to act solely as an inspector on
        behalf of the Employer under the directions of the Archi-
        tect/the Contract Administrator.

COMMENTARY ON CLAUSE 3.10

A competent clerk of works is a great asset on a building project and
he is often an essential member of the construction team. However,
the contract pays him scant regard. He acts on behalf of the
employer solely as an inspector under the directions of the architect.
By the wording of this clause, he has no power at all to give
directions to the contractor. He can only report on what he inspects.

Under JCT 80 (clause 12) it is contemplated that the clerk of
works might give directions to the contractor but these are stated to
be of no effect unless confirmed in writing by the architect within
two working days of being issued, and in any event such directions
must be in regard to a matter in respect of which the architect is
empowered under those conditions to issue an instruction. This
power under JCT 80 for the clerk of works to give such directions
appears to be at odds with the reference in clause 12 of that contract
to his acting solely as an inspector.

Under IFC 84 there is no reference at all to this power to give
directions, even subject to confirmation. However, whether under
JCT 80 or IFC 84, it is suggested that if directions are given and
complied with which amount to a variation in the contract works,
then the architect may sanction this in writing as a variation.

In the case of *Kensington and Chelsea and Westminster Area
Health Authority* v. *Wettern Composites and Others* (1984), the
employer was held vicariously liable for the negligence of a clerk of
works in his employ even though he was under the direction and
control of the architect. As a result the employer's damages claim for
negligent inspection on the part of the architect was reduced by 20%
on the basis of the contributory negligence of the clerk of works in
failing to detect inadequate wall fixings by sub-contractors.

Clause 3.11

Work not forming part of the Contract

3.11   Where the Contract Documents provide for work not

forming part of this Contract to be carried out by the Employer or by persons employed or engaged by the Employer, the Contractor shall permit the execution of such work on the site of the Works concurrent with[1] his execution of the Contract Works. Where the Contract Documents do not so provide the Employer may nevertheless with the consent of the Contractor (which consent shall not be unreasonably withheld) arrange for the execution of such work.

Every person so employed or engaged shall for the purposes of clauses 6.1 (*Injury and damage*) and 6.3 (*Insurance of the Works*) be deemed to be a person for whom the Employer is responsible and not a sub-contractor.

COMMENTARY ON CLAUSE 3.11

This clause deals with the position where the employer wants to carry out or have carried out on his behalf, certain work, concurrently with the contract works. It should be noted that work carried out under clause 3.11 can give rise to the making of an extension of time under clause 2.4.8 and reimbursement of loss and expense under clause 4.12.3. The clause provides for two different situations:

(1) *Where the contract documents identify work to be executed by the employer or by other persons engaged by the employer.*

In such a case, the contractor must permit the execution of such other work on the site. Depending on the information provided in the contract documents, the contractor will be expected to make allowance for any interface problems which could reasonably be foreseen so that he should not be granted an extension of time or loss and expense for disturbance to regular progress caused by the carrying out of such other works where the delays or the disturbance could be anticipated. The contractor should take this into account in his programme. On the other hand, if the extent of the delay or disturbance caused is more than could reasonably be foreseen, there will be an entitlement to an extension of time and loss and expense in appropriate circumstances.

(2) *Where the contract documents do not identify work to be carried out by the employer or persons engaged by the employer.*

Here, the employer may only have such other work carried out if the contractor consents. However, this consent cannot be unreasonably withheld. As the contractor will clearly be entitled to claim an extension of time in relation to delays to completion and reimbursement of loss and expense in relation to disturbance of regular progress as appropriate, it will, it is submitted, be exceptional for the contractor to have a reasonable ground for withholding consent, though examples can be imagined e.g. where the employer wishes to introduce on to the site non-union labour which will cause severe industrial relations problems and consequential dislocation to the contractor's organisation.

Work executed by a public authority employer's direct labour organisation could fall within this clause provided it did not form part of the contract works. If, however, the direct labour organisation was engaged by the contractor to carry out part of the works, this clause would be inapplicable. See the discussion of this point earlier in considering clause 2.4.8 (page 80).

NOTES TO CLAUSE 3.11

[1] '. . . concurrent with . . .'.
This makes it clear that the contractor's possession of the site may well not be exclusive and can, by the contract documents, be limited. This is dealt with in greater detail in chapter 4 dealing with possession and completion.

Clauses 3.12 to 3.14

Instructions as to inspection – tests

3.12 The Architect/The Contract Administrator may issue instructions requiring the Contractor to open up for inspection any work covered up or to arrange for or carry out any test[1] of any materials or goods (whether or not already incorporated in the Works) or of any executed work. The cost of such opening up or testing (together with the cost of making good in consequence thereof) shall be added to the Contract Sum unless provided for in the Specification/Schedules of Work/Contract Bills or the inspection or test shows that the materials, goods or work are not in accordance with this Contract.

Instructions following failure of work etc.

3.13.1 If a failure of work or of materials or goods to be in accordance with this Contract is discovered during the carrying-out of the Works,[2] the Contractor upon such discovery shall state in writing to the Architect/the Contract Administrator the action which the Contractor will immediately take[3] at no cost to the Employer to establish that there is no similar failure in work already executed or materials or goods already supplied (whether or not incorporated in the Works). If the Architect/the Contract Administrator:

– has not received such statement within 7 days of such discovery, or
– if he is not satisfied with the action proposed by the Contractor, or
– if because of considerations of safety or statutory obligations he is unable to wait for the written proposals for action from the Contractor,

the Architect/the Contract Administrator may issue instructions requiring the Contractor at no cost to the Employer to open up for inspection any work covered up or to arrange for or carry out any test of any materials or goods (whether or not already incorporated in the Works) or any executed work to establish that there is no similar failure and to make good in consequence thereof.

The Contractor shall forthwith[4] comply with any instruction under this clause 3.13.1.

3.13.2 If, within ten days of receipt of the instruction under clause 3.13.1, and without prejudice to his obligation to comply therewith, the Contractor objects to compliance stating his reasons in writing, and if within 7 days of receipt thereof the Architect/the Contract Administrator does not in writing withdraw the instruction or modify the instruction[5] to remove the Contractor's objection, then any dispute or difference as to whether the nature or the extent[6] of the opening up for inspection or testing instructed by the Architect/the Contract Administrator was reasonable in all the circumstances shall be and is hereby referred to an Arbitrator[7] in accordance with clause 9, including clause 9.6[8] whether or not that clause is stated in the Appendix to apply.

To the extent that the Arbitrator finds the same is not fair and reasonable as aforesaid[9] he shall decide the amount, if any, to be paid[10] by the Employer to the Contractor in respect of compliance (together with making good in consequence thereof) and the extensions of time,[11] if any, for completion of the Works to be made in respect of such compliance. Any amount so awarded by the Arbitrator shall be a debt due and payable to the Contractor by the Employer.

Instructions as to removal of work etc.

3.14   The Architect/The Contract Administrator may issue
        instructions in regard to the removal from the site of any
        work, materials or goods which are not in accordance
        with this Contract.

COMMENTARY ON CLAUSES 3.12 TO 3.14

Clauses 3.12, 3.13 and 3.14 deal with the uncovering of work for
inspection, the testing of materials or goods, and the removal of
defective work, materials or goods not in accordance with the
contract. There are equivalent clauses in JCT 80, namely clauses 8.3
and 8.4. However, the treatment of instructions for opening up and
testing following discovery of work or materials or goods which are
not in accordance with the contract is significantly different – see
clause 8.4.4 of JCT 80 and the code of practice referred to therein.
The JCT 80 approach to this problem is both more succinct and
more helpful than the approach adopted in clause 3.13 of IFC 84.

Furthermore, JCT 80 now has detailed provisions giving the
architect other extensive powers in respect of non-complying work,
e.g. the power to order consequential variations without cost to the
employer – see clauses 8.4.2 and 8.4.3. It seems a great pity that such
necessary and useful provisions have not as yet found their way into
IFC 84.

Both clauses 3.12 and 3.13 deal with opening up and testing etc.
The difference is that following a failure of work or materials or
goods to be in accordance with the contract, the architect can choose
to issue an instruction under clause 3.12 or 3.13, whereas in the
absence of such a failure the architect can only issue an instruction
for inspection or testing under clause 3.12. In addition, the financial
consequences can vary depending on which clause is used.

*(i)   Inspections and tests etc. where there has been no previous
        failure*

By clause 3.12 the architect can issue instructions requiring the
contractor to open up covered work for inspection and to test
materials or goods, whether or not incorporated in the works. Such
testing can therefore clearly cover materials and goods on site, the
ownership in which has passed to the employer under the provisions
of clause 1.10, and no doubt extends to off-site materials or goods in
which the ownership has passed under clause 1.11.

It may be that there is some specialist plant or equipment being manufactured off-site, the ownership in which has not passed but which the architect requires to be tested before being transported to site for installation. Provided such materials or goods are intended for incorporation into the works then, it is submitted, even if the ownership in them has not yet vested in the employer, an instruction for testing under this clause can validly be given provided it is within the power of the contractor to comply with it. However, in such a case, to avoid any practical problems arising, the requirement for such testing should be clearly set out in the contract documents so that the contractor can make the necessary arrangements when entering into the sub-contract for the supply of such materials or goods. Such a requirement may appear as an item in section I and/or section II of NAM/T.

Costs of opening up or testing and making good: extensions of time; loss and expense
The cost of opening up or testing and of making good in compliance with instructions given under clause 3.12 is to be added to the contract sum except in two situations:

(a) where the contract documents make provision for such cost;
(b) where the inspection of the uncovered work or the results of the testing show that the work or materials or goods do not comply with the contract. In this case the contractor bears the costs.

In (a) above therefore, on a literal interpretation, it appears that the employer pays even if the opening up or testing discloses a failure to meet the contractual requirements, whereas in (b) above the contractor pays.

This clause does not apply where the contractor has been required to price for specific testing within the contract sum and which does not stem from an architect's instruction. It appears therefore to relate to the expenditure of a provisional sum, in which case, it is submitted, that the wording attached to the description of the provisional sum needs to be carefully considered if the cost of the testing etc. (which shows the work or materials not to be in accordance with the contract) is not to be borne by the employer (always bearing in mind the provisions of clause 1.3 (priority of contract documents)).

Clause 2.4.6 provides for an extension of time for the contractor if the opening up or testing causes delay to the completion date unless the results disclose a failure to meet the contractual requirements. In making an extension of time the architect should take into account any requirement in the contract documents as to opening up or testing. A reasonably competent contractor should have foreseen that time would be required for testing, and the contractor should have incorporated such time in his programme, and it is only any excess time that should be allowed.

Clause 4.12.2 provides for reimbursement to the contractor of loss and expense due to compliance with an instruction under clause 3.12 unless the results show a failure to comply with the contractual requirements. Again, in determining the amount of loss and expense, account should be taken of the extent to which a reasonably competent contractor could foresee that any requirement in the contract documents would, if not adequately reflected in his programme, result in some disturbance to the regular progress of the works.

There is nothing in the contract to prevent the architect from issuing an instruction under clause 3.12 even where a failure of work or materials or goods to be in accordance with the contract has been discovered. Whether or not an instruction in such circumstances should be given under clause 3.12 or 3.13 depends on a number of factors and it is discussed in more detail below. In any event, if there has been such a failure the architect would be well advised to seek express instructions from the client if it is intended to issue instructions under clause 3.12 rather than 3.13.

(ii)   Inspections and tests etc. following a failure
Where it is discovered during the carrying out of the works that there is a failure of work, materials or goods to be in accordance with the contract, the architect may be able to issue instructions requiring the contractor, at no cost to the employer, to open up for inspection any other covered work or to arrange for, or carry out, tests on any other materials or goods to establish that there is no similar failure. However, unless owing to considerations of safety or because of statutory obligations it is necessary to issue instructions immediately, the architect must wait for up to seven days following the discovery for the contractor to state what action he proposes to take (immediately and at no cost to the employer) to establish that there is no similar failure. If the architect is not satisfied with the contractor's proposed action, or if the contractor fails to put forward a proposed course of action, the architect can proceed to issue instructions.

Whether the contractor wishes to object to the instruction or not, he is obliged to comply with it forthwith. However, within 10 days of receipt of the instruction he may object to compliance, and must state his reasons in writing. If within seven days of the architect's receipt of this objection he does not either withdraw the instruction or modify it so as to remove the contractor's objection, any resulting dispute as to whether the nature or extent of the opening up was reasonable in all the circumstances is referred to an arbitrator.

If the arbitrator finds that the nature or extent of the opening up for inspection or testing was not fair and reasonable in all the circumstances, he will decide the amount, if any, to be paid by the employer to the contractor. He also has power to deal with the extensions of time to be awarded.

The position therefore is that the contractor's compliance with the instructions issued under this clause is to be at no cost to the employer except where an arbitrator determines otherwise, unless of course the employer and contractor reach an agreement without proceeding to arbitration. The policy behind the clause appears to be reasonable. A failure of work, materials or goods to be in accordance with the contract will generally amount to a breach of contract (some may say only a temporary disconformity – see Lord Diplock's Speech in *Hosier and Dickinson Ltd* v. *Kaye (P & M) Ltd* (1972)) by the contractor. The architect may reasonably enough wish to find out if there are any further similar failures especially where there are structural or safety considerations involved. Even if the opening up etc. showed that there had been no further failure, it may reasonably be argued that the employer should not have to meet the cost associated with it and pay the contractor's loss and expense as he would under clause 3.12 (even if it was the contractor's breach of contract which necessitated the opening up etc.). Under clause 3.13 in such circumstances the contractor meets the cost subject to his right to object to compliance and to have that objection considered by an arbitrator if the architect is unwilling to accept it.

Unfortunately the drafting of the clause is not particularly clear. It appears to be too detailed. For instance, the whole of clause 3.13.2 could be removed by a requirement in clause 3.13.1 that the instruction must be reasonable in all the circumstances, taking into account considerations of health and safety, structural significance and cost. As such it compares unfavourably with JCT 80 clause 8.4.4. The rather involved wording may well stem from a suspicion on the contracting side of the industry, whether well founded or not, that this clause opens the door for the architect to exercise an unfettered licence to spend the contractor's money. However a requirement that the instruction should be reasonable in all the circumstances would normally ensure that the architect issued such instructions responsibly. Some of the difficulties raised by the wording of this clause are dealt with under the Notes [1] to [11] below.

Under which clause should architect issue instructions?
Should the architect issue instructions under clause 3.12 or 3.13? Of course, if there has been no failure of work or materials etc. to be in accordance with the contract the architect can only issue an instruction for opening up or testing etc. under clause 3.12. He only has a choice where a failure has been discovered and he wishes to ascertain whether or not there are further similar failures.

Depending on which clause is relied on, the financial consequences can vary considerably. For instance, if the instruction is given under clause 3.12 then, if the opening up or testing etc. shows that work or materials are not in accordance with the contract, the contractor must bear the cost of the opening up and testing, together with the costs of any disruption etc. caused. He may also be liable to pay liquidated damages if the opening up or testing etc. causes delay. This may be so even if the nature and extent of the opening up or testing required by the architect is over and above what might be strictly necessary. If the result of the opening up or testing etc. is that the inspection or test shows that the work, materials or goods are not in accordance with the contract, that fact in itself will mean that the contractor meets the cost.

In the same situation, if the instruction were given under clause 3.13.1, the contractor must have first been given the opportunity to put forward his own proposals, unless there were considerations of safety or statutory obligations involved. Furthermore, following the instruction, the contractor can object to compliance and if the architect does not accommodate the contractor's objection then the contractor can take the matter to an arbitrator for a decision. The arbitrator may decide that the nature and extent of the inspection or testing etc. required was unreasonable, even if it shows that the materials, goods or work were not in accordance with the contract, and may make an award requiring the employer to pay part of the cost of this, including, it is submitted, disruption costs of the contractor.

On the other hand, if compliance with an instruction under clause 3.13 reveals that the materials, goods or work are in accordance with the contract, provided the nature and extent of the testing required was reasonable in all the circumstances, compliance with the instruction will still be at the contractor's cost. Without the benefit of hindsight the architect may be in something of a dilemma. Generally however, provided the architect feels sure that the nature and extent of the testing is reasonable in all the circumstances, the employer is probably better served by the architect seeking the contractor's proposed course of action under clause 3.13.1 and, if dissatisfied with this, issuing an instruction under that clause accordingly.

In any event the architect should make it abundantly clear under which clause he is actually issuing the instruction.

Removal of work

Under clause 3.14 the architect may issue instructions requiring the removal of any work, materials or goods not in accordance with the contract. It is not sufficient for the architect to issue an instruction simply condemning the work, materials or goods even if that instruction also refers to clause 3.14. He must expressly require its removal. In *Holland Hannen and Cubitts (Northern) Ltd* v. *Welsh Health Technical Services Organisation and Others* (1981) Judge John Newey QC said:

> '. . . In my opinion, an architect's power [under clause 6(4) JCT 63] is simply to instruct the removal of work or materials from the site on the ground that they are not in accordance with the contract. A notice which does not require the removal of anything at all is not a valid notice under clause 6(4).'

An instruction under clause 3.14 is not of course a variation – see clause 3.6.1.

NOTES TO CLAUSES 3.12 TO 3.14

[1] '. . . *any test* . . .'.
Bearing in mind the important financial consequences which flow from the result of any inspection or test instructed under clause 3.12, it may be very important to determine whether or not an instruction refers to a single test or in fact to a series of different tests. For example, if the instruction calls for an identical test on six critical welded joints, two of which are discovered to be unsatis-

factory and four of which are satisfactory, are there six tests, two of which are at the contractor's expense and four at the employer's expense? Or is there just one single test which has disclosed defective work and which will therefore be wholly at the contractor's expense? While it must clearly be a question of degree in all the circumstances of any particular case, it is tentatively submitted that generally the determining factor will be the nature of the test i.e. critical welds in the above example rather than the number of times the same test is repeated.

[2]   '. . . during the carrying-out of the Works . . .'.
An instruction cannot be issued after the date of practical completion. While this must surely be right in relation to inspections and tests under clause 3.12 where there has been no failure (even though incidentally clause 3.12 is not expressly so limited), under clause 3.13 which operates following the discovery of a failure it might be a useful power for the architect to compel the contractor to investigate the extent of a problem which is discovered during the defects liability period. In complying with his obligations under clause 2.10 to make good defects at no cost to the employer, the contractor may well investigate the extent of any failures. If he does not, and it is not known for certain whether further work is similarly defective, the architect, if he has suspicions about the work, must presumably advise the employer to have any necessary inspection or tests etc. carried out at the employer's cost, leaving the employer to claim (whatever the result of the tests etc.) the cost from the contractor as special damages for breach of contract, if the expenditure was reasonably and properly incurred as a result of the contractor's breach.

[3]   '. . . the action which the Contractor will immediately take . . .'.
If the contractor wishes to put forward his own proposals to demonstrate that there is no similar failure he must do so before the architect issues an instruction under clause 3.13.1. Of course in practice there will no doubt often be an attempt to agree an appropriate course of action. If the contractor does not put forward any proposals in accordance with clause 3.13.1, he can always submit his proposals as part of his objection in writing under clause 3.13.2 after receiving the architect's instruction, and no doubt the architect will have to consider them accordingly.

[4] '. . . *forthwith* . . .'.

Whether or not an objection is made by the contractor, he must comply forthwith with the instruction. The exercise by the contractor of his right to object does not entitle him to suspend compliance with the instruction.

[5] '. . . *modify the instruction* . . .'.

This will enable the architect to alter the nature or extent of the opening up or testing required following receipt of the contractor's objection. If the nature of the contractor's objection is that he thinks it unreasonable that he should bear all the costs of compliance with such an instruction, it is submitted that the architect has no power to modify his instruction so as to place all or part of the costs of compliance on the employer. The architect only has power under clause 3.13.1 to issue an instruction requiring the contractor to open up for inspection or to test etc. at no cost to the employer. If the cost is to be shared or is to fall on the employer, this will require the express agreement of the employer outside the provisions of clause 3.13.

[6] '. . . *the nature or the extent* . . .'.

These are perhaps the most important words in clause 3.13. The objection by the contractor is almost certainly going to be based on the nature or extent of the opening up or testing required. Much will depend on the nature and extent of the failure discovered. In some circumstances the adoption of progressive sampling techniques may suffice.

[7] '. . . *shall be and is hereby referred to an Arbitrator* . . .'.

It is submitted that a written notice from the contractor to the employer requiring arbitration is still required under clause 9.1 despite the apparent automatic reference to arbitration. Once notice is given the architect then ceases to have any function in connection with the resulting dispute and there is an immediate confrontation between employer and contractor, all within as little as 17 days of the issue of the instruction.

[8] '. . . *in accordance with clause 9, including clause 9.6* . . .'.

The joinder provisions in dealing with arbitration are to apply whether or not clause 9.6 is stated in the appendix to the contract to apply. The deletion of the word 'applies' in the appendix against clause 9.6 does not therefore prevent it applying to this particular dispute or difference. (See also clause 5.6 of NAM/SC.)

[9] '. . . fair and reasonable as aforesaid . . .'.
This is odd and could be a defect in the drafting. It refers to a finding by the arbitrator that the nature and extent of the testing is not fair and reasonable 'as aforesaid' and yet there is no reference earlier in the clause to the question of fairness. The only reference is to whether or not the nature and extent of the opening up or testing was reasonable in all the circumstances. What then if the arbitrator finds that the nature and extent was reasonable although not fair? Without the introduction of the word 'fair', the arbitrator would doubtless have to look at all the circumstances objectively and not subjectively, so that any subjective peculiarities of the parties e.g. their respective financial positions should not be taken into account. However, if the arbitrator must in fact consider whether or not the nature and extent of the opening up or testing is fair in all the circumstances, perhaps this introduces a more subjective element. It seems inappropriate for an arbitrator to determine whether or not the nature and extent of the opening up or testing is fair rather than reasonable in all the circumstances. It is to be hoped that the arbitrator's consideration will be limited to what is reasonable only in all the circumstances.

The actual results of the inspection or tests should have no bearing on the arbitrator's determination of whether the nature and extent of the opening up for inspection or testing was reasonable in all the circumstances. Reasonableness must be determined at the time when the instruction is given, or at the latest at the end of seven days from the receipt by the architect of the contractor's objection.

[10] '. . . the amount, if any, to be paid . . .'.
This relates to the costs of any tests etc. including making good, and presumably also covers any other costs e.g. loss and expense suffered by the contractor. The arbitrator can no doubt apportion between the employer and the contractor the costs incurred in testing etc. together with any disruption costs incurred by the contractor.

[11] '. . . extensions of time . . .'.
By clause 2.4.6. the contractor is entitled to an extension of time for compliance with an instruction under clause 3.13.1 where the effect of compliance has been that the completion date has been delayed

thereby, except where the result of the opening up or testing reveals that the work, materials or goods were not in accordance with the contract. However, even where the results show that the contractual requirements have not been complied with, so that the architect has no power to make an extension of time, the arbitrator appears to be given such power. This is novel indeed. It means that the arbitrator can extend the completion date even though there is no event within clause 2.4 which permits this. Was this intended?

Clause 3.15

Instructions as to postponement

3.15 The Architect/The Contract Administrator may issue instructions in regard to the postponement of any work to be executed under the provisions of this Contract.

COMMENTARY ON CLAUSE 3.15

The architect can issue an instruction to postpone any or all of the work. If he does so, the contractor is entitled to an extension of time under clause 2.4.5 and to reimbursement of loss and expense under clause 4.12.5. Further, if the postponement results in the carrying out of the whole, or substantially the whole, of the uncompleted works being suspended for a continuous period of one month, then unless that postponement was caused by reason of some negligence or default of the contractor, the contractor will be entitled to determine his own employment under the contract and to make various financial claims against the employer – see clause 7.5.3(a) and clause 7.7.

It is unlikely that clause 3.15 can be used where there is a delay in giving initial possession of the site to the contractor. This view is supported by the inclusion in this contract of an express power for the employer to defer, for a limited period, the giving of possession of the site to the contractor, where clause 2.2 is stated in the appendix to apply.

An instruction to postpone may be implicit in the architect's request for the contractor to reprogramme all or part of his work even though clause 3.15 is not expressly referred to. In the case of *M. Harrison & Co. (Leeds) Ltd* v. *Leeds City Council* (1980) the main contractor was required to reprogramme part of his work by delaying it so as to enable a nominated sub-contractor to come on to the site earlier than had originally been planned. It was held that this amounted in effect to a postponement of the work, though no mention was made of that word or of clause 21(2) of JCT 63 which was the relevant contract in that case. (Clause 21(2) of JCT 63 is similar in terms to clause 3.15 of IFC 84.)

The acknowledgement by the architect of a *de facto* suspension of work may also apparently amount to an instruction to postpone – see the first instance decision in *Jarvis* v. *Rockdale Housing Association* (1987) where the contractor was unable to proceed with the works following the installation of defective piles by a nominated sub-contractor involving the need for redesign.

Chapter 7

Payment

CONTENT

This chapter deals with section 4 which concerns payment and related matters. It deals with the following:

The contract sum
Interim payments
The final certificate
Fluctuations
Disturbance of regular progress (including reimbursement of direct loss and/or expense).

SUMMARY OF GENERAL LAW

(A) Entire and severable contracts

In very many everyday contracts, payment by one party is due only on the complete fulfilment by the other party of his contractual obligations. It may involve contracts for the supply of materials or goods or the performance of a service. Such contracts are known as entire contracts. A classic example of this is to be found in the very old case of *Cutter* v. *Powell* (1795).

Facts:
A sailor agreed the following terms in his employment contract:

'Ten days after the ship *Governor Pary* . . . arrives at Liverpool I promise to pay to Mr T. Cutter the sum of 30 guineas, provided he proceeds, continues and does his duty as Second Mate in the said ship from here to the Port of Liverpool . . .'.

The sailor died before completion of the voyage and his personal representatives sought to recover a proportionate part of the agreed remuneration. It was held that they could not do so as this was an entire contract so that the sailor had to continue carrying out his duties until the ship arrived at Liverpool and failure to do this, even though as a result of his death, disentitled his personal representatives to any part of the remuneration.

On the other hand, some contracts, by their terms or by their nature, permit of a final payment against an interim valuation of work done or on the completion of a stage of the work having been reached. These are known as severable contracts.

Most standard forms of building contract will contain express provisions for payment by instalments, though this does not prevent them from being entire contracts as the provision for payment by instalments will usually be treated as being for payment on account of a final sum, so that the fundamental principle in relation to entire contracts i.e. payment in full on completion in full, will apply to the last instalment or the release of the final balance. A consequence of this is that instalment payments for work done or materials supplied create no estoppel against the employer if the work or materials are discovered subsequently to be defective. A later valuation can take this decrease in value into account. The mechanism by which interim instalments are paid is generally through the issue of interim payment certificates by the architect named in the contract. Subject always to the express terms of the contract, especially any arbitration clause, production of an interim certificate will generally be a condition precedent to the employer's obligation to make payment.

In contracts for work and materials which involve work being carried out over a significant period of time, the courts will readily imply a term, in the absence of an express term, to the effect that interim payments on account reflect the presumed intention of the parties – see e.g. the first instance decision in *Williams* v. *Roffey Bros and Nicholls (Contractors) Ltd* (31 January 1989).

On a strict application of the law relating to entire contracts, a contractor carrying out work on the employer's land which he fails to complete will not be entitled to payment. This could of course provide an unexpected and possibly unwarranted benefit to the employer. The rigours of the operation of this principle are qualified by the law in a number of ways. Firstly, if in such a case the employer sues the contractor for breach of contract to recover the increased cost of having the work completed, he will, in the assessment of damages, have to give some credit against what he would have had to pay had the contract been properly performed. Secondly, the doctrine of substantial performance may aid the contractor. The essence of this doctrine is that, provided the contractor has substantially performed his obligations, he will be permitted to sue for the price, giving credit for the outstanding work left incomplete.

Where the contract expressly provides for payment by instalments, provided any conditions required to be fulfilled before an instalment becomes due have been met, a debt is created. If the contract requires a certificate to be issued in respect of an instalment this may be a condition precedent to the debt coming into existence.

(B) The certification process

Most standard forms of building contract allow for interim payments by means of the certificate of the architect. This may relate to a stage which is reached in the work, in which case there is little or no act of valuation required. On the other hand, many contracts provide for interim certificates against valuations of work carried out and usually of goods and materials delivered to the site and sometimes those delivered off site but intended for incorporation into the works. The certificate will be issued by the architect who may be dependent on the valuation carried out by the quantity surveyor if one is appointed under the contract concerned. Depending on the wording of the contract it may well be that the issue of the certificate is a condition precedent to the employer's obligation to pay the contractor, so that no debt becomes due in favour of the contractor unless and until a certificate is issued for the amount being claimed – see e.g. *Lubenham Fidelities and Investment Ltd* v. *South Pembrokeshire District Council and Another* (1986). Most building contracts which provide for interim certificates state the period at which such certificates will be issued e.g. monthly.

Even if the issue of a certificate from the architect is a condition precedent to the employer's obligation to pay, on the basis that no debt becomes due in favour of the contractor unless and until a certificate is issued for the amount being claimed, it will generally be possible, provided the arbitration clause is worded sufficiently widely, to enable an arbitrator to ascertain and award any sum which ought to have been the subject of or included in any certificate. However, if instead of arbitrating the parties decide to litigate, the court will not have the same powers as the arbitrator and accordingly the contractor will not be able to challenge the absence of a certificate in the courts unless there is evidence that the architect had acted in bad faith or had clearly failed to fulfil the duties required of him under the building contract: *Northern Regional Health Authority* v. *Derek Crouch Construction Co. Ltd and Others* (1984) (dealt with in greater detail in the Commentary on clause 9 ('Settlement of disputes – arbitration', page 382). However, the court will substitute its own machinery for resolving a claim even in the absence of a certificate, where the contractual machinery has completely broken down and is incapable of operating. Certainly, if the failure of the architect to issue a certificate is due to some breach of contract on the part of the employer, the court is likely to provide a remedy – see *Croudace Ltd* v. *London Borough of Lambeth* (1986).

If the architect negligently undercertifies in an interim certificate, this will not be a breach of contract on the part of the employer. Further, even if the employer is fully aware that the sum certified is lower than it should be, possibly even if due to an arithmetical error, provided the employer pays the certified sum he will not be in breach, at any rate not if there exists an appropriate method within the contract by which the contractor can then challenge the certificate, for example an arbitration clause allowing immediate arbitration – see *Lubenham Fidelities and Investment Co. Ltd* v. *South Pembrokeshire District Council and Another* (1986). The situation is otherwise if the employer has sought to put pressure upon the architect to undercertify payment or the quantity surveyor to undervalue work.

Defective work
In relation to the issue of a payment certificate, it is undoubtedly the case that the certifier owes a duty of care to the employer. If there is included within the certificate payment for work which is defective and which the certifier ought to have known was defective and which thereby renders the valuation excessive, the certifier will be liable to the employer in damages. Further, there is no immunity from liability because of the apparent quasi-judicial role of the certifier i.e. his need to act independently using his own professional judgment in the valuation and certification process: *Sutcliffe* v. *Thackrah* (1974). In that particular case the defendants acted both as architect and quantity surveyor.

If the certifier negligently undervalues the work done, thereby causing the contractor loss, the question whether he may be liable to the contractor for a breach of duty of care under the tort of negligence is difficult to answer. On the one hand there is judicial support of considerable authority for the view that a duty of care is owed by the architect to the contractor in such circumstances to avoid negligently causing financial loss, e.g. see Lord Salmon in the case of *Arenson* v. *Casson, Beckman* (1977) who said:

'. . . The architect owed a duty to his client, the building owner, arising out of the contract between them to use reasonable care in issuing his certificate. He also, however, owed a similar duty of care to the contractor arising out of their proximity, see *Hedley-Byrne*. In *Sutcliffe* v. *Thackrah* the architect negligently certified that more money was due than was in fact due and he was successfully sued for the damage which this had caused his client. He might, however, have negligently certified that less money was payable than was in fact due, and thereby starved the contractor of money. In a trade in which cash flow is especially important, this might have caused the contractor serious damage for which the architect could have been successfully sued.'

This view was subsequently endorsed, firstly in the excellent judgment of Hunter J in the Hong Kong case of *Shui On Construction Co. Ltd* v. *Shui Kay Co. Ltd* (1985), and secondly in the English case of *Michael Salliss & Co. Ltd* v. *E.C.A. Calil and F.B. Calil and William F. Newman & Associates* (1987) a decision in which the Official Referee said:

'But it is self evident that a contractor who is a party to a JCT contract looks to the architect . . . to act fairly as between him and the building employer in matters such as certificates and extensions of time. Without a confident belief that that reliance will be justified, in an industry where cash flow is so important to the contractor, contracting could be a hazardous operation. If the architect unfairly promotes the building employer's interest by low certification or merely fails properly to exercise reasonable care and skill in his certification it is reasonable that the contractor should not only have the right as against the owner to have the certificate reviewed in arbitration but also should have the right to recover damages against the unfair architect.'

However, serious doubt has been cast over the *Sallis* case by the

Court of Appeal in the case of *Pacific Associates, Inc. and Another* v. *Baxter and Others* (1988) in which it was held that an engineer under the FIDIC form of international civil engineering contract did not owe the contractor or sub-contractor any duty of care in relation to his duty to act impartially in dealing with contractual claims. A decision that there was no duty of care was quite understandable on the facts of that case. It is perhaps unfortunate, however, that the judgments were not so qualified. Instead they extend to general principles so that at the present time it is unlikely that a duty will be owed except in the most exceptional circumstances. The prospect of an architect, in exercising functions under the building contract, in which he is clearly required to act impartially, owing a legal contractual and no doubt tortious duty to his employer client on the one hand but only at best some sort of moral duty to the contractor on the other, can only bring the law into contempt.

It is submitted that the existence of such a legal duty owed by an architect to the contractor would be sound in law and also accord with common sense.

Although in the *Sutcliffe* case there was no quantity surveyor acting, nevertheless the role of the quantity surveyor in valuing, and the relationship between the quantity surveyor and the architect, were considered in detail in the first instance judgment of Judge Stabb QC in the *Sutcliffe* case under the name of *Sutcliffe* v. *Chippendale and Edmondson* (1971). The judge said as follows:

'I readily acknowledge and accept that any prolonged or detailed inspection or measurement at an interim stage is impracticable, and not to be expected. On the other hand . . . the issuing of certificates is a continuing process, leaving each time a limited amount of work to be inspected and I should have thought that more than a glance round was to be expected . . . . The quality of the work was . . . the responsibility of the architect and never that of the quantity surveyor and since work properly executed is the work for which the progress payment is being recommended, I think that the architect is in duty bound to notify the quantity surveyor in advance of any work which he, the architect, classifies as not properly executed, so as to give the quantity surveyor the opportunity of excluding it.'

As a matter of practice it is advisable to have such notification in writing and also for the quantity surveyor, on his part, to notify the architect of any apparently defective work when carrying out his valuation.

(C) Retention

Standard forms of building contract which provide for payment by instalments will usually enable the employer to retain a certain percentage of the total value included in the interim certificate. This percentage is often fixed at three, five or 10 per cent, with the first half of it released at practical or substantial completion and the remaining half after the making good of any defects at the end of the defects liability period.

From the employer's point of view it is a useful system as it represents some protection against the inclusion of defective work in a valuation and which is therefore included in the amount of an interim certificate. It also provides security for the performance by the contractor of his obligations. Its main purpose, however, is to provide the employer with a fund during the defects liability period following practical or substantial completion, should the contractor fail to return and make good any defects of which he is notified.

Often, the express terms of the contract will give to the retention fund the status of trust money. In other words, the employer will hold the retention fund in a fiduciary capacity as trustee of it for the benefit of the contractor. In such a case, provided the retention money is kept as a separate identifiable fund (and it is the duty of a trustee to keep trust funds separate from other funds), a trustee in bankruptcy or a liquidator of the employer will not be able to include the retention fund as part of the employer's estate. It will have to be held for the benefit of the contractor and released to him at the appropriate time. A case in point is re *Tout and Finch* (1954).

Facts:

A sub-contract was entered into under the NFBTE/FASS ('green' form). Clause 11(h) provided that the main contractor's interest in retention money withheld under the sub-contract was held by him in a fiduciary capacity as trustee for the sub-contractor. The main contractor went into voluntary liquidation after completion but before the defects liability period had expired and before, therefore, the release of the final balance of the retention money.

Held:

The provisions of clause 11(h) created an equitable assignment of the appropriate part of the retention money in favour of the sub-contractor. In other words, the beneficial interest in the retention fund belonged in equity to the sub-contractor, and even though at the time that the contractor went into liquidation the retention money had not been released to the sub-contractor, the court regarded it, in equity, as having already been so transferred. Accordingly, the sub-contractor was entitled to it.

The protection of the trust money from the creditors of the employer (or the main contractor in the case of a sub-contract) is a very significant benefit to the party concerned. However, this benefit can be lost should the employer (or main contractor as the case may be) not place the retention money in a separate identifiable fund. If this is not done and the fund cannot be identified, then the other contracting party will rank only as an ordinary creditor without any priority in respect of the retention money due to him.

If the contract does not contain any express provisions dealing with the status of retention money, then it is submitted that it will not be treated as trust money in the hands of the employer.

(D) Deductions from payments: set-off and counterclaim

Authorised deductions
If a building contract provides for payments against interim certificates it will also usually entitle the employer to make certain deductions, generally in relation to specific ascertained amounts. The most important such entitlement will be the employer's right to deduct liquidated and ascertained damages. However, other deductions may also be authorised e.g. recovery of insurance premiums paid by the employer where the contractor has failed to take out any necessary insurance required by the contract; and the cost of employing another contractor to carry out work which the contractor has failed to do despite having received a valid instruction.

Set-off and counterclaim
An equitable set-off is a cross claim which is in the nature of a defence. Strictly so called, it is to be used only as a shield to an action and not as a cause of action in its own right. It is used only as a defence and not as a counterclaim as such, so that if the claim were abandoned, the set-off would not stand on its own in the same way as a counterclaim. It is not always easy to determine if a cross claim

is a set-off or a counterclaim or both. This subject received detailed consideration in the case of *Hanak* v. *Green* (1958).

Facts:

The plaintiff bought a house from the defendant who was a builder. He agreed to carry out certain work to the property. The plaintiff became dissatisfied with that work and sued the defendant for damages for breach of contract in the sum of £266. The defendant counterclaimed for, *inter alia*, damages due to the plaintiff's failure to admit one of his workmen into the house to carry out work and for trespass to his tools. The plaintiff was awarded £74 17s 6d and the defendant was awarded £84 19s 3d. The judge awarded the plaintiff her costs on the applicable county court scale for a judgment of £74 17s 6d and the defendant his costs on the applicable county court scale for a judgment of £84 19s 3d. The defendant claimed that as he had recovered net the sum of £10 1s 9d whereas the plaintiff had recovered nothing, the defendant's cross claim should be treated as a set-off i.e. as a defence so that he should have his costs met by the plaintiff but should not be obliged to pay her costs.

Held:

The defendant's contention with regard to costs was correct and the defendant was entitled to the costs of his defence on the scale appropriate for £266 being the amount of the plaintiff's original claim. Also on his counterclaim he should receive costs on the appropriate county court scale for £10 1s 9d. Lord Justice Morris said:

'A set-off . . . is a defence, but not every defence is a set-off . . . since the Judicature Acts there may be (i) a set-off of mutual debts; (ii) in certain cases a setting up of matters of complaint which, if established, reduce or even extinguish the claim, and (iii) reliance upon equitable set-off and reliance as a matter of defence upon matters of equity which formerly might have called for injunction or prohibition.'

Lord Morris further said that a cross claim was to be treated as a set-off if it arose

'. . . directly under and affected the contract on which the plaintiff herself relies . . . Some counterclaims might be quite incompatible with the plaintiff's claim, in no way connected with it and wholly unsuitable to be used as a set-off, but the present class of action involving building or repairs, extras and incidental work so often leads to cross claims for bad or unfinished work, delay or other breaches of contract that a set-off would normally prove just and convenient . . .'.

A cross claim may not amount to a set-off but may nevertheless be the subject of a counterclaim e.g. where it arises out of a different and separate contract between the same parties, not being part of a series of similar contracts.

A vital question concerning cross claims is the extent to which one party can withhold payment to the other by reliance on a cross claim, even to the extent of a refusal to pay certified sums. The terms of the contract in question may expressly deal with such rights e.g. as does clause 21 of NAM/SC, the JCT Standard Form of Sub-Contract Conditions for Sub-Contractors named under IFC 84, in which case the contractual terms will override what might otherwise be the position under common law. If there are no express provisions covering the point, then a right of set-off will, it is submitted, exist even against certified sums. Where the contract in question permits some deduction by way of set-off, it is a difficult question as to whether this, by implication, excludes an equitable right to set-off. A detailed consideration of this subject is to be found in an article by Mr I. N. Duncan Wallace in the *Law Quarterly Review* (LQR 36 page 89) in which most of the cases are considered and which leads to the conclusion that the equitable right of set-off is excluded only if the contract between the parties expressly excludes it or does so by necessary implication. For a contrasting view as to how this applies to JCT 80 – see John Parris: *The Standard Form of Building Contract: JCT 80* (second edition, pages 149 to 154 inclusive).

(E) Contractor's claims arising under the contract and for breach of contract

Most standard forms of building contract expressly entitle the contractor to recover sums in addition to the value of the work done when he is involved in extra cost, expenditure or losses due to certain matters, some of which may also amount to a breach of contract by the employer. Frequently words are used such as 'direct loss and/or expense', 'direct loss and/or damage', 'increased cost' etc. The extent of recovery by the contractor will depend on the precise words used.

However, most standard forms of contract also preserve, either expressly or by their silence on the point, the contractor's common law rights to claim damages for breach of contract.

The general legal principles on which damages are awarded for breach of contract are considered later in chapter 10, dealing with determination of the employment of the contractor. The application of these general principles is likely to overlap to a considerable degree with any discussion of the particular principles on which recovery can be claimed under the express wording of the contract itself. Accordingly, these particular principles are discussed in detail in relation to the specific relevant clauses in IFC 84 and any possible difference between recovery under the express terms of the contract and at common law will, where appropriate, be highlighted.

CONSIDERATION OF THE RELEVANT CLAUSES OF IFC 84

Clause 4.1

Contract Sum

4.1   The Contract Sum shall not be adjusted or altered in any way otherwise than in accordance with the express provisions of the Conditions and subject to clause 1.4 (*Inconsistencies*) any error or omission, whether of arithmetic or not, in the computation of the Contract Sum shall be deemed[1] to have been accepted by the parties hereto.

COMMENTARY ON CLAUSE 4.1

The contract sum is referred to in article 2 of the agreement attached to the conditions of contract. It is expressed as being exclusive of value added tax and it can only be varied in the manner specified in the conditions (see clause 4.5). Clause 4.1 expressly provides that the contract sum is not to be adjusted or altered in any way otherwise than in accordance with express provisions of the conditions of contract so that there can be no adjustment of the contract sum by implication.

Subject to the exception contained in clause 1.4 (see page 37) the parties are bound by any errors or omissions incorporated in the contract sum unless there exists sufficient ground to persuade a court to grant the equitable remedy of rectification. This remedy is discretionary and the courts will have to be satisfied that the written contract fails in some way to express what was the clear intention of both parties. It does this firstly in order that the written document properly reflects the agreement actually made between the parties, and secondly to prevent a party unfairly holding the other to a written contract which he knows does not accurately reflect the agreement reached. Cases on the rectification of building contracts are rare but one such a case is: *A. Roberts and Co. Ltd* v. *Leicestershire County Council* (1961).

Facts:
The contractor's revised tender contained a completion period of 18 months. The county council decided that the period for completion should be 30 months, that is, the same date but one year later than the date put forward by the contractor. The county council did not refer to any date in its letter of acceptance. Instead, the formal contract, when drawn up, contained a completion date one year later than that put forward by the contractor. The contractor did not notice the change of year and sealed and returned the contract. Before the county council itself sealed the formal contract, it held a meeting with the contractor during which the contractor referred to his plans to complete in 18 months. The county council's officers did not mention the later date inserted in the formal contract. The county council subsequently sealed the formal contract.

Held:
Rectification would be ordered. A contracting party is entitled to rectification of a contract if he can prove that he believed a particular term to be included and the other party concluded the contract without that term being included in the knowledge that the first party believed that it was included.

The exception in clause 1.4 referred to in this clause deals with inconsistencies, errors and omissions in and between contract documents and other documents issued under clause 1.7 and clause 3.9, and departures from the Standard Method of Measurement referred to in clause 1.5. This is a significant exception and the reader is referred to those clauses and the commentary on them.

It is important to distinguish the contract sum from the contractor's tender sum. The sum put forward by the contractor in his tender does not become the contract sum until the tender is accepted. Depending on the tendering procedures adopted, the tender sum may be adjusted by agreement between the parties e.g. on the discovery of an error in the make-up of the tender sum.

Alternatively, the contractor may be asked to elect to maintain the tender sum notwithstanding the error or to withdraw his tender.

NOTES TO CLAUSE 4.1

[1] '*. . . shall be deemed . . .*'
In other words, the parties shall be regarded as having been aware of the errors or omissions and, subject of course to clause 1.4, to have accepted them.

Interim and final payments and the final certificate

Clauses 4.2 to 4.8

Interim payments

4.2 Subject to any agreement between the parties as to stage payments, the Architect/the Contract Administrator shall, at intervals of one month, unless a different interval is stated in the Appendix, calculated from the Date of Possession stated in the Appendix,[1] certify the amount of interim payments to be made by the Employer to the Contractor within 14 days of the date of the certificate.

Interim valuations shall be made by the Quantity Surveyor whenever the Architect/the Contract Administrator considers them to be necessary for the purpose of ascertaining the amount to be stated as due in an interim payment.

The amount of the interim payment to be certified shall be the total of the amounts in clauses 4.2.1 and 4.2.2, at a date not more than 7 days before the date of the certificate, less any sums previously certified for payment.[2] These amounts are:

4.2.1 95% of

(a) the total value of the work properly executed[3] by the Contractor, including any items valued in accordance with clause 3.7 *(Valuation of Variations)*, together with, where applicable, any adjustment of the value under clause 4.9(b) *(Formulae adjustment)* but excluding any restoration, replacement or repair of loss or damage or removal and disposal of debris which in clauses 6.3B.3.5 and 6.3C.4.4 are treated as if they were a Variation;

(b) the total value of the materials and goods which have been reasonably and properly[4] and not prematurely delivered to or adjacent to the Works for incorporation

therein and which are adequately protected against
weather and other casualties;[5]

(c) at the discretion of the Architect/the Contract
Administrator, the value of any off-site goods and
materials.

[j]

4.2.2   100% of

any amounts payable to the Contractor or to be added to
the Contract Sum under clauses
2.1   *(Insurance premium)*
3.12   *(Inspection)*
4.9(a)   *(Tax etc. fluctuations)*
4.10   *(Fluctuations: named persons)*
4.11   *(Disturbance of progress)*
5.1   *(Statutory obligations)*
6.2.4   *(Insurance: liability etc. of Employer)*
6.3A.4   *(Insurance monies)*
6.3B.2   *(Insurance premiums)*
6.3B.3.5   *(Restoration etc. of loss or damage)*
6.3C.3   *(Insurance premiums)*
6.3C.4.4   *(Restoration etc. of loss or damage)*,
to the extent that such amounts have been ascertained,
together with any deduction made under clause 3.9 *(Levels)* or 4.9(a) or 4.10.

Interim payment on Practical Completion

4.3   The Architect/The Contract Administrator shall, within
14 days after the date of Practical Completion, certify payment to be made by the Employer to the Contractor of
97½% of the total value referred to in clauses 4.2.1(a)
and 100% of any amounts payable pursuant to clause
4.2.2 together with any deduction under clause 3.9 *(Levels)* or 4.9(a) *(Tax etc. fluctuations)* or 4.10 *(Fluctuations:
named persons)* less any sums previously certified for
payment. The amount so certified shall be paid to the
Contractor within 14 days of the date of the certificate.

Interest in percentage withheld

4.4   Where the Employer is not a local authority[6] the Employer's interest in the percentage of the total value not
included in the amounts of the interim payments to be
certified under clauses 4.2 and 4.3 shall be fiduciary as
trustee for the Contractor (but without obligation to
invest) and the Contractor's beneficial interest therein
shall be subject only[7] to the right of the Employer to

---

[j] See also clause 1.11

have recourse thereto from time to time for payment of any amount which he is entitled under the provisions of this Contract to deduct from any sum due or to become due to the Contractor.

Computation of adjusted Contract Sum

4.5   Not later than 6 months after Practical Completion[8] of the Works the Contractor shall provide the Architect/the Contract Administrator, or if so instructed by the Architect/the Contract Administrator, the Quantity Surveyor with all documents reasonably required for the purposes of the adjustment of the Contract Sum. Not later than 3 months after receipt by the Architect/the Contract Administrator or the Quantity Surveyor as the case may be of the aforesaid documents a statement of all the final Valuations under clause 3.7 *(Valuation of Variations)* shall be prepared by the Quantity Surveyor and a copy of such statement and a copy of the computations of the adjusted Contract Sum shall be sent forthwith to the Contractor.

The adjustment of the Contract Sum shall be in accordance with clause 3.7 and, where applicable, clause 4.9(b), and with the amounts, together with any deductions, referred to in clause 4.2.2 as finally ascertained less all provisional sums and the value of any work for which an Approximate Quantity is included in the Contract Documents and any amount to be deducted under clause 2.10 or under clause 3.9.

Issue of final certificate

4.6   The Architect/The Contract Administrator shall, within 28 days of the sending of such computations of the adjusted Contract Sum to the Contractor or of the certificate issued by the Architect/the Contract Administrator under clause 2.10 *(Defects liability)*, whichever is the later, issue a final certificate certifying the amount due to the Contractor or to the Employer as the case may be. The amount to be certified shall be the Contract Sum adjusted as stated in clause 4.5 less any sums previously certified for payment. The amount so certified shall as from the twenty-eighth day after the date of the final certificate be a debt payable as the case may be by the Employer to the Contractor or by the Contractor to the Employer,[9] subject to any amounts properly deductible by the Employer.

Effect of final certificate

4.7   The final certificate for payment shall be conclusive, except for any matter which is the subject of proceedings

commenced before or within 28 days after the date of the final certificate for payment,
- that the quality of materials and the standard of workmanship is, where the proviso to clause 1.1 applies, to the reasonable satisfaction of the Architect/the Contract Administrator, and
- that any necessary effect has been given to all the terms of this Contract that require additions or adjustments or deductions from the Contract Sum, save in regard to any accidental inclusion or exclusion of any item[10] or any arithmetical error in any computation, and
- that all and only such extensions of time, if any, as are due under clause 2.3 have been given and
- that the reimbursement of direct loss and/or expense, if any, to the Contractor pursuant to clause 4.11 is in final settlement of all and any claims which the Contractor has or may have arising out of the occurrence of any of the matters referred to in clause 4.12 whether such claim be for breach of contract, duty of care, statutory duty or otherwise.

Effect of certificates other than final

4.8   Save as provided in clause 4.7, no certificate of the Architect/the Contract Administrator shall of itself be conclusive evidence that any work, materials or goods to which it relates are in accordance with this Contract.

COMMENTARY ON CLAUSES 4.2 TO 4.8

Clause 4.2 commences with reference to the possibility of stage payments, although no detailed machinery is provided as it is in the case of interim certificates, dependent on valuations. For the simple type of contract for which IFC 84 is envisaged, it might have been anticipated that machinery for stage payments as a basis for interim payment would have been given equal weight to detailed valuations.

The machinery for interim payments

Except where stage payments have been agreed, interim certificates regulate the interim payments to be made to the contractor by the employer. Unless otherwise stated in the appendix, certificates of the amount of interim payments are to be issued by the architect at intervals of one month starting from the date of possession stated in the appendix. The employer must pay the contractor such amount within 14 days of the date of the certificate, subject to any right he may have to deduct e.g. liquidated damages.

The amount to be included in interim payment certificates

The amount to be included in an interim certificate will usually be the product of interim valuations made by the quantity surveyor, together with the ascertainment of any reimbursement of loss and expense to which the contractor may be entitled (see clauses 4.11 and 4.12). While the contract states that it is the quantity surveyor who shall carry out the valuation, the second paragraph of clause 4.2 goes on to state that he shall only do so whenever the architect considers such valuations to be necessary for the purpose of ascertaining the amount to be stated as due in an interim payment. In other words, the architect need not use the quantity surveyor for carrying out interim valuations if such valuations are not necessary in determining the amount of the interim payment certificate. Only rarely will an interim payment not involve an interim valuation e.g. where the interim payment deals only with reimbursement to the contractor of loss and expense.

By virtue of the third paragraph of clause 4.2, a valuation, if any, must be carried out not more than seven days before the date of the certificate. The amount included in the interim payment certificate is to be the total of the following:

(1) *95 per cent of:*
    (a) the value of the work (including varied work) properly executed by the contractor, together with, where applicable, any adjustment of the value under the formulae adjustment for fluctuations (see 4.9(b) and supplemental condition D) *(per clause 4.2.1(a))*
    (b) the value of materials and goods reasonably and properly (and not prematurely) delivered to or adjacent to the works for incorporation therein, provided they are adequately protected against weather and other casualties *(per clause 4.2.1(b))*
    (c) at the discretion of the architect, the value of any off-site goods and materials *(per clause 4.2.1(c))*.

Clause 4.2.1(c), dealing with the value of any off-site goods and materials, perhaps understandably gives little guidance to the architect as to either the circumstances in which he might exercise his discretion or the precautions which might be taken to protect the employer in relation to the passing of property in the goods and materials if the discretion is to be exercised. Clearly, depending on the circumstances, checks should be made and precautions taken (see Commentary and Notes to clause 1.11, page 50, and see JCT 80 clause 30.3 for possible steps which the architect may wish to consider taking).

(2)   *100 per cent of any amounts payable to the contractor to be added to the contract sum under the following clauses:*

3.12	(inspection)
4.9(a)	(tax etc. fluctuations)
4.10	(fluctuations: named persons)
4.11	(disturbance of progress)
5.1	(statutory obligations)
6.2.4	(insurance: liability etc. of employer)
6.3A.4	(insurance monies)
6.3B.2	(insurance premiums)
6.3B.3.5	(restoration etc. of loss or damage)
6.3C.3	(insurance premiums)
6.3C.4.4	(restoration etc. of loss or damage)

to the extent ascertained, together with deductions under clause 3.9 (levels) or 4.9(a) (tax etc. fluctuations) or 4.10 (fluctuations: named persons) (per clause 4.2.2).

Interim payment certificate on practical completion

Within 14 days of practical completion the architect must certify payments of 97.5 per cent of the total value referred to in (1) above: in other words, half of the 5 per cent withheld under clause 4.2.1 is released at this time. Further, 100 per cent of the amounts referred to in (2) above (adjusted in the same way as stated there) must be certified, less of course any sums previously certified. This is in effect an updating exercise by the architect or the quantity surveyor including adjustments of the contract sum in respect of reimbursement of loss and expense under clause 4.11. The employer must pay the contractor within 14 days of the date of the certificate (per clause 4.3).

Retention money

For some general comments on this subject see under the heading 'Summary of General Law', page 205.

The status of retention money is dealt with in clause 4.4. The first point to note is that the contract treats the status of retention money differently depending on whether the employer is, or is not, a local authority.

If the employer is not a local authority, then the employer's interest in the retention money is fiduciary as trustee for the contractor. There is however an express provision to the effect that the employer has no obligation to invest the retention fund. It has been strongly argued (see John Parris: *The Standard Form of Building Contract JCT 80*, Second Edition pages 143 to 144) that despite this express provision (which is also to be found in JCT 80 clause 30.5.1) the employer must account to the contractor for any interest earned by the retention money on the basis that it is contrary to the general law that a trustee should benefit from the trust fund. However, it is submitted that, although the legal position is not beyond doubt, the terms of the contract will suffice to prevent the application of the ordinary rules relating to the duties of a trustee to account to the beneficiary for profits received out of the trust fund. It has been said that the rule:

'is not that reward . . . is repugnant to the fiduciary duty, but that he who has the duty shall not take any secret remuneration or financial benefit not authorised by the law, or by his contract, or by the trust deed under which he acts, as the case may be' – per Lord Norman: *Dale* v. *I.R. Commissioners* (1953).

A trustee may therefore be able to establish his right to remuneration by virtue of the provisions of the trust instrument, in this case the conditions of contract.

Where the employer is a local authority, there are no express provisions dealing with the employer's interest in the retention fund. It is submitted that in such a case there is no obligation on the employer either to keep the retention fund in a separate account or to account for interest earned thereon.

The 5 per cent withheld is released as to one half following the certificate of practical completion and the balance in the final certificate issued under clause 4.6.

Deductions by employer

The contract permits the employer to deduct certain sums from the amount shown in interim payment certificates. The circumstances where he may do so are as follows:

—liquidated damages (clauses 2.7);
—the cost of employing others following the contractor's failure to comply with an instruction (clause 3.5.1);
—the contractor's default in insuring against injury to persons and property (clause 6.2.3);
—the failure of the contractor to insure the works (clause 6.3A.2);
—following determination of the contractor's employment under clause 7.1 to 7.3 (clause 7.4(d)).

The contract makes no specific reference to any other rights of cross claim, whether of ascertained sums or otherwise. Whether the employer can set off any other amounts against payment certificates will depend therefore on a construction of the contract as a whole in the light of the general law on this topic (see 'Summary of General Law' page 206). It is submitted that on balance it is likely that the general common law right of set-off will be available to the employer in appropriate cases.

The adjusted contract sum and the final payment certificate

The contractor must send to the architect or, if so instructed by him, to the quantity surveyor, all documents reasonably required for the purposes of the adjustment of the contract sum. The contractor can do this before practical completion of the works or within six months thereafter. The quantity surveyor must prepare a statement of all the final valuations under clause 3.7 (valuations of variations), and a copy of the computations of the adjusted contract sum must be sent to the contractor not later than three months after the contractor has provided the required documents (clause 4.5).

The final adjustment of the contract sum is generally in line with that for interim payments except that there is no retention percentage withheld and any valuation and ascertainment is final. The same deductions as are permitted from interim valuations, namely under clause 3.9 (levels) or 4.9(a) (fluctuations) or clause 4.10 (fluctuations: named persons) are to be made, but in addition, as this is the final adjustment of the contract sum, there will also be a deduction of all provisional sums as well as the value of any approximate quantity work included in the contract documents, and finally any amount to be deducted under clause 2.10 (defects liability) where the architect has instructed that a defect shall not be made good by the contractor with an appropriate deduction being made from the contract sum in respect thereof.

The final adjustment of the contract sum should not under clause 4.5 exceed nine months and may of course be very much earlier.

Having sent to the contractor a copy of the statement and computation of the adjusted contract sum, the architect must, within 28 days of such date or of the certificate under clause 2.10 (certificate of making good defects), whichever is the later, issue a final certificate certifying the amount due to the contractor or the employer, as the case may be.

The amount certified must include 100 per cent of the total value referred to in (1) above (see page 215): in other words, the release of the second half of the retention, together with the amounts referred to in (2) above as finally ascertained and adjusted as therein required, less any sums previously certified. The amount certified is payable either to the contractor or to the employer, as the case may be, from the twenty eighth day after the date of the final certificate. If the balancing payment is due from the employer to the contractor, the employer may deduct from this such sums as the contract permits (*per clause 4.6*).

There is no specific provision relating to agreement on any final account between employer and contractor. The only stipulation is that a certificate be issued within 28 days of the computations of the adjusted contract sum having been sent to the contractor. Clearly, in practice, there will be an attempt by the contractor and the architect or the quantity surveyor to reach agreement. As the amount to be included in the final certificate is of course governed by what is authorised under the conditions of contract, it will be unlikely that the architect will issue a final payment certificate containing a sum other than that reflected in the quantity surveyor's final adjustment of the contract sum, unless there is an obvious error in it, even if the contractor disagrees with the quantity surveyor's computations. The contractor's remedy will be to await the certificate and then attack this as provided in clause 4.7 (see below).

Effect of final certificate (clause 4.7)

To a limited extent only, the issue of the final certificate for payment is conclusive evidence in any proceedings, whether by arbitration or litigation, that:

(a)   the quality of materials or the standard of workmanship is to the reasonable satisfaction of the architect;

(b)   all terms of the contract that require additions or adjustments to, or deductions from, the contract sum, with the exception of any accidental inclusion or exclusion of any items or any arithmetical error in any computation, have been complied with.

The limits to its conclusiveness result firstly from the matters to which its application is expressly restricted. Secondly the employer or the contractor may take steps to restrict its effect.

Matters covered by conclusiveness of final certificate if no steps taken by employer or contractor to avoid its effect

(a)   Quality of materials and standards of workmanship
The final certificate is conclusive as to the quality of materials or the standard of workmanship in the limited sense that where, and to the extent that, such approval of quality or of materials or standards of workmanship is a matter for the opinion of the architect, such quality and standards are to his reasonable satisfaction (see clause 1.1). Provided that the reference to the architect's approval in clause 1.1 is restricted to situations where the contract documents do not refer to an objective criterion, but leave the question of compliance with specification solely for the subjective opinion and judgment of the architect (and this is surely the intention behind the proviso to clause 1.1), then the effect will be that, in respect of all other matters of quality and standards, the final certificate will not be conclusive. See Note [3] to clause 1.1, page 32, for a further discussion.

Even in relation to those matters of quality and standards where the final certificate is conclusive evidence, it may not extend to consequential losses e.g. loss of use suffered by the employer owing to defects appearing after practical completion but before the issue of the final certificate. On this basis, the final certificate is conclusive that, as at its date, the quality and standards referred to are satisfactory and this can include previously defective work which has been made good. But the certificate will not, it is argued, prevent the employer from recovering his reasonably foreseeable losses resulting from the defect. This view of the effect of the final certificate on consequential losses is given some support by the judgment of Lord Diplock in the House of Lords' case of *Hosier and Dickinson Ltd* v. *Kaye (P and M) Ltd* (1972).

(b) That any necessary effect has been given to all the terms of the contract that require adjustments to the contract sum

It should be noted firstly that it is only those terms of the contract which affect the contract sum to which the conclusiveness relates under this inset to clause 4.7. Secondly, it does not apply to the accidental inclusion or exclusion of any item or arithmetical error in any computation.

Under this head therefore as the conclusiveness of the final certificate refers to any necessary effect having been given to all the terms of the contract requiring additions or adjustments to, or deductions from, the contract sum, its issue would not prevent the contractor from challenging any decision of the architect in relation to extensions of time or any failure to make extensions of time, by way of proceedings at any time after the issue of the final certificate, subject always to the general limitation period under the Limitation Act 1980. Such proceedings, concerned as they would be with extensions of time and therefore the employer's entitlement to liquidated damages, could not affect the contract sum. It is for this reason that the third inset to clause 4.7 was introduced by Amendment 3 to IFC 84 dated July 1988 (see (c) on page 222).

On the wording of this inset of clause 4.7, and indeed on similar wording in JCT 80 (clause 30.9.1.2), it appeared that as the provisions of clause 4.11, dealing with disturbance of regular progress, are without prejudice to other rights and remedies which the contractor may possess, he could where appropriate claim damages for breach of contract after the issue of the final certificate as this would not require any adjustment to the contract sum. Accordingly, in order to ensure that this could not be done the fourth inset to clause 4.7 was introduced in Amendment 3 to IFC 84 dated July 1988 (see (d) on page 222).

The final certificate is not conclusive so as to prevent the employer claiming any liquidated damages to which he was entitled prior to the issue of the final certificate, provided of course that the architect has issued the appropriate certificate of non-completion under clause 2.6, and the employer has complied with the written notice required under clause 2.7.

However, if the architect purports to issue a certificate or further certificate under clause 2.6 after the issue of the final certificate, the clause 2.6 certificate is likely to be invalid. The reason for this is that it is arguable that after the issue of the final certificate, the architect becomes *functus officio* (without office) so that the clause 2.6 certificate may be of no effect and, being a condition precedent to the employer's right to deduct liquidated damages under clause 2.7, the employer could not then deduct these even if the employer had sent the appropriate written notice under that clause. The employer may still be able, however, to claim damages at common law for any delay on the part of the contractor in completion where this amounts to a breach of contract. The issue of the final certificate does not appear to prevent a claim for unliquidated damages for delay.

In the case of *Fairweather* v. *Asden Securities Ltd* (1978) Judge Stabb held that, once the final certificate was issued, the architect was *functus officio* so that he could not then issue a certificate under clause 22 of the 1963 JCT contract (for this purpose similar to clause 2.6 of IFC 84), which certificate was a condition precedent to the employer's right to claim liquidated damages. Although the final certificate under that contract (clause 30(7).JCT 63 1972 revision) was significantly different to that in later revisions of JCT 63 and in JCT 80 and IFC 84, the same point could be argued. The decision in this case has been criticised (e.g. by the learned editors of *Building Law Reports* – see 12 BLR, page 41 *et seq.*). However, in the case of *A. Bell & Son (Paddington) Ltd* v. *CBF Residential Care and Housing Association* (1989) it was assumed without argument that the architect could not issue such a certificate of non-completion after the final certificate had been issued.

(c)   *That all and only such extensions of time, if any, as are due under clause 2.3 have been given.*
This is intended to ensure that the final certificate is conclusive as to any claim by the contractor for further extensions of time or any claim by the employer that extensions of time given should be .educed.

(d)   *That the reimbursement of direct loss and/or expense, if any, to the Contractor pursuant to clause 4.11 is in final settlement of all and any claims which the contractor has or may have arising out of the occurrence of any of the matters referred to in clause 4.12 whether such claim be for breach of contract, duty of care, statutory duty or otherwise.*

The effect of the second inset to clause 4.7 prevents loss and expense claims being reopened after the issue of the final certificate as to do so would be to further adjust the contract sum and this the final certificate prevents. However, this did not prevent a claim by the contractor for damages for breach of contract where a clause 4.12 matter entitling the contractor to reimbursement of loss and expense also amounted to a breach of contract. This could apply in a number of instances e.g. late instructions (clause 4.12.1); failure to give ingress to or egress from the site over adjoining or connected land (clause 4.12.6). The fourth inset to clause 4.7 prevents such claims.

Further, the possibility of framing a claim for loss and expense in the form of an action in negligence or for breach of statutory duty is also prevented by this inset.

The overall effect of the amendments to this clause introduced in Amendment 3 to IFC 84 dated July 1988 is to render the final certificate much more conclusive in terms of extensions of time and actions by the contractor for breach of contract where the factual background would have justified a loss and expense claim under clauses 4.11 and 4.12.

Steps which may be taken by the employer or the contractor to avoid the conclusiveness of the final certificate.

Even the limited matters on which the final certificate can be conclusive can be saved from such a fate if they are the subject of proceedings (whether by way of arbitration or litigation) commenced either before the date of the final certificate or within 28 days thereafter. If any matter is to be the subject of proceedings it is vital that the issues in the proceedings, i.e. in the notice requesting the concurrence of the other party in the appointment of an arbitrator or the endorsement on the writ of summons, as the case may be, are drafted sufficiently widely to include all the possible areas of dispute which it is sought to resolve by the proceedings. If the issues are too narrowly defined it could result in an important and relevant matter being inadmissible.

NOTES TO CLAUSES 4.2 TO 4.8

[1] '. . . *the Date of Possession stated in the Appendix* . . .'.
If the date of possession stated in the appendix is deferred under clause 2.2 for, say, four weeks, then the first certificate is likely to be

of nil amount unless any loss and expense attributable to the deferment has already been ascertained.

[2] '. . . less any sums previously certified for payment.'
Interim payment certificates are issued on account of the final certificate. This means that interim payment certificates should contain the total value of work and material executed together with other sums required to be included in interim payment certificates, with a deduction from this running total of any amounts previously certified. In this way, adjustments· can be made in the current valuation to take into account errors in previous certificates e.g. fluctuations wrongly included or the value of work or materials included in earlier certificates and then found to be defective.

The inclusion in an interim payment certificate of the value of work is not of course conclusive evidence that the work concerned has been carried out in accordance with the contract: see clause 4.8.

[3] '. . . properly executed . . .'.
Clearly, the architect's duty to the employer will include proper routine visits between interim payment certificates to ensure that no obviously defective work is mistakenly included in the quantity surveyor's valuations (discussed in more detail under the heading 'Defective work', page 202). For further consideration of the architect's duties in relation to inspection of work, the reader is referred to standard works covering this subject e.g. *Hudson's Building and Engineering Contracts, Building Contracts* by Donald Keating and to the numerous relevant cases e.g.:

> *Sutcliffe* v. *Chippendale and Edmondson* (1971);
> *East Ham Corporation* v. *Bernard Sunley & Sons Ltd* (1965);
> *Moresk Cleaners* v. *Hicks* (1966);
> *London Borough of Merton* v. *Lowe* (1981);
> *Brickfield Properties Ltd* v. *Newton* (1971);
> *Holland Hannen and Cubbitts (Northern) Ltd* v. *Welsh Health Technical Services Organisation* (1981);
> *Kensington and Chelsea and Westminster Area Health Authority* v. *Wettern Composites and Others* (1984);
> *EDAC* v. *William Moss Group Ltd and Others* (1984).

[4] '. . . reasonably and properly . . .'.
What is reasonable and proper in relation to the delivery of materials

and goods to the site will depend on a number of factors and all the relevant circumstances need to be taken into account e.g. it may be reasonable for a contractor to stack up materials on site at a time of shortage. Also, where there is a particularly tight building programme it may well be reasonable for the contractor to hold greater stocks of materials on site than he would otherwise do if there were greater leeway in the overall construction period. As a matter of practice, if the architect intends to disallow any of the materials or goods on site, he should notify not only the contractor but also the quantity surveyor so that they can be excluded from his valuation.

[5] '. . . adequately protected against weather and other casualties.'
The architect must decide whether such reasonable protection has been afforded. However, if the quantity surveyor, in carrying out his valuation, is of the view that materials or goods have not been adequately protected, then as a matter of practice he should notify the architect.

If the materials or goods are delivered to, placed on or adjacent to the works, and are intended for incorporation therein, they are site materials within the meaning of those words in IFC 84 – see clause 6.3.2. Accordingly, if they are lost or damaged due to an event within the all risks insurance cover for the works required under clauses 6.3A, 6.3B or 6.3C, the policy should pick up the cost of replacement or rectification with the contractor getting an extension of time if appropriate and, where 6.3B or 6.3C applies, the replacement will be treated as an instruction requiring a variation and will attract therefore reimbursement of direct loss or expense where suffered or incurred by the contractor, though only in relation to the period following that point in time at which the contractor has the obligation to replace lost or damaged site materials (see page 323). This will be the case whether or not the site materials are adequately protected against weather or other casualties, even where this is instrumental to the loss or damage occurring, unless the absence of protection causes the site materials to suffer from deterioration, rust or mildew which is outside the all risks cover required under the contract. If the loss or damage is outside the scope of the policy then the risk will, it is submitted, fall upon the contractor who will have to replace or rectify at his own cost – see clause 1.10.

[6] 'Where the Employer is not a local authority . . .'.
The restriction to local authorities is too narrow. Bearing in mind

that the imposition of a trust status on retention money is primarily to protect the contractor's interest should the employer become insolvent, it would have been reasonable to extend the scope of the non-trust retention money to public bodies generally e.g. health authorities and nationalised industries.

[7]   '. . . subject only . . .'.
These words may be sufficient to prevent the non-local authority employer from claiming a common law right of set-off against the retention money, unless the words '. . . provisions of this Contract . . .' extend to implied as well as express provisions. This will not prevent the employer from claiming a set-off in the event that the contractor goes into liquidation or becomes bankrupt. This is because Rule 4.90 of the Insolvency Rules 1986 (in respect of winding-up) and section 323 Insolvency Act 1986 (in the case of bankruptcy) provides that where there have been mutual credits, mutual debts or other mutual dealings between a debtor and any person proving a debt in the winding-up or bankruptcy, an account must be taken of what is due from the one party to the other in respect of such mutual dealings and the balance of the account, and no more, can be claimed or paid on either side respectively (note also clause 7.4(d)).

[8]   '. . . not later than six months after Practical Completion . . .'
While it is unlikely that the contractor can supply all the documents reasonably required for a final adjustment of the contract sum before practical completion, he can nevertheless submit what relevant documentation he has at any particular time. Indeed, it is often desirable that the preparation of the final account be a continuous operation from the outset of the contract, especially where agreement can be reached with the contractor at each stage. It makes the issue of the final payment certificate very much easier for the architect.

[9]   '. . . by the Contractor to the Employer . . .'.
Interim payments are on account of the final amount payable in accordance with the conditions of contract. Accordingly, it is possible that an over-valuation or other mistake could result in an over-payment during the course of the contract. This would be corrected in the final account by a payment being required from the contractor to the employer. This adjustment in respect of an over-

payment due to an over-valuation can be subsequently adjusted even if the contractor has thereby over-paid a named or domestic sub-contractor and is unable, as a result of the sub-contract works having been completed, to deduct it from further sums due to the sub-contractor: see *John Laing Construction Ltd* v. *County and District Properties Ltd* (1982).

[10] '. . . *accidental inclusion or exclusion of any item* . . .'.
Does this mean an item which is, or perhaps should be, included in the final account or does it relate to items of work or materials? It is submitted that it covers both matters.

Clause 4.9

Fluctuations

4.9 The Contract Sum, less any amount included therein for work to be executed by a named person as a sub-contractor under clause 3.3, shall be adjusted in accordance with

(a) Supplemental Condition C (*Tax etc. fluctuations*), unless

(b) Supplemental Condition D (*Formulae fluctuations*) is stated in the Appendix to apply and Contract Bills are included in the Contract Documents.

[*Note:* Supplemental conditions C and D are printed in a separate booklet called 'JCT Fluctuations Clauses for use with the JCT Intermediate Form of Building Contract IFC 84'. The Formula Rules in relation to Supplemental Condition D are to be found in a booklet called 'Consolidated Main Contract Formula Rules' (October 1987) for use, *inter alia*, with IFC 84. Neither booklet has been reprinted in this book.]

Clause 4.10

Fluctuations: named persons

4.10 In respect of any amount included in the Contract Sum for work to be executed by a named person as mentioned in clause 4.9, the Contract Sum shall be adjusted by the net amount which is payable to or allowable by the named person under clause 33 or 34 (as applicable) of the Sub-Contract Conditions NAM/SC.

Provided that there shall be excluded from the net
amount referred to above any sum which would have
been excluded under clause 33.4.7 or 34.7.1 of the Sub-
Contract Conditions NAM/SC had not the period or
periods for completion of the Sub-Contract Works been
extended by reason of an act, omission or default of the
Contractor as referred to in clause 12.2.1 of the said Sub-
Contract Conditions.

COMMENTARY ON CLAUSES 4.9 AND 4.10

It is not proposed in this book to deal in detail with the application
of the fluctuations provisions set out in supplemental conditions C
and D of IFC 84. This commentary is limited therefore to a brief
outline only.

A discussion of supplemental conditions C and D must commence
with a discussion of the fluctuations clauses in JCT 80 on which the
supplemental conditions are closely based.

The fluctuations clauses in JCT 80 (and in its predecessor JCT 63)
are now very familiar to those whose task it is to deal with building
contracts. Where, under JCT 80, it is decided that full fluctuations
should apply, it is possible to choose one of two possible methods of
price adjustment. The first method in JCT 80 is covered by clause 39
and allows adjustment of the contract sum based on actual changes
in rates of wages and in the prices of materials payable by the
contractor. The second method is covered by clause 40 and provides
for adjustment of the contract sum by reference to published
indices, this latter method being known generally as the formula
method. Where neither of these two methods of price adjustment is
required and the contract sum is based on what is generally called
(not altogether accurately) a 'fixed' or 'firm' price basis for price
adjustment, JCT 80 in clause 38 permits an adjustment of the
contract sum by reference to changes in contributions, levies and
taxes imposed under or by virtue of any Act of Parliament.

Turning to IFC 84 the first point to appreciate is that there is no
labour and materials fluctuations clause for this contract, the only
options being either the fixed or firm price basis in supplemental
condition C (the equivalent of clause 38 of JCT 80) or the formula
method in supplemental condition D (equivalent to clause 40 of JCT
80). It is somewhat surprising that the materials and wages
fluctuations option has been omitted. Although rapidly losing

popularity in favour of the formula method, it would have been a useful inclusion in the contract, particularly for refurbishment schemes and for those contracts where the employer particularly wishes to limit the list of materials which are to be subject to adjustment. Clause 4.9(b) is so drafted that if bills of quantities are not a contract document then the contract must be on the fixed or firm price basis with adjustments to the contract sum being restricted to changes in taxes, levies etc. and this must place a serious limitation on the use of IFC 84. However, it would be possible to incorporate formula fluctuations even where there are no contract bills, provided a contract sum analysis is used and provided careful drafting is employed to adapt Amendment 3 to JCT 80 Without Quantities Form – March 1987, which enables formula fluctuations to be used in that contract. See also page 16.

Where the formula method is used, the contract sum must be adjusted in accordance with the provisions of supplemental condition D and the formula rules referred to earlier. Briefly, this method of adjustment is based on the 49 'work categories', see Rule 3 and Appendix A to the formula rules, the work included in the bills of quantities being broken down as far as possible into the separate categories. The items in the bills of quantities are annotated so that all items are as far as possible divided into a series of work categories or, alternatively, into broader work groups. Indices which broadly reflect the changes in the cost of materials and the rates of wages within each category are published monthly by the Property Services Agency of the Department of the Environment. Valuations are then prepared at monthly intervals and the valuation divided into the various work categories. The difference between the indices for the various work categories for that month and those applicable during the month on which the tender was based is then applied to the value of the work completed within each work category during that month in order to arrive at the amount due in respect of fluctuations. The figure thus produced is then adjusted to take into account those items introduced in the bills which, although subject to formula adjustment, cannot be allocated to a particular work category e.g. preliminaries or lump sum adjustments on the tender summary. These items, grouped under the heading of 'balance of adjustable work' are adjusted in accordance with rule 38 or 26 of the formula rules. Where the contract involves a local authority, a further adjustment is made by deducting a percentage for what is termed the 'non-adjustable element'. This abatement, which is stated as a percentage, must be indicated in the appendix to IFC 84 and must not exceed 10 per cent.

The formula rules require the contract bills to indicate whether part I or part II of section 2 of the rules shall apply. The contract bills must also specify the base month. This information must be included in the appendix to IFC 84. The base month identifies the calendar month the index numbers of which are deemed to equate to the price levels represented by the rates and prices contained in the contract bills. They are index numbers on which the contractor will have based his tender. If tenderers have not been informed of the base month, it will normally be the month prior to that in which the tender is due to be returned.

Part 1 of section 2 of the rules applies where it is intended to use the work category method of adjustment. Part II of section 2 applies where what is known as the work groups method is to apply. This latter method in effect condenses the work categories by combining them into groups. The contract bills must state the method of grouping to be adopted.

The need to measure work in detail for each valuation is paramount where the formula method is used. Work which has been completed between valuations must be accurately measured in each separate category in order that the appropriate indices can be applied. However, should the amount of any adjustment included in a previous certificate require correction, then the correction shall be given effect in the next certificate – see supplemental condition D5 and rule 5 of the formula rules. Should there be a delay in the publication of the monthly bulletin containing the indices, adjustment of the contract sum must be made in each interim certificate during such period of delay on a fair and reasonable basis and can be adjusted later on the recommencement of the publication – see supplemental condition D10.

Interim valuations must be made before the issue of each interim certificate, and accordingly the words 'whenever the Architect/the Contract Administrator considers them to be necessary' in clause 4.2 shall be deemed to have been deleted – see supplemental condition D6. Both supplemental conditions C and D provide (see C4.7 and C4.8, and D12 respectively) that adjustment of the contract sum will not continue after the date for completion or any extended time for completion of the works. In other words, the operation of the fluctuations provision is frozen at that date. However, it is important to appreciate that the freezing of these provisions is subject to clause 2.4 remaining unamended and to the architect having, in respect of every written notification by the contractor under clause 2.3, fixed or confirmed in writing such date for completion as he considers to be in accordance with that clause. Accordingly, if the employer wishes to amend the printed text of clause 2.4 without this having the effect of unfreezing the application of the fluctuations provisions, he must also ensure that the relevant supplemental condition is suitably amended. Further, it is of course of the utmost importance that the architect properly fulfils his duties under clause 2.3. Clearly, a failure to properly amend the supplemental condition or a failure by the architect to properly exercise his duties could involve the employer in substantial additional costs in paying fluctuations during a period of over-run by the contractor.

Even where the application of the fluctuation provisions is frozen, it is prudent for the quantity surveyor to continue to separate work completed after that time so that adjustment can readily be made should the architect subsequently make further extensions of time for completion of the works under clause 2.3.

It should be appreciated that both types of fluctuations contained in IFC 84 are 'rise and fall' provisions. In other words the contract sum can be adjusted downwards as well as upwards in appropriate circumstances.

It should be noted that, in relation to supplemental condition C (fluctuations for tax etc.), the contractor is required to give written notice to the architect of the occurrence of any of the events which trigger off an adjustment of the contract sum – see C4.1. The notice is required to be given within a reasonable time after the occurrence of the event to which the notice relates, and it is expressly stated that the giving of a written notice in that time is a condition precedent to any payment being made to the contractor in respect of the event in question: see C4.2. Supplemental condition C4.3 states that the quantity surveyor and the contractor may agree what shall be deemed for all purposes of the contract to be the net amount payable or allowable by the contractor in respect of the occurrence of any of the listed events. This does not give the quantity surveyor authority to so agree where the contractor has failed to serve the written notice within a reasonable time after the occurrence of the event in question: see *John Laing Construction Ltd* v. *County and District Properties Ltd* (1982).

Clause 4.10 deals with fluctuations for named persons as sub-contractors. It states that the contract sum shall be adjusted by the net amount payable to or allowable by the named person under clause 33 or 34 (as applicable) of the sub-contract conditions NAM/SC. Named persons are treated separately as the sub-contract fluctuations may be on a different basis to the main contract fluctuations. The contractor will be reimbursed for the actual amount of fluctuations payable by him to the named sub-contractor. However, there is a proviso in clause 4.10 to the effect that the employer is not required to reimburse the contractor the fluctuations which he (the contractor) has paid to the named sub-contractor to the extent that those fluctuations are payable by reason of the period or periods for completion of the sub-contract works having been extended by any act, omission or default of the contractor as referred to in clause 12.2.1 of NAM/SC.

Clauses 4.11 and 4.12

Disturbance of regular progress

4.11   If, upon written application being made to him by the Contractor within a reasonable time of it becoming apparent, the Architect/the Contract Administrator is of the opinion[1] that the Contractor has incurred or is likely to incur direct loss and/or expense, for which he would not be reimbursed by a payment under any other provision of this Contract, due to

(a) the deferment of the Employer giving possession of the site under clauses 2.2 where that clause is stated in the Appendix to be applicable; or

(b) the regular progress[2] of the Works or part of the Works being materially affected by any one or more of the matters referred to in clause 4.12,

then the Architect/the Contract Administrator shall ascertain,[3] or shall instruct the Quantity Surveyor to ascertain, such loss and expense incurred and the amount thereof shall be added to the Contract Sum provided that the Contractor shall in support of his application submit such information required by the Architect/the Contract Administrator or the Quantity Surveyor as is reasonably necessary for the purposes of this clause.

The provisions of this clause 4.11 are without prejudice to any other rights or remedies which the contractor may possess.

Matters referred to in clause 4.11

4.12 The following are the matters referred to in clause 4.11:

4.12.1 the Contractor not having received in due time necessary instructions (including those for or in regard to the expenditure of provisional sums), drawings, details or levels from the Architect/the Contract Administrator for which he specifically applied in writing provided that such application was made on a date which having regard to the Date for Completion[4] stated in the Appendix or any extended time then fixed was neither unreasonably distant from nor unreasonably close to the date on which it was necessary for him to receive the same;

4.12.2 the opening up for inspection of any work covered up or the testing of any of the work, materials or goods in accordance with clause 3.12 (including making good in consequence of such opening up or testing), unless the inspection or test showed that the work, materials or goods were not in accordance with this Contract;

4.12.3 the execution of work not forming part of this Contract by the Employer himself or by persons employed or otherwise engaged by the Employer as referred to in clause 3.11 or the failure to execute such work;

4.12.4 the supply by the Employer of materials and goods which the Employer has agreed to supply for the Works or the failure so to supply;

4.12.5 the Architect's/the Contract Administrator's instructions under clause 3.15 issued in regard to the postponement of any work to be executed under the provisions of this Contract;

4.12.6 failure of the Employer to give in due time ingress to or egress from the site of the Works, or any part thereof through or over any land, buildings, way or passage adjoining or connected with the site and in the possession and control of the Employer, in accordance with the Contract Documents after receipt by the Architect/the Contract Administrator of such notice, if any, as the Contractor is required to give, or failure of the Employer to give such ingress or egress as otherwise agreed between the Architect/the Contract Administrator and the Contractor;

4.12.7 the Architect's/the Contract Administrator's instructions issued under clauses

1.4 *(Inconsistencies)* or
3.6 *(Variations)* or
3.8 *(Provisional sums)*
except, where the Contract Documents include bills
of quantities, for the expenditure of a provisional sum
for defined work* included in such bills;
or, to the extent provided therein, under clause
3.3 *(Named sub-contractors)*.

4.12.8   the execution of work for which an Approximate Quantity
is included in the Contract Documents which is not a
reasonably accurate forecast of the quantity of work
required.

---

* See footnote to clause 8.3 (Definitions).

COMMENTARY ON CLAUSES 4.11 AND 4.12

Clause 4.11 gives the contractor an entitlement to recover from the employer direct loss and/or expense arising from the employer deferring giving possession or where the regular progress of the works is materially affected by any of the matters itemised in clause 4.12.1 to 4.12.8. These are all matters which, if they cause disturbance to the contractor, will directly or indirectly be the responsibility of the employer.

This commentary begins with a general discussion of the scope of the contractor's ability to recover any increased expenditure or losses suffered by him as a result of the deferment of possession or of one of the matters in clause 4.12 occurring, and it will then deal with the machinery and operation of the two clauses.

Scope of recovery
The governing words limiting the extent of recovery are 'direct loss and/or expense'. These words are the same as appear in JCT 63 and JCT 80 and they are unquestionably intended to have the same meaning. The first point to appreciate is that the right of recovery does not extend to indirect, or what may be called consequential, loss or expense. The case of *Croudace Construction Ltd* v. *Cawoods Concrete Products Ltd* (1978) gives some assistance in drawing this distinction.

Facts:

The plaintiffs were main contractors for the erection of a school. They entered into a sub-contract with the defendants for the supply and delivery of masonry blocks. The sub-contract contained a clause which included the following words:

'. . . if any materials or goods supplied by us should be defective or not of the correct quality or specification ordered our liability shall be limited to free replacement of any materials or goods shown to be unsatisfactory. We are not under any circumstances to be liable for any consequential loss or damage caused or arising by reason of late supply or any fault, failure or defect in any materials or goods supplied by us or by reason of the same not being of the quality or specification ordered or by reason of any other matters whatsoever.'

The main contractor sued the sub-contractor for losses alleged to have arisen because of late delivery and defects in the materials and goods supplied, seeking as part of the claim to recover loss of productivity and additional costs of delay in executing the main contract works and also the cost to them of meeting other sub-contractors' claims which were brought about by the delays of the defendant sub-contractor. A preliminary issue was ordered to be tried as to:

'Whether on a proper construction of such Contract or Contracts, including the Defendants' Standard Conditions of Trading, the Plaintiffs are entitled to recover damages under any, and if so, which, of the Heads of Damage which they had pleaded.'

Held:

Consequential loss or damage meant that loss or damage which did not result directly and naturally from the breach of contract complained of. The Court of Appeal held that the meaning of the words 'consequential loss or damage' had already been decided in the case of *Millar's Machinery Co. Ltd* v. *David Way & Son* (1935) and that they were bound by that decision. They also agreed with its reasoning. The plaintiffs could therefore recover the losses which they had pleaded.

It is clear therefore that 'consequential' will be treated as meaning indirect and will be interpreted quite restrictively as including only such heads of loss or expense which do not flow naturally from the breach without other intervening cause and independently of special circumstances.

This interpretation of the words 'direct loss and/or expense' brings them very close to the general common law position as to the measure of damages for a breach of contract. Some of the matters listed in clause 4.12 will arise owing to a breach of contract by the employer. At common law it is well settled that, while the governing purpose of damages is to put the party whose rights have been violated in the same position, so far as money can do so, as if his rights had been observed, this overall purpose is qualified in that the aggrieved party is only entitled to recover such part of the loss actually resulting from the breach as was at the time of the contract reasonably foreseeable as liable to result from it. This will depend firstly on imputed knowledge and secondly on actual knowledge.

So far as imputed knowledge is concerned, a reasonable person is taken to know that in the ordinary course of things certain losses are liable to result from a breach of the contract. This is known as the first limb of the rule in the leading case of *Hadley* v. *Baxendale* (1854) (dealt with in more detail later in chapter 10, page 337). Secondly, the contracting parties may have particular knowledge which would lead them to the conclusion that a breach of contract would result in losses being suffered over and above those which might be thought to flow naturally from the breach in the ordinary course of things: *Victoria Laundry Ltd* v. *Newman Industries Ltd* (1949). In a claim for damages for breach of contract both types of damages will be recovered although in relation to the second, i.e. that depending on the actual knowledge of the parties, specific evidence must be adduced to demonstrate this. The measure of damages for breach of contract at common law is dealt with briefly later in this book (see page 339). However, as the second type of damage is arguably indirect in nature it could be said to be outside the meaning of 'direct loss and/or expense' and so irrecoverable under clause 4.11.

Further, as has been stated above the words 'direct loss and/or expense' are interpreted in line with the first limb of the rule in *Hadley* v. *Baxendale* (1854). There is further reason for giving these words a restricted scope. In the case of *Wraight Ltd* v. *P.H. and T. (Holdings) Ltd* (1968) Mr Justice Megaw said:

'In my judgment, there are no grounds for giving the words 'direct

loss and/or damage caused to the contractor by the determination'
any other meaning than that which they have, for example, in a
case of breach of contract.'

It is therefore apparent that the words 'direct loss and/or damage'
mean the same as damages at common law for breach of contract.
IFC 84 includes, as does JCT 63, and JCT 80, the phrase 'direct loss
and/or damage' as being recoverable by the contractor in the event
of his employment being determined for certain stated reasons (see
IFC clause 7.7(b); JCT 63 clause 26(2)(b)(vi); JCT 80 clause
28.2.2.6). It is a general principle in construing the meaning of
words in a contract to presume that the same words have the same
meaning and that different words are intended to have a different
meaning: in other words, that the words 'direct loss and/or expense'
mean something different to the words 'direct loss and/or damage'.
If this is so, then the former words appear to be more restrictive in
their scope than do the latter words. Despite the present trend
therefore of allowing under these words such items as the loss of
profit which could have been earned on another contract had the
disruption and delay not occurred, it may not yet be beyond doubt
that such claims, whilst recoverable where the word 'damage' is
used (as in the case of *Wraight* v. *P.H. and T. (Holdings) Ltd*
above), will not be recoverable where the word 'expense' is used
instead. However, it is equally arguable that it is the common
interpretation of the word 'loss' which appears in both phrases, and
which therefore can be interpreted in both places as being consistent
with common law damages for breach of contract.

Heads of typical loss and expense claims
(1) Loss of productivity
Delay and disruption can lead to additional expenditure on labour
and plant. While clearly recoverable, it is not always easy to
establish the correct figures. This head of claim may refer to the
contractor's own plant as well as that which is on hire. As is the case
under every head of damage, the contractor will be in a much
stronger position to substantiate his claim if he has adequate
records. It cannot be too often emphasised just how important the
keeping of detailed and efficient records can be to substantiate a
valid claim whether to the satisfaction of the quantity surveyor, who

is attempting to ascertain its value, or, should it be necessary, in arbitration or other proceedings.

(2)   Site-overheads
This head of claim will cover such matters as extra involvement of supervisory and administrative staff engaged on the site; the cost of accommodation, services etc.

(3)   Head office overheads
Delay or disruption will almost certainly mean increased involvement of staff from head office dealing with problems caused by disruption. Further, the pricing of the contract in question will have included a contribution from the earnings on that contract to cover general head office overheads such as rent, rates, light, office equipment etc. This contribution will usually be related to the contract period. In other words, over a fixed period of time, it is anticipated that a certain level of contribution will be achieved. If there is delay and disruption, the same period of time will produce a lower contribution and therefore a loss will be incurred.

The measure of that loss is not always easy to determine and has led to the use of formulae. The two most common ones are that set out in *Hudson's Building and Engineering Contracts*, 10th edition, known as the 'Hudson Formula', and that set out in *Emden's Building Contracts and Practice*, 8th edition. It is not possible in a book of this kind to debate in detail the validity or usefulness of such formulae. Suffice it to say that while their use, and in particular that of the Hudson Formula, has been criticised from time to time, they are simply an attempt at an approximation of the losses where the actual losses are extremely difficult, if not impossible, to quantify precisely. The key to their use is that they can only ever be regarded as the next best thing to actual proof, but so regarded, they may each in their own right have some merit.

In the English case of *Finnegan Ltd* v. *Sheffield City Council* (1988) and the Canadian case of *Ellis-Don Ltd* v. *The Parking Authority of Toronto* (1978) the use of formula received some judicial support. However, the blind use of formula where it is likely to produce an unrealistic result has been trenchantly attacked in the American case of *Berley Industries, Inc.* v. *City of New York* (1978) in which Judge Fuchsberg in giving judgment said:

'So far as the record shows, all expenditure for office overhead may have been completed when the 87% level was reached... Above all, there was no claim by the plaintiff that proof was not available. It fell back, instead, on the formula...

The mathematical formula did not fill the void. At not a single point in the equation which it set up was there a component which represented an actual item of increased costs, whether attributable to the delay on the city's job or not. The computation it essayed was therefore no less speculative because it was cast in a mathematical milieu. And, insofar as it was offered as a substitute for direct evidence of overhead damage, there was no accompanying foundation from which it could be found that, because of the character of Berley's business, increased overhead attributable to delay was impossible of proof without the aid of the formula. Nor was there any attempt to prove that the formula was logically calculated to produce a fair estimate of actual damages. Absent these preconditions, it was but an unsupported opinion. It did not serve to expand Berley's ability to recover any more than the reasonable value of any additional home office expenses it might have been able to prove by other means.

In rejecting the formula we, of course, are aware of the fact that, under the Eichleay nomenclature, it has been applied in a series of home office cases, which for the most part accept the formula largely without analysis and almost as a matter of administrative convenience.

The case before us readily reveals how the mechanical imposition of a formula akin to the one advanced by the plaintiff can all too easily bring a harsh daily penalty when only compensatory damages are warranted or even when the doctrine of *damnum absque injuria* is in order. For all practical purposes, it would completely ignore the safeguards against overreaching and arbitrariness to which the law of evidence has long been committed. As Justice (now Presiding Justice) Murphy pointed out in his dissent below, "The damages computed under the 'Eichleay formula' would be the same in this case whether the plaintiff had completed only 1% or 99% of the job on the scheduled completion date of May 7, 1971. This rather bizarre result is caused by the fact that the 'Eichleay formula' focuses on the length of the delay to the exclusion of many other important factors bearing on actual damages. If, on May 7, 1971, the plaintiff was merely required to spend $100 to complete the job, the 'Eichleay formula' would still require that the defendant pay $19,262 for 335-day delay . . .

"I can only conclude that the mathematical computations under the 'Eichleay formula' produce a figure with, at best, a chance relationship to actual damages, and at worst, no relationship at all." Neither at argument nor in its brief did plaintiff deny the implications of this critique.'

Finally, it should be mentioned that the decision of the High Court in *Tate and Lyle Food and Distribution Co. Ltd* v. *Greater London Council* (1982) may appear at first sight to have thrown into doubt, in the recovery of damages for breach of contract, the use of a percentage of head office expenses to represent the managerial resources expended as a result of delays and disruption on contracts.

However, it is submitted that it should not be so regarded. Firstly, it would seem that no actual evidence of any kind was produced in substantiation of the claim. It may have been different if some method of time recording had been employed so that a reasonable percentage could have been derived from such records. Secondly, the case concerned plaintiffs not in the business of construction so that they would not normally expect to seek recovery of their head office managerial time through construction contracts. Such recovery would no doubt be achieved through product sales. In the case of a construction company the head office managerial time can be recovered only through income from construction contracts, and if such contracts run into delay or disruption this has a direct bearing on the company's ability to recover its head office managerial cost.

(4)   Interest and financing charges

When the contractor incurs loss and expense, this has to be financed by him either from his own capital resources or alternatively by increased borrowing. In either case, it is self-evident that the use of money costs money. Depending on the level of inflation, the cost of being stood out of money which would otherwise be available for other uses, or the cost of borrowing which would not otherwise have occurred can at best be a source of irritation to the contractor and at worst can be damaging to the point of bankruptcy or liquidation. It is therefore a matter of great importance as to how such interest or financing charges should be recovered.

Applying basic legal principles as to the measure of damages and ignoring existing case law and the tradition of the common law in relation to interest, there would appear to be little doubt that it would in the ordinary course of things be obvious that a breach of contract by the employer involving the contractor in the financing of losses and the incurring of additional expenditure would involve him in direct costs of financing and interest charges in which he would not otherwise have been involved. Unfortunately, the hostility of the traditional common law approach to interest, culminating in the House of Lords decision in *London Chatham and Dover Railway Co.* v. *S. E. Railway Co.* (1893), has meant that this head of damages at common law has been irrecoverable, unless perhaps it can be demonstrated as being within the actual and particular knowledge of the contracting parties, enabling it to be claimed as special damages. For example, in the case of *Wadsworth* v. *Lydall* (1981) the parties entered into an informal partnership agreement which granted an agricultural tenancy to Wadsworth to live in the farmhouse and run the farm. Some years later, on the dissolution of the partnership, Lydall was to pay £10,000 to Wadsworth on being given vacant possession. On the strength of this Wadsworth entered into an agreement with someone else. However, when he vacated the farm the £10,000 was not paid. Some time later he was paid £7,200 which he used to meet part of his obligations in favour of the third party. He raised the balance of the money due to the third party by taking out a mortgage. The question arose as to the special damages of £335 which Wadsworth had to pay in respect of interest to the third party for late completion of a purchase. The trial judge had disallowed this interest. Wadsworth appealed to the Court of Appeal who allowed Wadsworth's claim for interest. The principle was stated by Lord Justice Brightman as follows:

'In my view the damage claimed by the plaintiff was not too remote. It is clearly to be inferred from the evidence that the defendant well knew at the time of the negotiation of the contract ... that the plaintiff would need to acquire another farm or smallholding as his home and his business, and that he would be dependent on the £10,000 payable under the contract in order to finance that purchase.

. . . In *London, Chatham and Dover Railway Company* v. *South Eastern Railway Company* (1893) the House of Lords was not concerned with a claim for special damages. The action was an action for an account . . . If a plaintiff pleads and can prove that he has suffered special damages as a result of the defendant's failure to perform his obligation under a contract and such damage is not too remote . . . I can see no logical reason why such special damage should be irrecoverable merely because the obligation on which the defendant defaulted was an obligation to pay money and not some other type of obligation.'

This type of approach has been considerably widened. See such cases as the *President of India* v. *Lips Maritime Corp* (1987) in the Court of Appeal in which Lord Justice Neill stated the current principles as follows:

'What then is the present law as to the recovery of damages at common law for a breach of contract which consists of the late payment of money? I would venture to state the position as follows:

(1) A payee cannot recover damages by way of interest merely because the money has been paid late. The basis for this principle appears to be that the court will decline to impute to the parties the knowledge that in the ordinary course of things the late payment of money will result in loss. I would express my respectful agreement with the way in which Hobhouse J. explained the surviving principle in *International Minerals and Chemical Corp.* v. *Karl O Helm AG* . . .

'In my judgment, the surviving principle of legal policy is that it is a legal presumption that in the ordinary course of things a person does not suffer any loss by reason of the late payment of money. This is an artificial presumption, but is justified by the fact that the usual loss is an interest loss and that compensation for this has been provided for and limited by statute.'

(2) In order to recover damages for late payment it is therefore necessary for the payee to establish facts which bring the case within the second part of the rule in *Hadley* v. *Baxendale*. In *Knibb*

v. *National Coal Board* (1986) . . . Sir John Donaldson MR stated the effect of the decision in *President of India* v. *La Pintada Compania Navigacion SA* in these terms:

> "From this (the decision) it emerges that there is no general common law power which entitles courts to award interest . . . but that if a claimant could bring himself within the second part of the rule in *Hadley* v. *Baxendale* (1854) . . . he could claim special damages, notwithstanding that the breach of contract alleged consisted in the nonpayment of a debt."

It may be said that the line between the two parts of the rule in *Hadley* v. *Baxendale* has become blurred so that the division has lost much of its utility . . . But it is clear, as Staughton J. recognised in the instant case, that the court must find the dividing line because it is only if the claim falls within the second part of the rule that the loss can be recovered.

(3) It is important to keep in mind that the question in each case is to determine what loss was reasonably within the contemplation of the parties at the time when the contract was made. As I understand the matter, the principle in *London and Dover Railway Co.* v. *South Eastern Railway Co.* in its modern and restricted form, goes no further than to bar the recovery of claims for interest by way of general damages. Thus, in the case of a claim for damages for the late payment of money the court will not determine in favour of the plaintiff that damages flow from such delay "naturally, that is, according to the usual course of things" (*Hadley* v. *Baxendale*). But a plaintiff will be able to recover damages in respect of a special loss if it is proved that the parties had knowledge of facts or circumstances from which it was reasonable to infer that delay in payment would lead to that loss.

Moreover, I do not understand that, provided that knowledge of the facts and circumstances from which such an inference can be drawn can be proved, it is necessary further to prove that the facts or circumstances were unusual, let alone unique to the particular contract.

As I have stated earlier, the question in each case is to determine what loss was reasonably within the contemplation of the parties at the time when the contract was made. In dealing with this question the court will not impute to the parties the knowledge that damages flow "naturally" from a delay in payment. But where there is evidence of what the parties knew or ought to have known the court is in a position to determine what was in their reasonable contemplation. For this purpose, the court is entitled to take account of the terms of the contract between the parties and of the surrounding circumstances, and to draw inferences. In drawing inferences as to the parties' actual or imputed knowledge, the court is not obliged to ignore facts or circumstances of which other people doing similar business might have been aware.'

In the case of *Holbeach Plant Hire Ltd* v. *Anglian Water Authority* (1988) it was held that financing costs (for example, in running an overdraft) associated with the hiring of plant and equipment and the purchase of materials and the payment of wages, were capable, if they satisfied the criteria set out in the *Lips Maritime* case referred to above, of being special damages and therefore recoverable provided there is evidence that the parties had such losses in contemplation or would have had they addressed the point at the time of entering into the contract.

In any event a claim for financing charges in respect of loss and expense is different in nature to a claim for interest for late payment of a debt which is the situation in which the common law displays such hostility. Until the architect has issued his certificate including the ascertainment of the loss and expense, there is no debt due and accordingly the strict common law approach has no obvious application. However, as regards any delay by the employer in paying an interim certificate containing an ascertainment of loss and expense, the common law does apply so that, provided payment is in fact made before proceedings are instituted by the contractor, he cannot claim interest in respect of late payment of the ascertained sum unless it can be squeezed into the category of special damage under the second limb of the rule in *Hadley* v. *Baxendale* (1854). Similarly, there can be no claim at common law for interest where the employer (in circumstances where he is vicariously liable for the actions of his architect) is in breach of contract following late certification by the architect. However, in this latter situation it may be possible, depending on the circumstances, to include such interest or financing charges as part of a claim for loss and expense: *Rees and Kirby Ltd* v. *Swansea City Council* (1985).

So far as the actual claim for loss and expense is concerned, it is now clear that financing charges, or the cost of being stood out of one's money, are a recoverable head of claim and rightly so. This is established beyond doubt in the Court of Appeal case of *F.G. Minter Ltd* v. *Welsh Health Technical Services Organisation* (1980). The Court of Appeal held that the historic common law hostility towards awarding interest was no bar to the recovery of financing charges because there should be no presumption in favour of an anomaly and anachronism. Subject therefore to the particular contractual machinery for recovering loss and expense, in principle interest and financing charges are properly recoverable. The only question therefore is the period for which they are recoverable and this will depend on the wording of the contractual provisions concerned.

So far as IFC 84 is concerned, the important words in clause 4.11 are '. . . the Contractor has incurred or is likely to incur direct loss and/or expense'. Interest and financing charges will therefore be recoverable, subject to the machinery of the clause, up until at least the date of ascertainment by the quantity surveyor. The contractor would therefore be well advised to state a daily interest rate or some other method by which the cost can be calculated.

Under JCT 63 clause 11(6), the written application must be made within 'a reasonable time' of the loss and expense having been incurred. Under JCT 63 clause 24 it must be made within a reasonable time of it becoming apparent that the progress of the works has been affected. The application under both clauses relates to loss and expense having already been incurred. In this connection, what is a reasonable time with regard to loss and expense due to financing charges could well be different from what is a reasonable time in relation to other immediate loss and expense suffered. The latter is what might be called a primary loss e.g. increased site overheads and so on, whereas the former is a secondary loss. The secondary loss is a continuing one and may not have been fully incurred until the primary loss is both quantified and paid. It may only be at that point in time that the period of 'reasonable time' commences so far as the application for financing charges is concerned – see *Rees and Kirby Ltd* v. *Swansea City Council* (1985).

Under IFC 84 clause 4.11, it appears (the drafting is not as clear as it might be) that the 'reasonable time' relates to loss and expense which the contractor has incurred *or is likely to incur*. The use of a different measure of 'reasonable time' for the so-called secondary loss by way of financing charges is not therefore, it is submitted, applicable under IFC 84. The measure of what is a reasonable time should therefore be the same, whether the loss is already suffered or is a continuing loss. A consideration of the judgments delivered in the Court of Appeal in the *Minter* and *Rees* and *Kirby* cases prompts the following suggested guidelines for the recovery of interest or financing charges as part of a claim for reimbursement of direct loss or expense.

(i) Timing of application
(a)   The application should be made 'within a reasonable time' of it becoming apparent that direct loss or expense has been or is likely to be incurred.

(b)   If the direct loss or expense is of a continuing nature it is unwise and possibly fatal, under JCT 63, for the contractor to wait until the loss or expense has stopped before making the application. It is suggested that the first application should be no later than one month after the loss or expense starts to run and that, in the case of JCT 63, there should be successive monthly applications thereafter.

Under the provisions of IFC 84 and JCT 80 successive applications are probabaly not required as the initial application is likely to cover continuing losses, including financing charges.

(c)   However, even if the contractor has failed to make his application within a reasonable time, unless the employer takes the point at an early stage, he may find himself estopped from challenging the contractor's application on this ground. This will be particularly so if the employer has treated the application as being valid and the contractor has relied upon this to his detriment.

(ii) Contents of application
Generally, a liberal approach is to be adopted in relation to the contents of the written application. However, the cost of being stood out of money i.e. the financing or interest charges should be referred to in the application though a specific reference to financing charges or any other continuing loss may not strictly be necessary under IFC 84 or JCT 80.

(iii) Period for which financing charges can be recovered
Financing charges can be recovered from the date of their being incurred up to the date of the primary loss being reimbursed with the exception of any period during that time where the financing charge is being incurred for a reason not directly attributable to the incurring of the primary loss to which the finance charge initially

related. This could for example relate to the situation where the contractor takes an unreasonable time to provide necessary details of his loss or expense (but note that in the *Rees* case itself a period of almost 12 months to produce a claim was held not to be unreasonable).

There is no automatic cut off point for recovery of financing charges at practical completion.

(iv) Compound and simple interest

During the period for which the financing charges are recoverable (see above) compound interest is payable based upon periodic rests. It is suggested that usually the rate of interest and frequency of rests will be either those which applied to the contractor or those which accord with normal banking practice, whichever produces the lower figure. If the contractor pays particularly high rates of interest in circumstances where this was not reasonably foreseeable by the employer then it is unlikely that the employer will be responsible for other than normal commercial rates.

If this compounded figure is not paid on the due date by the employer and proceedings are issued before payment has been made, the contractor can claim simple interest only upon the compounded financing charge, though it has to be remembered that the payment of this interest and the rate is ultimately within the discretion of the court.

In conclusion, there appears to be a paradox. On the one hand, at common law, interest or financing charges cannot be recovered as general damages for breach of contract under the first limb of the rule in *Hadley* v. *Baxendale* (1854). If recoverable at all it must be as special damages under the second limb of the rule as being in the actual contemplation of the parties, even if not flowing naturally in the ordinary course of things from the breach. On the other hand, interest or financing charges are clearly now recoverable as part of direct loss and expense and yet, as we have seen, direct loss and/or expense has been construed as falling within only the first limb of the rule in *Hadley* v. *Baxendale* (1854) under which interest or financing charges are not recoverable.

(5) Loss of profit

If the words 'direct loss and/or expense' are to be interpreted in line with the measure of damages at common law for breach of contract, then there is no doubt at all that the contractor can claim for a loss of profit which he would have earned on other contracts had there been no delay and disruption to the current contract, provided such loss is foreseeable. The contractor's entitlement to claim loss of profit for breach of contract at common law is perfectly reasonable and just, and provided there is clear evidence the contractor can properly include this head of claim in his claim for loss and expense under the contract. Again, the use of formulae is commonplace. The comments relating to formulae under heading (3), *Head office over-heads* (see page 238), are also relevant here.

(6) The costs of the claim

Often contractors will include a sum in their claims for the cost of preparing the claim itself. Where the preparation of the claim is undertaken by in-house staff of the contractor, this cost will no doubt be reflected somewhere in the claim for increased head office managerial time. If the preparation of the claim is undertaken by independent consultants a specific sum in respect of their fees for the preparation of the claim may be included. However, in either case, the architect or quantity surveyor may well object to such a claim. As the loss and expense must be due either to a deferment of possession or to the regular progress of the works being affected by one of the matters referred to in clause 4.12, it is certainly arguable that the cost of preparation of a claim is due not to any of the matters referred to, but to the contractor's decision to claim reimbursement of loss and expense. It can perhaps be mentioned in passing that there is authority for the proposition that professional fees paid to a claims consultant for work done as an expert witness in the preparation of a building case for arbitration are allowable in the taxation of legal costs: *James Longley & Co. Ltd* v. *South West Thames Regional Health Authority* (1983).

It should be pointed out, and should be borne in mind by the architect and the employer, that the contractor's entitlement under clause 4.11 is expressly without prejudice to any other rights or remedies which he may possess. Where the cause of the disturbance of regular progress is a breach of contract by the employer, the contractor may therefore claim damages at common law for breach of contract and thereby include in his claim any matters which the architect or the employer may seek to reject as part of a claim for reimbursement of loss and expense. For example, in the case of

London Borough of Merton v. *Stanley Hugh Leach* (1985) 32 BLR at pages 107–108, Mr Justice Vinelott in considering clause 24 of JCT 63 (similar in many respects to clause 4.11 and 4.12 of IFC 84) said:

'Moreover there is a clear indication in the contract that the draftsman contemplated that the contractor might have parallel rights to claim compensation under the express terms of the contract and to pursue claims for damages. That arises under clause 24(2) which I have already read and which, of course, expressly provides that the provisions of the conditions are to be without prejudice to other rights and remedies of the contractor. The effect of clause 24(2) (as I understand it) is this. Clause 24(1) specifies grounds upon which the contractor is entitled to make a claim for reimbursement of direct loss or expense for which he would not otherwise be reimbursed by a payment made under the other provisions of the contract. The grounds specified may or may not result from a breach by the architect of his duties under the contract; a claim by the contractor under sub-paragraph (a) will normally, though not perhaps invariably, arise from a failure by the architect to answer with due diligence a proper application by the contractor for instructions, drawings and the like, while a claim by the contractor under sub-clause (b) following a proper instruction requiring the opening up of works under clause 6(3) normally (though not perhaps invariably) will not involve any breach by the architect of any obligation under the contract. In either case the contractor can call on the architect to ascertain the direct loss or expense suffered and to add the loss or expense when ascertained to the contract sum. The contractor will then receive reimbursement promptly and without the expense and delay of a claim for damages. But the contractor is not bound to make an application under clause 24(1). He may prefer to wait until completion of the work and join the claim for damages for breach of the obligation to provide instructions, drawings and the like in good time with other claims for damages for breach of obligations under the contract. Alternatively he can, as I see it, make a claim under clause 24(1) in order to obtain prompt reimbursement and later claim damages for breach of contract, bringing the amount awarded under clause 24(1) into account.'

Global or rolled-up claims

Loss and expense, in order to be reimbursible, must have been caused by a deferment of possession or by one of the clause 4.12 matters. The link between cause and effect must be established. It is sometimes not practicable to relate loss or expense to one specific instance of one specific matter under clause 4.12. If this can be done then it should be. However, there could be a series of events, the interaction of which, prevents this approach from working. In the case of *Crosby* v. *Portland Urban District Council* (1967) 5 BLR page 126, a case which concerned the ICE Conditions of Contract, Mr Justice Donaldson (as he then was) said at 5 BLR pages 135–6:

'. . . Since, however, the extent of the extra cost incurred depends upon an extremely complex interaction between the consequences of the various denials, suspensions and variations, it may well be difficult or even impossible to make an accurate apportionment of the total extra cost between the several causative events.

. . . so long as the Arbitrator . . . ensures that there is no duplication, I can see no reason why he should not recognise the realities of the situation and make individual awards in respect of those parts of individual items of the claim which can be dealt with in isolation and a supplementary award in respect of the remainder of these claims as a composite whole . . .'

In suitable cases therefore a global or rolled up claim may be permissible. It is, however, to be regarded as the exception rather than the rule. In practice by far the majority of contractor's claims for reimbursement of direct loss and expense are framed, at any rate initially, on this global basis.

In the more recent case of *London Borough of Merton* v. *Stanley Hugh Leach* (1985) this issue was considered again in relation to JCT 63 by Mr Justice Vinelott who said at 32 BLR page 102:

'In *Crosby* the arbitrator rolled up several heads of claim arising under different heads and indeed claims for which the contract provided different bases of assessment. The question accordingly is whether I should follow that decision. I need hardly say that I would be reluctant to differ from a judge of Donaldson J's experience in matters of this kind unless I was convinced that the question had not been fully argued before him or that he had overlooked some material provisions of the contract or some relevant authority. Far from being so convinced, I find his reasoning compelling. The position in the instant case is, I think as follows. If application is made (under clause 11(6) or 24(1) or under both sub-clauses) for reimbursement of direct loss or expense attributable to more than one head of claim and at the time when the loss or expense comes to be ascertained it is impracticable to disentangle or disintegrate the part directly attributable to each head of claim, then, provided of course that the contractor has not unreasonably delayed in making the claim and so has himself created the difficulty the architect must ascertain the global loss directly attributable to the two causes, disregarding, as in *Crosby*, any loss or expense which would have been recoverable if the claim had been made under one head in isolation and which would not have been recoverable under the other head taken in isolation. To this extent the law supplements the contractual machinery which no longer works in a way in which it was intended to work so as to ensure that the contractor is not unfairly deprived of the benefit which the parties clearly intend he should have.

. . . a rolled up award can only be made in a case where the loss or expense attributable to each head of claim cannot in reality be separated and . . . where apart from that practical impossibility the conditions which have to be satisfied before an award can be made have been satisfied in relation to each head of claim.'

It is not possible in a book dealing with IFC 84 generally to treat the topic of claims for loss and expense in any greater detail. The reader is referred to specialist works on this topic e.g. Vincent Powell-Smith and John Sims: *Building Contract Claims*.

Other limits to right of recovery
    (i)   Loss too remote
As at common law, if the loss is too remote from the cause of the disruption it will not be sufficiently 'direct' to be allowable. For example, the architect may issue an instruction requiring a variation by substituting one specified material for another. The effect of this could be to disturb the regular progress of the works e.g. a delivery date for the new materials which takes their delivery beyond the date on which the original materials were intended by the contractor to be incorporated into the works. This will properly entitle the contractor to claim.

Supposing, however, that when they reach the site they are discovered to be defective. This in turn causes a further period of disturbance to the regular progress of the works while other materials are ordered and obtained. While the loss and expense caused by the original disturbance of progress due to the variation, namely that caused by the delay in delivery, would be reimbursable, the further disturbance of regular progress brought about by reason of the materials being defective would not be sufficiently direct, either at common law or under this clause, to entitle the contractor to recover loss and expense resulting therefrom. While in one sense the contractor can argue that, had it not been for the instruction requiring a variation being issued, those materials would not have been part of the contract and therefore regular progress could not have been affected by them, it is not the *causa causans* (the actual and direct cause) but is only what is called a *causa sine qua non*. The real cause of the disturbance of regular progress is the defective materials and not the issue of the instruction requiring a variation. The line is sometimes difficult to draw, but the principle must be kept firmly in mind. See also as to remoteness in relation to prolongation costs between the contractor's programme completion date and the contractual completion date under heading *(iv) Restrictions on period for recovery of prolongation costs* (page 253).

(ii) The duty to mitigate

At common law, following a breach of contract, it is the duty of the aggrieved party to take reasonable steps to mitigate his loss. Subject to what is said below concerning notification from the contractor being required when in his opinion he is likely to incur loss and expense, there is no express requirement in the contract to mitigate the loss caused by one of the matters referred to in clause 4.12, either in relation to the extent of disturbance of regular progress or the financial consequences of it. Even so, it is submitted that there must be a general duty on the contractor to take reasonable steps to mitigate. This may be by reducing the extent to which regular progress is disturbed by the matter in question, or alternatively by limiting the loss and expense which flows from it. If the contractor fails to mitigate the former, then it can be argued that, to that extent, it is not the matter referred to in clause 4.12 which caused the disturbance of regular progress, and if he fails in relation to the latter then it can be argued that, to that extent, not all of loss and expense is attributable to the disturbance of regular progress as part of it is the result directly of the contractor's failure to mitigate his loss. There are of course limits to the steps which the aggrieved party must take in order to mitigate his loss. It must be a reasonable step to take in all the circumstances. It will certainly not include the expenditure of substantial sums of money.

While the architect or the employer may in appropriate circumstances be able to challenge part of the contractor's claim on the basis of his failure to mitigate, it should also be borne in mind that if the contractor does take reasonable steps to mitigate his loss, and thereby incurs expenditure, then, even if the effect of this is to inadvertently aggravate the loss, this is nevertheless recoverable by him.

(iii) The regular progress of the works must be materially affected
If the disturbance of regular progress is minimal and insignificant, this will not warrant reimbursement of loss and expense.

It is submitted that regular progress of the works can be materially affected without any overall delay occurring. It is not therefore necessary to establish delay or prolongation in order to found a claim for reimbursement of loss and expense.

The words 'regular progress' are not easy to define.

Although disturbance of progress will often result in delay to completion, this is not necessarily so. It could happen that the disturbance to progress is the result of out-of-sequence working brought about by one of the matters referred to in clause 4.12. In this way, for example, the labour force may be less effectively deployed. Certain skilled craftsmen may have to spend longer on a particular task than had been anticipated. The contractor will be involved in additional wage payments. There may be no overall delay to the completion date. Progress, while being disturbed, may not result in delay to completion e.g. if the activity is not on the critical path.

Another aspect of the words 'regular progress' is considered below (see page 261).

(iv) Restrictions on period of recovery for prolongation costs
The contractor may have programmed to complete before the contractual completion date. His pricing will be based on a construction period shorter than the contractual period. If the regular progress of the works is materially affected by a clause 4.12 matter which causes the contractor to overrun his programme, even though not the contract period, he will suffer site-wide preliminary costs, e.g. hutting, supervision, security and general site services which he other-

wise would not have incurred. Can he recover these as loss and expense?

If the claim for reimbursement is based upon matters falling within clause 4.12.1, he cannot recover them as the architect's obligations under that clause relate to the provision of necessary instructions, drawings, details etc. in relation to the contractor's need for such information having regard to the contract completion date – see *Glenlion Construction Ltd* v. *The Guinness Trust* (1987).

If the claim for reimbursement of these prolongation costs arises from any other clause 4.12 matter, the situation may well be different. However, it has been argued that these costs are also irrecoverable in the sense that they are not additional costs at all. It is argued that in tendering against a known contract period it must be assumed that the contractor has priced his preliminaries on the assumption that he may be on site for the whole contract period – see for example the case of *Finnegan Ltd* v. *Sheffield City Council* (1988) and in particular the editorial commentary thereon at 43 BLR page 126.

It is submitted that it cannot be 'deemed' that the contractor's tender includes for all time-related costs for the entire contract period. This would give the employer the best of both worlds. On the one hand he will be looking towards the lowest tender which in practice will generally be at least partly due to the contractor shortening the construction period. On the other hand, if the employer is responsible for causing the contractor to remain on site beyond that shorter period, he ought not to be able to contend that nothing is due for the period before the contract period has expired.

The correct approach, it is submitted, is to be found in the ordinary principles of remoteness of damage. What could be said at the date of entering into the contract as the likely result of such disturbance to progress of the kind set out in clause 4.12. The knowledge of the employer at the time of entering into the contract is crucial. For instance, if the contractor has provided, with his tender, a realistic programme, showing a construction period shorter than the contract period, it is likely to be foreseeable that prolongation costs will be suffered by an overrun of that shorter period. Whereas, if the information available at the time of entering into the contract, indicates that the contractor intends or expects to be on site for the whole contract period, clearly no prolongation costs will be suffered during that period. It has to be said, however, that if this approach is the correct one, the situation where a clause 4.12.1

matter runs concurrently with one of the other matters in clause 4.12 produces a very difficult situation. If it means that the contractor would not be entitled to recover his prolongation costs between the expiry of his programmed period for completion and the contractual completion date, it might become very tempting for employers, where the contractor looks as though he could have a claim for prolongation costs under a matter other than that in clause 4.12.1, to deprive the architect of information or instructions so as to ensure that the flow of information from the architect is geared to the full contract period plus a short overrun so as to wipe out the contractor's prolongation claim for the period between the contractor's realistic programmed completion date and the contract completion date.

The machinery and operation of the clause
The first point to appreciate is that applications for reimbursement of loss and expense are completely unconnected with the granting of extensions of time for completion under clause 2.3. Indeed, IFC 84 usefully separates the clauses whereas their equivalents in both JCT 63 and JCT 80 are consecutive. It is true that all the matters listed in clause 4.12 are mirrored in clause 2.4, although clause 2.4 also contains other matters. Clauses 2.3 and 2.4 deal with the contractor's right to an extension of time for completion of the works, which relieves him of the payment of liquidated damages. The fact that he is so relieved because of a delay, even if arising out of an event which has a mirror provision in clause 4.12, does not entitle him automatically in itself to a claim for reimbursement of loss and expense. The delay may not cause a disturbance to regular progress. In the other direction, it is possible for a claim for reimbursement of loss and expense under clause 4.11 to be made where regular progress is disturbed even though no delay at all is caused and therefore there is no question of an extension of time being granted. Of course, in practice, if there is an extension of time granted owing to delay to completion caused by an event which mirrors one of those listed in clause 4.12, then as a delay to completion almost always involves the contractor in loss and expense, he can use the fact of the grant of an extension of time as compelling evidence that some loss and expense must have been suffered simply by the fact of a delay being involved.

A further point which should be made is that there can be no reimbursement of direct loss and expense where the cause of it falls outside those matters listed in clause 4.12. Simply because the contractor finds that the work is more difficult or more time-consuming, and therefore more expensive, than he had envisaged is not of itself a ground for him to claim.

Clause 4.11 is clearly a mixture of JCT 63 and JCT 80. This cross-fertilisation has not been carried out without leaving in its wake some difficulties of construction brought about by the drafting. These difficulties are mentioned in passing during this discussion of the operation of these clauses and also in the Notes to the clause which follow.

The first requirement before clause 4.11 can operate is that the contractor must make a written application. It is a condition precedent to a claim for reimbursement of loss and expense – see *London Borough of Merton* v. *Stanley Hugh Leach* (1985).

The application is the subject of the first paragraph of clause 4.11. Sub-paragraphs (a) and (b) are also relevant. It is in relation to this first paragraph of clause 4.11 that the greatest difficulties in interpretation arise. A comparison of the provisions of JCT 63 (clause 24) and JCT 80 (clause 26) in regard to the contractor's written application helps to illustrate the difficulties.

Under JCT 63 clause 24(1), the written application must be made within a reasonable time of it becoming apparent that the progress of the works has been materially affected.

Under JCT 80 clause 26.1, the written application must be made as soon as it has become, or should reasonably have become, apparent that the regular progress of the works has been or is likely to be materially affected.

Under IFC 84, the words 'within a reasonable time of it becoming apparent' do not appear to refer to the regular progress being materially affected, especially having regard to paragraph (a) of this clause. Indeed, it is difficult to establish to what the word 'it' relates at the end of the second line of the first paragraph of clause 4.11. It probably relates to the occurrence or likely occurrence of loss and expense. Further, JCT 80 goes on to state that the notice is also required where the regular progress of the works is likely to be affected even though it has not been at that date. This extra provision in JCT 80 can clearly be of considerable assistance to the architect in considering ways of preventing or limiting the disturbance of regular progress. It would have been helpful therefore to have such a requirement of advance notice in IFC 84.

Sub-paragraph (a) of clause 4.11 is treated separately rather than being included in one of the matters listed in clause 4.12. One advantage of this is that it avoids the argument which could otherwise be raised that a deferment of possession cannot result in the regular progress of the works being materially affected. To affect progress it may be argued that progress of some sort must have started. If the employer fails to give possession then it could be said that this does not disturb progress at all if the contractor has not even begun work on site. (See also page 261.)

The next matter to be considered is the contents of the written application made by the contractor. Under JCT 63 there is no express requirement as to the contents of the written application. Under JCT 80, the written application must state that the contractor has incurred, or is likely to incur, loss and expense. In IFC 84, the situation is not clear but appears to be as in JCT 80, there being a requirement that the contractor should state that he has incurred, or is likely to incur, loss and expense, if the word 'it' at the end of the second line of the clause relates to 'has incurred or is likely to incur'. The clause is certainly not as clear as it could be. Generally the courts will take a liberal approach in relation to the contents of the written application – see *Rees and Kirby* v. *Swansea City Council* (1985).

Having received the notice within the time stated, the architect must form an opinion. Under JCT 63 clause 24(1), the architect's opinion relates to whether or not the loss and expense *has been* incurred. The ascertainment relates back to such loss and expense. Under JCT 80 clause 26.1, the architect's opinion relates to whether the regular progress of the works has been, or is likely to be, materially affected and the ascertainment relates to the loss and expense which has been or *is being* incurred by the contractor. Under IFC 84 the architect's opinion relates to whether the contractor has incurred or *is likely to* incur loss and expense. This is rather odd in that, while the architect's opinion extends to loss and expense which is likely to be incurred by the contractor, the ascertainment carried out by the architect or the quantity surveyor is in relation to the loss and expense incurred. Literally, therefore, if the architect is of the opinion that the contractor is likely to incur loss and expense, he can only ascertain or instruct the quantity surveyor to ascertain such loss and expense which has been incurred. Sense can perhaps be made of these words if they are interpreted to mean that the architect, having formed an opinion that loss and expense is likely to be incurred, can then ascertain or instruct the quantity surveyor to ascertain actual loss and expense as and when incurred without the necessity for further written applications by the contractor.

Under IFC 84 (as in JCT 80 but unlike JCT 63) the contractor must support his application with such information required by the architect or the quantity surveyor as is reasonably necessary for the purpose for the operation of the clause. JCT 80 goes on to state (per clause 26.1.3) that the contractor shall also submit on request such details of loss and expense as are reasonably necessary for the ascertainment to be made. Such an express requirement is not to be found in IFC 84 although it is indirectly covered by the requirement referred to above for the contractor to provide such information on request as is reasonably necessary for the purposes of the clause. The degree of information required of the contractor will depend on many factors, including the state of the architect and quantity surveyor's knowledge of relevant matters, as well as the ease or difficulty faced by the contractor in obtaining the information. The words of Mr Justice Vinelott in *London Borough of Merton* v. *Stanley Hugh Leach* (1985) though relating to the written application under clause 24 of JCT 63 are it is submitted of some relevance to clause 4.11 of IFC 84. He said at 32 BLR page 97:

'The contractor must act reasonably: his application must be framed with sufficient particularity to enable the architect to do what he is required to do. He must make his application within a reasonable time: it must not be made so late that, for instance, the architect can no longer form a competent opinion on the matters on which he is required to form an opinion or satisfy himself that the contractor has suffered the loss or expense claimed. But in considering whether the contractor has acted reasonably and with reasonable expedition it must be borne in mind that the architect is not a stranger to the work and may in some cases have a very detailed knowledge of the progress of the work and of the contractor's planning. Moreover, it is always open to the architect to call for further information either before or in the course of investigating a claim. It is possible to imagine circumstances where the briefest and most uninformative notification of a claim would suffice: a case, for instance, where the architect was well aware of the contractor's plans and of a delay in progress caused by a requirement that works be opened up for inspection but where a dispute whether the contractor had suffered direct loss or expense in consequence of the delay had already emerged. In such case the contractor might give a purely formal notice solely in order to ensure that the issue would in due course be determined by an arbitrator when the discretion would be exercised by the arbitrator in the place of the architect.'

The contractor's written application must be made 'within a reasonable time of it becoming apparent'. The difficulties of determining exactly to what 'it' refers have been discussed above. Once that question has been answered, the contractor has a reasonable time to submit his written application to the architect. Failure to notify the architect within a reasonable time may, particularly if the clause can be construed to include an obligation on the contractor to include in his application advance warning of likely future loss and expense, prejudice the architect and therefore the employer in his ability to take avoiding action. However, what is a reasonable time will depend on all the circumstances. In the case of *Tersons Ltd* v. *Stevenage Development Corporation* (1965) Lord Justice Willmer said in relation to the requirement of notice under an early edition of the ICE Conditions of Contract that:

'. . . The contractors must at least be allowed a reasonable time in which to make up their minds. Here the contractors are a limited company and that involves that, in the matter of such importance as that raised by the present case, the relevant intention must be that of the board of management.'

In that case the notice had to be given as soon as practicable but similar considerations will apply to the words 'within a reasonable time'. In the event of the contractor clearly failing to make the written application within a reasonable time, the architect is not empowered to operate the clause. Of course, if this deprives the contractor of his right to claim reimbursement, he may nevertheless pursue an action for damages for breach of contract at common law if the disruption is due to a breach of contract on the part of the employer.

Having received the written application and obtained any further information reasonably required, the architect or the quantity surveyor on his behalf must carry out the ascertainment. If the quantity surveyor is instructed to ascertain the loss and expense it will not be possible subsequently for the architect to ignore that ascertainment. The contract provides that the quantity surveyor, having been instructed by the architect, *shall* carry out the ascertainment of the loss and expense which therefore ceases to be the function of the architect. Once ascertained it shall be added to the contract sum.

The contractor's programme

The contractor's programme of works may be of some assistance both to the contractor in establishing his claim and to the architect in forming an opinion as to whether or not regular progress of the works has been materially affected. However, an agreed programme, while of some evidence, is by no means conclusive as to what regular progress should have been under the contract.

The loss and expense, to be recoverable under clause 4.11, must not be reimbursable by payment under any other provision of the contract. This will avoid the possibility of double payment e.g. where part of the increased cost of materials or labour is already covered under the fluctuations provisions and therefore paid for in the manner provided elsewhere in the contract.

The matters listed in clause 4.12.1 to 4.12.8

The wording of the listed matters mirrors the wording of certain of the events listed in clause 2.4. Many of the observations made in relation to the wording of those events in the Commentary and Notes to clause 2.4 are therefore relevant in considering these listed matters. Listed below are the particular events in clause 2.4 to which the reader is referred for a consideration of the words used.

Clause 4.12 matters		*Clause 2.4 events*
4.12.1	(late instructions etc.)	*see* 2.4.7
4.12.2	(opening up and testing)	*see* 2.4.6
4.12.3	(work not forming part of this contract carried out by the employer or persons engaged by him)	*see* 2.4.8
4.12.4	(employer supplying materials or goods)	*see* 2.4.9
4.12.5	(instructions as to postponement)	*see* 2.4.5 and 3.15
4.12.6	(failure to give ingress or egress)	*see* 2.4.12

| 4.12.7 | (instructions under clauses 1.4, 3.6, 3.3, 3.8) | *see* 2.4.5 |
| 4.12.8 | (approximate quantities not a reasonable forecast) | *see* 2.4.15 |

Apart from the Commentary and Notes on clause 2.4 to which the reader is referred in relation to the specific matters, one or two further points are made as follows.

Clause 4.12.2 refers to opening up and testing in accordance with clause 3.12. The corresponding event in clause 2.4.6 refers also to clause 3.13.1 dealing with opening up and testing following the discovery of a failure of work or materials to be in accordance with the contract. The question of whether or not any loss and expense is recoverable in respect of testing in such circumstances is discussed in the Commentary to clause 3.13.

In JCT 63 clause 24(1)(d), the approximate equivalent of what is now contained in clause 4.12.3 of IFC 84 related to *delay* on the part of direct contractors. This reference to the word delay was removed when the equivalent of this provision was inserted into JCT 80 (see clause 26.2.4). It is proper that loss and expense incurred as a result of disruption to the work carried out by the employer or his direct contractors should be covered even if there was no delay on their part.

The listed matters in clauses 4.12.1 to 4.12.8 follow quite closely those in clause 26.2 of JCT 80 and represent an improvement over those contained in JCT 63.

NOTES TO CLAUSES 4.11 AND 4.12

[1] '. . . opinion . . . '.
This opinion is reviewable in arbitration proceedings although not if any difference or dispute is dealt with in the courts – see *Northern Regional Health Authority* v. *Crouch* (1984).

[2] '. . . the regular progress . . . '.
As the claiming of loss and expense is tied to the regular progress of the works being materially affected, it is necessary to show that there has been a disturbance of the regular progress. Before there can be disturbance there must be some form of progress so that if any of the matters listed cause the contractor to be unable to commence progress on the works, this may not be covered by the clause e.g. failure of the employer to give ingress to or egress from the

site of the works over any land, buildings etc. in the possession and control of the employer, in such circumstances that the contractor cannot even commence progress of the works. Although there is a certain logic in this argument, it is considered that it is unlikely to succeed before an arbitrator or the courts. Further, the employer should appreciate that preventing the contractor from even commencing progress will often be a serious breach of contract so that the employer could well be faced with a substantial claim for damages for breach of contract at common law. If the breach of contract is serious enough, the contractor will also have the right to regard the employer's failure as a repudiation of his obligations under the contract and may then regard the contract as at an end.

In a similar way, loss and expense caused by one of the listed matters after progress has apparently ended could cause difficulties to the contractor in seeking to claim under this clause. It may be that the works have been completed and are awaiting a certificate of practical completion, but that before such a certificate is issued certain tests have to be carried out e.g. in relation to a heating and ventilation system. If the testing is delayed owing to one of the matters listed in clause 4.12, can it be argued that there has been no disturbance of regular progress as there is nothing left for the contractor to do except await the results of the testing? Again it is submitted that this argument would not succeed. It is submitted that progress can be disturbed by any of the listed matters (with the possible exception of that in clause 4.12.5 (postponement of work)) occurring between the date of possession given in the appendix, or any deferred date allowed under clause 2.2, and the date of practical completion. This uncertainty could have been avoided by including within clause 4.11(b) a reference to the commencement and regular progress of the works being materially affected. (See for example clause 14.1 of NAM/SC.)

[3] '. . . ascertain . . .'.
This may be regarded as an unfortunate choice of word. Dictionary definitions refer to e.g. 'finding out for certain'. Some employers, and particularly auditors on their behalf, take the view that if they cannot know for certain that every penny of the contractor's claim for loss and expense has been suffered or incurred, then nothing is due. In other words, if the contractor cannot prove every penny of his claim, he is entitled to nothing in respect thereof as it is not possible under the clause to make estimates. If it is reasonable for

the contractor to produce evidence of actual losses and expenses, this evidence can be required of him. If it is not possible to itemise the figures in precise detail, the making-up of a reasonable estimate based on what actual information is available is, it is submitted, sufficient.

[4]   '. . . having regard to the Date for Completion . . .'.
It is not automatically the chosen sequence of operations of the contractor which determines when it is reasonably necessary for the contractor to receive instructions, drawings, details or levels from the architect. Depending on the contractor's actual methods of working and sequence of operations, instructions issued may disturb the regular progress of the works of one contractor whereas another contractor, carrying out precisely the same contract works but with a different sequence of operations, may find that his regular progress is not disturbed. It is submitted therefore that the architect must be aware of this sequence of working e.g. through an agreed realistic programme, before this sequence can be related to the date for completion and thereby become relevant.

Chapter 8

Statutory obligations etc.

CONTENT

This chapter considers section 5 which deals with statutory obligations, notices, fees and charges together with value added tax and the statutory tax deduction scheme under the Finance (No. 2) Act 1975 as amended by the Finance Act 1980.

SUMMARY OF GENERAL LAW

Statutory requirements in relation to contract works

A building contractor, apart from his contractual obligations to the employer, will also sometimes owe a statutory duty to him and to others under various statutes e.g. the Defective Premises Act 1972 and possibly the Building Act 1984 (and the Building Regulations enacted pursuant to the powers vested in the Secretary of State under section 1 and schedule 1 thereof). In this chapter we are primarily concerned with duties owed under statute by the contractor to the building owner/employer. The duty may be owed to the employer as an owner or as an occupier or user of the building.

It is not every breach of an obligation imposed by statute which will give rise to a civil claim for damages by someone who has been injured or suffered loss or damage by reason of the non-compliance. The injured part must show firstly that the injury suffered was within the ambit of the statute; secondly that the statutory duty imposes a liability to civil action; thirdly that the statutory duty was not fulfilled; and fourthly that the breach of the statutory duty caused the injury.

Where the statute is silent as to a remedy, there is a presumption that the injured party will have a right of action for breach of statutory duty. For a detailed discussion of this topic the reader is referred to leading text books e.g. *Clerk and Lindsell on Torts*, 16th edition.

An issue of particular importance is whether a building contractor owes an independent statutory duty to the employer to comply with the Building Regulations 1985. The building regulations include many important matters e.g. relating to design. Do they impose a duty on the builder to comply with the regulations even though this may be the employer's responsibility under the building contract?

The interaction between the contractual terms and any statutory duty falling on the contractor, arising under the building regulations, raises interesting and difficult questions. There have been a number of cases under earlier building regulations dealing with liability arising from a breach of duty owed under building regulations. Most, though not all, deal with the liability of a local authority for its building inspector's negligent performance of his duties to ensure compliance with building regulations. Two cases of particular interest in the current context are *Acrecrest Ltd* v. *W. S. Hattrell & Partners and The London Borough of Harrow* (1983) and *Street and Another* v. *Sibbabridge Ltd and Another* (1980).

In the *Acrecrest* case, Acrecrest Ltd were the owners and developers of land in the London Borough of Harrow. In 1971 they employed the first defendants, a firm of architects, for the development of a site including flats. The flats were defective owing to defective foundations which were not constructed to a required uniform depth of five feet, having instead been constructed to depths which varied over the site (but nevertheless in accordance with the instructions of the local building inspector). The building owner therefore sued his architects for damages for negligence and breach of contract. The architects brought in as third party the London Borough of Harrow and the building owner subsequently brought in the borough as a second defendant. The building contract was governed by the JCT 63 (July 1971 revision). The building contractors were not joined because the building owner did not think them worth suing.

The local authority was held to owe a duty of care to the developers even though they were not occupiers. This part of the decision has since been emphatically overruled in the case of *Governors of the Peabody Donation Fund* v. *Sir Lindsay Parkinson & Co. Ltd* (1984) by limiting the scope of a local authority's duty of care to

occupiers and users. Further, the question of whether a duty of care
is owed at all by a local authority in favour of occupiers and users
has not been finally determined – see *Murphy* v. *Brentwood District
Council* (1990). Even if such a duty is owed, it is confined to actual
personal injury or physical damage to separate property. It does not
extend to liability in respect of the cost of averting damage to health
or safety of person or property.

The employment by building owners of 'approved inspectors'
under the 1984 Act on contractual terms is likely to enhance
significantly the building owner's position by providing non-
occupying building owners with a contractual remedy where none
existed before.

However, the Court of Appeal made a number of interesting
observations in relation to the duties owed by the builder under the
building regulations.

Per Lord Justice Stevenson:

'. . . the builder, whether he is the owner or a contractor employed
by the owner, is beyond doubt under a duty, both statutory and
contractual, to comply with the regulations. . . .

The contractual duty of the builders is contained in clause 4(1)
[of JCT 63], subject to which they agree to carry out the works . . .
clause 18 [indemnity provisions] contains the builders' agreement
to indemnify the plaintiffs in very comprehensive terms. . . .'

So far as the duty owed by the builder to the building owner was
concerned, this was covered by clause 4 of JCT 63 in its July 1971
revision. It is important to appreciate that following the July 1976
revision of JCT 63 (and subsequently repeated in JCT 80 and in IFC
84), a provision was added to the contract which seeks to absolve the
contractor from liability to the employer under the contract where
the failure of the works to comply with statutory requirements
results from the contractor having carried out the works in accordance with the contract documents or any instruction of the architect. This provision must therefore be carefully considered in
determining the liability of the contractor. This question is considered in detail in the Commentary to clause 5.3 of IFC 84, page 273.

In *Street* v. *Sibbabridge* it was held that there was an implied term
in a building contract that the contractor would comply with build-

ing regulations which overrode the contractor's particular obligation under the contract to comply with the architect's design and instructions. This meant that, despite the contractor's compliance with the architect's instructions as to depth of foundations, the contractor was still liable to the building owner. Judge Fay QC held that the express term of the contract as to the depths of foundations did not prevail over the implied term to comply with building regulations.

The House of Lords decisions in the *Murphy* case and in *The Department of the Environment* v. *Thomas Bates and Sons Ltd and Others* (1990) make it abundantly clear that the duty, if any, owed by local authorities is based upon ordinary principles of common law negligence in performance by them of their statutory functions. It does not give rise to a separate action or breach of statutory duty.

Further, in the case of *Perry* v. *Tendring District Council and Others* (1984) His Honour Judge John Newey QC held, after considering full argument on the issue, that the builder did not owe a separate statutory duty to the building owner though he could be liable in negligence if he negligently breached the building regulations.

An interesting question is whether a contractor, if held liable to the building owner for a breach of building regulations when he has simply carried out the design and instructions of the architect without being in any way culpable, can claim an indemnity or contribution from the architect, who is himself likely to owe a duty of care to the building owner. There seems no reason why the contractor should not be able to claim a full indemnity, or where he is partly at fault, a contribution from the architect.

The appropriate course for the contractor to take is to join in the proceedings the architect and to seek an indemnity or a contribution by virtue of the Civil Liability (Contribution) Act 1978. The claim to contribution would most probably be based upon the architect's breach of a contractual duty to take care in favour of the building owner. It may also, however, be based upon a breach of a concurrent tortious duty to take care although this is somewhat less certain.

In the case of *Townsends Ltd* v. *Cinema News etc. Ltd* (1958) the contractor was in breach of his statutory duty to serve notice on the sanitary authority before executing certain work. It was established that there was a clear practice in such a situation for the contractor to rely on the architect to give the notices and see that the regulations were complied with. The contractor relied on the architect serving the proper notice but the architect failed to carry out this task correctly. There was a breach of the byelaws causing loss to the employer which he was entitled to recover from the contractor. In turn, the contractor recovered his loss from the architect who, it was held, owed a duty of care to the contractor and who had been negligent in the performance of that duty even though this was undertaken gratuitously.

Further, it may often be the case that if the builder's breach of contract resulting from a breach of statutory duty in failing to comply with the building regulations is the inevitable result of compliance with the architect's design or specification or instructions, as the architect acts as the agent for the building owner, any claim by the building owner against the builder can be defeated by the builder by applying the doctrine of circuity of action. If the builder could claim an indemnity from the architect, he could equally claim an indemnity from the building owner so that the very same set of facts which give rise to the building owner's claim against the builder also gives the builder a claim for indemnity against the building owner – hence the circuity of action: see *Equitable Debenture Assets Corporation Ltd* v. *William Moss Group Ltd and Others* (1984) 2 CLR 1 in which His Honour Judge John Newey QC at page 35 said:

'To construe condition 4(1) as imposing upon Moss liability for failing to comply with the regulation at times when reasonably they were totally unaware of the breach seems decidedly harsh. However, because of the clear words of condition 4(1) and in view of Townsends' case I have no alternative but to reach that conclusion. But that is not the end of the matter. Although Moss acted in breach of condition 4(1) by constructing by their sub-contractor, Alpine, curtain walling which did not comply with Building Regulation C8, they were at the time acting in accordance with condition 1(1). It was Alpine's drawings issued by Morgan as architects under the head contract which caused Moss to act in breach of condition 4(1).

... I think that up to the time when Moss must have discovered the lack of buildability in aspects of the design of curtain walling, Moss probably would be entitled to an indemnity from Morgan if found liable to EDAC. However, Morgan were, as architects under the head contract, EDAC's agents: therefore Moss would equally well be entitled to indemnity from EDAC. In the circumstances it seems to me that the doctrine of circuity must apply to defeat EDAC's claim. A plea of circuity of action has been recognised in recent years in actions in negligence and for breach of statutory duty, see *Ginty* v. *Belmont Building Supplies Ltd* (1959) 1 All ER 414 and *Post Office* v. *Hampshire County Council* (1980) 1 QB 124 CA, but if I understand aright the pre-1975 precedent in Bullen & Leake's *Pleading and Practice*, they indicate that it can apply generally.'

Defective dwellings

The Defective Premises Act 1972 imposes duties on persons taking on work for, or in connection with, the provision of dwellings. It may therefore have an application to certain contracts let under IFC 84. Under the Act, the contractor will owe a duty to the employer in the terms of section 1 which says:

'A person taking on work for or in connection with the provision of a dwelling (whether the dwelling is provided by the erection or the conversion or enlargement of a building) owes a duty:

(a) if the dwelling is provided to the order of any person, to that person; and

(b) without prejudice to paragraph (a) above, to every person who acquires an interest (whether legal or equitable) in the dwelling;

to see that the work which he takes on is done in a workmanlike or, as the case may be, professional manner with proper materials and so that as regards that work the dwelling will be fit for habitation when completed.'

A person taking on work will ordinarily include the contractor and also any professional persons e.g. architects, engineers and quantity surveyors. Sub-contractors will also be covered.

It is not possible by contract to exclude or restrict the operation of any of the provisions of the Act or any liability arising by virtue of such provisions.

However, the Act itself provides a defence. Section 1(2) provides as follows:

'A person who takes on any such work for another on terms that he is to do it in accordance with instructions given by or on behalf of that other shall, to the extent to which he does it properly in accordance with those instructions, be treated for the purposes of this section as discharging the duty imposed on him by sub-section (1) above except where he owes a duty to that other to warn him of any defects in the instructions and fails to discharge that duty.'

While the Act will therefore provide the employer with an additional remedy in respect of defective materials or workmanship, it will not apply where materials or goods, which have been specified by the employer, are discovered to be of a kind unsuitable for the purpose for which they are being incorporated into the works. In other words, where the failure is of design or fitness for purpose, then provided the contractor has complied with his obligations under the contract documents, he will not be liable under the Act except where he owes a duty to the employer to warn him of any defects in the instructions and fails to discharge that duty. The duty to warn may well arise in a situation where a reasonably competent and experienced contractor would appreciate that the contract documents or the instructions of the architect were in some way defective.

As this statutory defence is similar in some respects to the operation of clause 5.3 of IFC 84, it is unlikely that that clause will run foul of the prohibition against restricting or excluding liability contained in the Act. In so far as clause 5.3 (discussed in detail later on page 273) is compatible with the statutory defence, it is not an attempt to restrict or exclude liability under the Act. However, clause 5.3 does not contain the proviso, as does the Act, relating to the duty to warn, and if a duty to warn arises, the contractor's failure to warn in appropriate circumstances will render clause 5.3 void so far as the employer's rights under the Act are concerned.

Any cause of action in connection with a breach of the Act is deemed to have accrued at the time when the dwelling was completed, or where defects in the dwelling have been rectified, from the time when such further rectification work was finished so far as that rectified work is concerned – see section 1(5) of the Act. The cause of action for breach of the statutory duty imposed by section 1 of the Act will expire six years from when it is deemed to have accrued.

CONSIDERATION OF THE RELEVANT CLAUSES OF IFC 84

Clauses 5.1 to 5.4

Statutory obligations, notices, fees and charges

5.1 The Contractor shall comply with, and give all notices required by, any statute, any statutory instrument, rule or order or any regulation or byelaw applicable to the Works (hereinafter called 'the Statutory Requirements') and shall pay all fees and charges in respect of the Works legally recoverable from him.

The amount of any such fees or charges (including any rates or taxes other than value added tax) shall be added to the Contract Sum unless they are required by the Specification/Schedules of Work/Contract Bills to be included in the Contract Sum.

Notice of divergence from Statutory Requirements

5.2 If the Contractor finds any divergence between the Statutory Requirements and the Contract Documents or between the Statutory Requirements and any instruction of the Architect/the Contract Administrator he shall immediately give to the Architect/the Contract Administrator a written notice specifying the divergence.

Extent of Contractor's liability for non-compliance

5.3 Subject to clause 5.2 the Contractor shall not be liable to the Employer under this Contract if the Works do not comply with the Statutory Requirements where and to the extent that such non-compliance of the Works results from the Contractor having carried out work in accordance with the Contract Documents or any instruction of the Architect/the Contract Administrator.

Emergency compliance

5.4 If in any emergency compliance with clause 5.1 requires the Contractor to supply materials or execute work before receiving instructions from the Architect/the Contract Administrator then:

5.4.1 the Contractor shall supply such limited materials and execute such limited work as are reasonably necessary to

secure immediate compliance with the Statutory
Requirements;

5.4.2   the Contractor shall forthwith inform the Architect/the
         Contract Administrator thereof; and

5.4.3   the work and materials shall be treated as if they had
         been executed and supplied pursuant to an Architect's/
         Contract Administrator's instruction requiring a Variation
         issued in accordance with clause 3.6 *(Variations)*, pro-
         vided that the Contractor has informed the Architect/the
         Contract Administrator in accordance with clause 5.4.2
         and the emergency arose because of a divergence
         between the Statutory Requirements and all or any of the
         following documents namely: the Contract Drawings or
         the Specification or the Schedules of Work or the Con-
         tract Bills, or any instruction or any drawing or document
         issued by the Architect/the Contract Administrator under
         clauses
         1.7   *(Further drawings or details)*,
         3.5   *(Architect's instructions)* or
         3.9   *(Levels)*.

COMMENTARY ON CLAUSES 5.1 TO 5.4

Clause 5.1 imposes an obligation on the contractor to comply with
what are called 'the statutory requirements' and to pay all fees and
charges in respect of the works legally recoverable from him.

The question of fees is dealt with differently in IFC 84 when
compared with JCT 63 and JCT 80. Under these latter forms the
contractor pays the fee or indemnifies the employer in respect of
any liability for fees. The amount of these is added to the contract
sum unless they have already been included in the contract
documents. Under IFC 84 there is no reference to indemnifying the
employer. The contractor simply pays those fees which are legally
recoverable from him, with the employer paying those legally
recoverable from the employer. Those paid by the contractor are
recoverable by an addition to the contract sum unless already
included for in it.

For a discussion of breaches of statutory requirements involving
non-compliance with building regulations see earlier under 'Summary of general law' page 264.

Clause 5.2 deals with divergences between the statutory requirements and the contract documents or an instruction of the architect.
For some reason, perhaps an oversight in drafting, there is no

reference to divergences between the statutory requirements and further drawings or details issued under clause 1.7 or levels under clause 3.9. If the contractor finds any such divergence, he is under a duty to immediately inform the architect by written notice specifying the divergence. From the contractual point of view, the contractor is under no obligation to look for divergences. However, as he has a potential liability for breach of statutory duty, apart from contract, he should not disregard matters to which such statutory requirements apply e.g. the building regulations (see earlier under 'Summary of general law' page 264).

Under JCT 63 and JCT 80, following notification of a divergence (clause 4(1)(c) and 6.1.3 respectively), the architect is expressly required to issue instructions in relation to the divergence. If it requires the works to be varied, any such instruction is to be treated as if it were a variation issued in accordance with the contract. However, under IFC 84 there is no such express requirement on the architect to issue instructions. This may be the result of an oversight in the drafting. In any event it is submitted that, on receipt of a notice from the contractor specifying the divergence, the architect must issue any necessary instructions, no doubt after consultation with the employer where appropriate, to remove the divergence. Failure to do so would be a breach of contract by the employer and would also entitle the contractor to an extension of time under clause 2.4.7 and to reimbursement of direct loss and/or expense under clause 4.12.1, provided that the contractor, in his notice specifying the divergence, also requested instructions.

Clause 5.3 seeks to exclude the liability of the contractor to the employer 'under this Contract' if the works do not comply with the statutory requirements provided that the non-compliance results from the contractor having carried out the work in accordance with the contract documents or any instruction of the architect. This exclusion of liability is expressed not to apply where the contractor has become aware of a divergence and has failed under clause 5.2 to notify the architect.

This exclusion raises two issues. Firstly, is it effective to reduce or extinguish the contractor's liability to the employer under the contract for breach of the statutory requirements? Secondly, can it affect any independent duty owed by the contractor to the employer to comply with statutory requirements e.g. under the building regulations?

As to the first issue, clause 5.3 clearly negatives the express

obligation imposed by clause 5.1. However, if, as has been held in
the case of *Street* v. *Sibbabridge* (see page 266), there is an implied
term in the contract involving the contractor in an absolute
obligation to comply with building regulations, the exclusion clause
is required to satisfy the requirement of reasonableness under the
Unfair Contract Terms Act – see section 7(1) and 7(3) and the
'Guidelines' for the application of the reasonableness test in
schedule 2 to the Act. It is submitted that the exclusion in clause 5.3
is reasonable as the contractor is doing no more than constructing
the works in accordance with the particular requirements of the
employer (c.f. by analogy paragraph (e) of the Guidelines in
schedule 2 of the Act which requires that attention must be given in
particular as to 'whether the goods were manufactured, processed
or adapted to the special order of the customer'). It is submitted that
this exclusion in the clause in favour of the contractor is reasonable
despite the fact that from the case of *Acrecrest* v. *Hattrell* (page 265)
it would appear that negligence or breach of statutory duty of the
architect in failing to design the works to comply with building
regulations is not attributable to the employer.

In regard to the second issue, clause 5.3 does not by its terms
attempt to exclude liability for any independent duty owed apart
from contract. It refers to liability 'under this Contract' being
excluded and not any liability apart from the contract itself. If this
independent obligation is also independent of fault on the part of
the contractor (the position is not clear – see earlier under
'Summary of general law' page 265) it would seem particularly harsh
on the contractor to incur a liability to the employer in circumstances where he has done no more than fulfil his contractual
obligations.

If any loss or damage is suffered by the employer owing to a
breach of such an independent duty owed by the contractor,
whether depending on fault or not, the indemnity provisions in
clause 6.1 will, subject to the provisos contained in that clause,
take effect except in relation to damage to the contract works
themselves – see clause 6.1.3 and note the case of *Tozer Kemsley
Millbourn (Holdings) Ltd* v. *Cavendish Land (Metropolitan) Ltd and
Others* (1983).

Clause 5.4 deals with compliance by the contractor with statutory
requirements in an emergency. This is a very useful provision
enabling the contractor in an emergency to take such limited steps by
way of the supply of materials or the execution of work as are

necessary to secure immediate compliance with any statutory requirements. The contractor is contractually entitled to do this in the absence of instructions from the architect provided the emergency situation makes compliance necessary before such instructions can be obtained. The contractor must then forthwith inform the architect and the emergency work carried out will be treated as having been done pursuant to an architect's instruction requiring a variation, provided always that the emergency arose because of a divergence between the statutory requirements and the contract documents or any instruction, drawing or document issued by the architect under clauses 1.7; 3.5 or 3.9.

Clause 5.5

Value Added Tax: Supplemental Condition A

5.5   The sum or sums due to the Contractor under article 2 of this Contract shall be exclusive of any value added tax and the Employer shall pay to the Contractor any value added tax properly chargeable by the Commissioners of Customs and Excise on the supply to the Employer of any goods and services by the Contractor under this Contract in the manner set out in Supplemental Condition A. [j.2] Clause A1.1 of Supplemental Condition A shall only apply where so stated in the Appendix.

To the extent that after the Base Date the supply of goods and services to the Employer becomes exempt from value added tax there shall be paid to the Contractor an amount equal to the loss of credit (input tax) on the supply to the Contractor of goods and services which contribute exclusively to the Works.

[j.2] Clause A1.1 can only apply where the Contractor is satisfied at the date the Contract is entered into that his output tax on all supplies to the Employer under the Contract will be at either a positive or a zero rate of tax.

On and from 1 April 1989 the supply in respect of a building designed for a 'relevant residential purpose' or for a 'relevant charitable purpose' (as defined in the legislation which gives statutory effect to the VAT changes operative from 1 April 1989) is only zero rated if the person to whom the supply is made has given to the Contractor a certificate in statutory form: see the VAT leaflet 708 revised 1989. Where a contract supply is zero rated by certificate only the person holding the certificate (usually the Contractor) may zero rate his supply.

A   Value Added Tax
           [q]   Clause 5.5

Treatment of VAT

A1   In this Supplemental Condition 'tax' means the value
            added tax introduced by the Finance Act 1972 which is
            under the care and management of the Commissioners of
            Customs and Excise (hereinafter called 'the Commission-
            ers'). Supplies of goods and services under this Contract
            are supplies under a contract providing for periodical pay-
            ment for such supplies within the meaning of Regulation
            26 of the Value Added Tax (General) Regulations 1985 or
            any amendment or re-enactment thereof.

Alternative provisions to clauses A2 to A3.2 inclusive

A1.1   .1 Where it is stated in the Appendix pursuant to clause
            5.5 of the Conditions that clause A1.1 of this Agreement
            applies clauses A2 to A3.2 inclusive hereof shall not
            apply unless and until any notice issued under clause
            A1.4 hereof becomes effective or unless the Contractor
            fails to give the written notice required under clause
            A1.2. Where Clause A1.1 applies clauses A1 and A4 to
            A12 of this Agreement remain in full force and effect.

A1.1   .2 Not later than 7 days before the date for the issue of
            the first certificate for interim payment the Contractor
            shall give written notice to the Employer, with a copy to
            the Architect/the Contract Administrator, of the rate of
            tax chargeable on the supply of goods and services for
            which certificates are to be issued. If the rate of tax so
            notified is varied under statute the Contractor shall, not
            later than 7 days after the date when such varied rate
            comes into effect, send to the Employer, with a copy to
            the Architect/the Contract Administrator, the necessary
            amendment to the rate given in his written notice and
            that notice shall then take effect as so amended.

A1.1   .3 For the purpose of complying with clause 5.5 for the
            recovery by the Contractor from the Employer of tax prop-
            erly chargeable by the Commissioners on the Contractor,
            an amount calculated at the rate given in the aforesaid
            written notice (or, where relevant, amended written
            notice) shall be shown on each certificate for interim pay-
            ment issued by the Architect/the Contract Administrator

---

[q] The application of Value Added Tax to supplies under
the Intermediate Form and other JCT Forms is explained
in Practice Note 6.

---

and, unless the procedure set out in clause A4 hereof shall have been completed, on the final certificate issued by the Architect/the Contract Administrator. Such amount shall be paid by the Employer to the Contractor or by the Contractor to the Employer as the case may be within the period for payment of certificates set out in clause 4.2 *(certificates for interim payment)* or clause 4.6 *(final certificate)* as applicable.

A1.1   .4 Either the Employer or the Contractor may give written notice to the other with a copy to the Architect/the Contract Administrator, stating that with effect from the date of the notice clause A1.1 shall no longer apply. From that date the provisions of clauses A2 to A3.2 inclusive hereof shall apply in place of clause A1.1 hereof.

Written assessment by Contractor

A2   Unless clause A1.1 applies the Contractor shall not later than the date for the issue of each certificate of interim payment and, unless the procedure set out in clause A4 shall have completed, for the issue of the final certificate for payment give to the Employer a written provisional assessment of the respective values (less 5% or 2½% applicable thereto: see clauses 4.2 and 4.3) of those supplies of goods and services for which the certificate is being issued and which will be chargeable, at the relevant time of supply under Regulation 26 of the Value Added Tax (General) Regulations 1985 on the Contractor at:

A2.1   a zero rate of tax (Category (i)) and

A2.2   any rate or rates of tax other than zero (Category (ii)).

The Contractor shall also specify the rate or rates of tax which are chargeable on those supplies included in Category (ii), and shall state the grounds on which he considers such supplies are so chargeable.

Employer to calculate amount of tax due – Employer's right of reasonable objection

A3.1   Upon receipt of such written provisional assessment the Employer, unless he has reasonable grounds for objection to that assessment, shall calculate the amount of tax due by applying the rate or rates of tax specified by the Contractor to the amount of the assessed value of those supplies included in Category (ii) of such assessment, and remit the calculated amount of such tax, together with the amount of the certificate issued by the Architect/the Contract Administrator, to the Contractor within the period for payment of certificates set out in clause 4.2, 4.3 and 4.6.

A3.2   If the Employer has reasonable grounds for objection to
        the provisional assessment he shall within 3 working
        days of receipt of that assessment so notify the
        Contractor in writing setting out those grounds. The
        Contractor shall within 3 working days of receipt of the
        written notification of the Employer reply in writing to
        the Employer either that he withdraws the assessment
        in which case the Employer is released from his
        obligation under A3.1 or that he confirms the assess-
        ment. If the Contractor so confirms then the Contractor
        may treat any amount received from the Employer in
        respect of the value which the Contractor has stated to
        be chargeable on him at a rate or rates of tax other than
        zero as being inclusive of tax and issue an authenticated
        receipt under clause A5.

**Written final statement – VAT liability of Contractor –
recovery from Employer**

A4.1   Where clause A1.1 is operated clause 4 only applies if no
        amount of tax pursuant to clause A1.1.3 has been shown
        on the final certificate issued by the Architect/the Con-
        tract Administrator. After the issue of the certificate
        under clause 2.10 *(discharge of obligations of Contractor
        on defects)* the Contractor shall as soon as he can finally
        so ascertain prepare a written final statement of the
        respective values of all supplies of goods and services for
        which certificates have been or will be issued which are
        chargeable on the Contractor at:

A4.1   .1 a zero rate (Category (i)) and

A4.1   .2 any rate or rates of tax other than zero (Category (ii))

        and shall issue such final statement to the Employer.

        The Contractor shall also specify the rate or rates of tax
        which are chargeable on the value of those supplies
        included in Category (ii) and shall state the grounds on
        which he considers such supplies are so chargeable.

        The Contractor shall also state the total amount of tax
        already received by the Contractor for which a receipt or
        receipts under clause A5 have been issued.

A4.2   The statement under clause A4.1 may be issued either
        before or after the issue of the final certificate for
        payment under clause 4.6.

A4.3   Upon receipt of the written final statement the
        Employer shall, subject to clause A7, calculate the final
        amount of tax due by applying the rate or rates of tax
        specified by the Contractor to the value of those supplies

included in Category (ii) of the statement and deducting therefrom the total amount of tax already received by the Contractor specified in the statement and shall pay the balance of such tax to the Contractor within 28 days from receipt of the statement.

A4.4   If the Employer finds that the total amount of tax specified in the final statement as already paid by him exceeds the amount of tax calculated under clause A4.3 the Employer shall so notify the Contractor who shall refund such excess to the Employer within 28 days of receipt of the notification, together with a receipt under clause A5 showing the correction of the amounts for which a receipt or receipts have previously been issued by the Contractor.

Contractor to issue receipt as tax invoice

A5   Upon receipt of any amount paid under certificates of the Architect/the Contract Administrator and any tax properly paid under the provisions of clause A2 or clause A1.1 the Contractor shall issue to the Employer a receipt of the kind referred to in Regulation 12(4) of the Value Added Tax (General) Regulations 1985 containing the particulars required under Regulation 13(1) of the aforesaid Regulations or any amendment or re-enactment thereof to be contained in a tax invoice.

Value of supply-liquidated damages to be disregarded

A6.1   If, when the Employer is obliged to make payment under clause A3 or A4 he is empowered under clause 2.7 to deduct any sum calculated at the rate stated in the Appendix as liquidated damages from sums due or to become due to the Contractor under this Contract he shall disregard any such deduction in calculating the tax due on the value of goods and services supplied to which he is obliged to add tax under clause A3 or A4.

A6.2   The Contractor when ascertaining the respective values of any supplies of goods and services for which certificates have been or will be issued under the Conditions in order to prepare the final statement referred to in clause A4 shall disregard when stating such values any deduction by the Employer of any sum calculated at the rate stated in the Appendix as liquidated damages under clause 2.7.

A6.3   Where clause A1.1 is operated the Employer shall pay the tax to which that clause refers notwithstanding any deduction which the Employer may be empowered to make under clause 2.7 from the amount certified by the Architect/the Contract Administrator in a certificate for interim payment or from any amount certified by the Architect/the Contract Administrator as due to the Contractor under the final certificate.

Employer's right to challenge tax claimed by Contractor

A7.1   If the Employer disagrees with the final statement issued by the Contractor under clause A4 he may but before any payment or refund becomes due under clause A4.3 or A4.4 request the Contractor to obtain the decision of the Commissioners on the tax properly chargeable on the Contractor for all supplies of goods and services under this Contract and the Contractor shall forthwith request the Commissioners for such decision. If the Employer disagrees with such decision then, provided the Employer indemnifies and at the option of the Contractor secures the Contractor against all costs and other expenses, the Contractor shall in accordance with the instructions of the Employer make all such appeals against the decision of the Commissioners as the Employer shall request. The Contractor shall account for any costs awarded in his favour in any appeals for which clause A4 applies.

A7.2   Where, before any appeal from the decision of the Commissioners can proceed, the full amount of the tax alleged to be chargeable on the Contractor on the supply of goods and services under the Conditions must be paid or accounted for by the Contractor, the Employer shall pay to the Contractor the full amount of tax needed to comply with any such obligation.

A7.3   Within 28 days of the final adjudication of an appeal (or of the date of the decision of the Commissioners if the Employer does not request the Contractor to refer such decision to appeal) the Employer or the Contractor, as the case may be, shall pay or refund to the other in accordance with such final adjudication any tax underpaid or overpaid, as the case may be, under the provisions of this Supplemental Condition and the provisions of clause A4.4 shall apply in regard to the provision of authenticated receipts.

Discharge of Employer from liability to pay tax to the Contractor

A8   Upon receipt by the Contractor from the Employer or by the Employer from the Contractor as the case may be, of any payment under clause A4.3 or A4.4 or where clause A1.1 of this Agreement is operated of any payment of the amount of tax shown upon the final certificate issued by the Architect/the Contract Administrator or upon final adjudication of any appeal made in accordance with the provisions of clause A4 and any resultant payment or refund under clause A4.3 or A4.4, the Employer shall be discharged from any further liability to pay tax to the Contractor in accordance with this Supplemental Condition. Provided always that if after the date of discharge under clause A8 the Commissioners decide to correct the tax due from the Contractor on the supply to the Employer of any goods and services by the Contractor under this Contract the amount of any such correction shall be an additional payment by the Employer to the Contractor or by the Contractor to the Employer, as the case may be. The provisions of clause A4 in regard to disagreement with any decision of the Commissioners shall apply to any decision referred to in this proviso.

Awards by arbitrator or court

A9   If any dispute or difference is referred to an Arbitrator appointed under clause 9 or to a court then insofar as any payment awarded in such arbitration or court proceedings varies the amount certified for payment for goods or services supplied by the Contractor to the Employer under this Contract or is an amount which ought to have been so certified but was not so certified then the provisions of this Agreement shall so far as relevant and applicable apply to any such payments.

Arbitration provision excluded

A10   The provisions of article 5 *(Arbitration)* shall not apply to any matters to be dealt with under clause A4.

Employer's right where receipt not provided

A11   Notwithstanding any provisions to the contrary elsewhere in the Conditions the Employer shall not be obliged to make any further payment to the Contractor under the Conditions if the Contractor is in default in providing the receipt referred to in clause A5: provided that clause A11 shall only apply where:

A11.1   the Employer can show that he requires such receipt to validate any claim for credit for tax paid or payable under this Supplemental Condition which the Employer is entitled to make to the Commissioners, and

A11.2   the Employer has

paid tax in accordance with the provisional assessment of

the Contractor under clause A2 unless he has sustained
a reasonable objection under clause A3.2 of this Agreement; or

paid tax in accordance with clause A1.1 of this
Agreement.

A12   Where clause 7.4 becomes operative there shall be added
to the amount allowable or payable to the Employer in
addition to the amounts certified by the Architect/the
Contract Administrator any additional tax that the
Employer has had to pay by reason of determination
under clause 7.1, 7.2 or 7.3 as compared with the tax the
Employer would have paid if the determination had not
occurred.

COMMENTARY ON CLAUSE 5.5

It is not proposed in this book to deal in any detail with the question
of Value Added Tax. The tax was originally introduced under the
Finance Act 1972. It has since been significantly amended on a
number of occasions, the most recent changes coming into effect on
1 April 1989.

For the most part new buildings as well as works of alteration will
attract value added tax at the standard rate. There will however still
be some occasions on which the appropriate value added tax is at the
zero rate e.g. domestic residences and buildings for charitable purposes. Accordingly there will also on occasions be situations where
part of the work is standard rated and part zero rated. This is
considered again below.

The contract sum is exclusive of value added tax and the employer
must pay to the contractor any value added tax properly chargeable
by the Commissioners of Customs and Excise on the supply to the
employer of any goods or services by the contractor under the
contract and in the manner set out in Supplemental Condition A.

In the event that after the base date the supply of goods and
services to the employer becomes exempt from value added tax, this
would prejudice the contractor's position in that he would lose the
credit of the input tax on the supply to him of goods and services. In
such a case the employer must pay the contractor a sum equal to the
loss of any such input tax credit.

The changes introduced on 1 April 1989

Prior to 1 April 1989, Supplemental Condition A provided a system of provisional assessment by the contractor whereby each month he had to analyse the supply to ascertain what proportion of the supply was chargeable on him at the standard rate and what proportion was chargeable on him at the zero rate.

With new buildings of a non-residential nature being standard rated, the split between standard and zero rate will now be less frequent though it will still be necessary from time to time e.g. an extension to a school building which includes a caretaker's flat.

Because it is now likely in many instances that the whole of the supply is either standard rated or zero rated, an alternative mechanism has been introduced into Supplemental Condition A. This does not replace the existing system which remains for use where applicable. The alternative system will apply where at the outset the contractor can be confident that the whole of the supply is either standard rated or zero rated. Where there is or may be a combination of these, the previous system of monthly assessments by the contractor will apply with the contractor having to supply the employer, once he is aware of the amount of the certificate, with a provisional assessment of the value upon which value added tax is due with the employer having himself to calculate the amount of value added tax thereon.

Whichever system is used there is still provision for issue by the contractor, when he receives the VAT, of a receipt in statutory form (the authenticated receipt) for that VAT for use by the employer as an input credit voucher.

The alternative system
As stated above the alternative mechanism is for use where the contractor is sure that the whole of the supply will either be standard rated or zero rated. In such a case, the new appendix entry for clause 5.5 can be completed so that it provides for clause A1.1 of Supplemental Condition A to apply, by deleting 'does not apply' from the appendix entry. The effect of the new clause A1.1 of Supplemental Condition A can be summarised as follows:

(1) The contractor must notify the employer with a copy to the architect before the issue of the first certificate for interim payment of the appropriate rate of VAT to apply to the tax exclusive amounts shown in payment certificates, both interim and final. If the rate of tax given in this notification is changed, currently either 15% (standard rate) or nil % (zero rate) then a further notification is required;

(2) Each payment certificate will have shown on it an amount calculated at the rate of tax notified by the contractor. By this means the employer receives certificates showing the VAT exclusive amount as certified by the architect together with an amount of VAT chargeable thereon. He will pay one total sum;

(3) Either party by written notice to the other can revoke the use of clause A1.1 of Supplemental Condition A and thereafter the existing provisional assessment provisions apply. This could happen for instance if variations were introduced resulting in part of the contract works being supplied at the standard rate and part at the zero rate;

(4) The architect is not involved in certifying what rate of VAT is due from the employer since the architect's role is limited to certification of the VAT exclusive amount to which the architect then simply by a matter of arithmetic applies the percentage VAT notified by the contractor.

Clause 5.6

Statutory tax deduction scheme: Supplemental Condition B

5.6 Where at the Base Date the Employer was a 'contractor', or where at any time up to the issue and payment of the final certificate for payment the Employer becomes a 'contractor' for the purposes of the statutory tax deduction scheme referred to in Supplemental Condition B, that Condition shall be operated.

B   Statutory tax deduction scheme

[r]   Clause 5.6

B1.1   In this clause 'the Act' means the Finance (No.2) Act 1975 amended by the Finance Act 1980; 'the Regulations' means the Income Tax (Sub-Contractors in the Construction Industry) Regulations 1975 S.I. No. 1960, S.I. 1980 No. 1135 and the Income Tax (Construction Operations) Order 1980 S.I. No. 171; 'contractor' means a person who is a contractor for the purposes of the Act

---

[r] The application of the Tax Deduction Scheme and these provisions is explained in JCT Practice Note 8.

and the Regulations; 'evidence' means such evidence as is required by the Regulations to be produced to a 'contractor' for the verification of a 'sub-contractor's' tax certificate; 'statutory deduction' means the deduction referred to in section 69(4) of the Act or such other deduction as may be in force at the relevant time; 'sub-contractor' means a person who is a sub-contractor for the purposes of the Act and the Regulations; 'tax certificate' is a certificate issuable under section 70 of the Act.

Provision of evidence – tax certificate

B2.1 Not later than 21 days before the first payment becomes due under clause 4.2 or after the Employer becomes a 'contractor' as referred to in clause 5.6 the Contractor shall:

either
B2.1 .1 provide the Employer with the evidence that the Contractor is entitled to be paid without the statutory deduction;

or
B2.1 .2 inform the Employer in writing, and send a duplicate copy to the Architect/the Contract Administrator, that he is not entitled to be paid without the statutory deduction.

B2.2 If the employer is not satisfied with the validity of the evidence submitted in accordance with clause B2.1.1 hereof, he shall within 14 days of the Contractor submitting such evidence notify the Contractor in writing that he intends to make the statutory deduction from payments due under this Contract to the Contractor who is 'a sub-contractor' and give his reasons for that decision. The Employer shall at the same time comply with clause B5.1.

Uncertificated Contractor obtains tax certificate

B3.1 Where clause B2.1.2 applies, the Contractor shall immediately inform the Employer if he obtains a tax certificate and thereupon clause B2.1.1 shall apply.

Expiry of tax certificate

B3.2 If the period for which the tax certificate has been issued to the Contractor expires before the final payment is made to the Contractor under this Contract the Contractor shall, not later than 28 days before the date of expiry:

either
B3.2   .1 provide the Employer with evidence that the
Contractor from the said date of expiry is entitled to be
paid for a further period without the statutory deduction in which case the provisions of clause B2.2 hereof
shall apply if the Employer is not satisfied with the
evidence;

or
B3.2   .2 Inform the Employer in writing that he will not be
entitled to be paid without the statutory deduction after
the said date of expiry.

Cancellation of tax certificate

B3.3   The Contractor shall immediately inform the Employer
in writing if his current tax certificate is cancelled and
give the date of such cancellation.

Vouchers

B4   The Employer shall as a 'contractor', in accordance with
the Regulations send promptly to the Inland Revenue
any voucher which, in compliance with the Contractor's
obligations as a 'sub-contractor' under the Regulations,
the Contractor gives to the Employer.

Statutory deduction – direct costs of materials

B5.1   If at any time the Employer is of the opinion (whether
because of the information given under clause B2.1.2 of
this clause or of the expiry or cancellation of the
Contractor's tax certificate or otherwise) that he will be
required by the Act to make a statutory deduction from
any payment due to be made the Employer shall
immediately so notify the Contractor in writing and
require the Contractor to state not later than 7 days
before each future payment becomes due (or within 10
days of such notification if that is later) the amount to
be included in such payment which represents the
direct cost to the Contractor and any other person of
materials used in carrying out the Works.

B5.2   Where the Contractor complies with clause B5.1 he
shall indemnify the Employer against loss or expense
caused to the Employer by any incorrect statement of
the amount of direct cost referred to in that clause.

B5.3   Where the Contractor does not comply with clause B5.1
the Employer shall be entitled to make a fair estimate of
the amount of direct cost referred to in that clause.

Correction of errors

B6 Where any error or omission has occurred in calculating or making the statutory deduction the Employer
shall correct that error or omission by repayment to, or
by deduction from payments to, the Contractor as the
case may be subject only to any statutory obligation on
the Employer not to make such correction.

Relation to other Clauses of the Contract

B7 If compliance with this Supplemental Condition involves the Employer or the Contractor in not complying with any other provisions of the Contract, then
the provisions of this Supplemental Condition shall
prevail.

Application of arbitration agreement

B8 The provisions of Article 5 (arbitration) shall apply to
any dispute or difference between the Employer or the
Architect/the Contract Administrator on his behalf and
the Contractor as to the operation of this Supplemental
Condition except where the Act or the Regulations or
any other Act or Parliament or statutory instrument
rule or order made under an Act of Parliament provide
for some other method of resolving such dispute or
difference.

COMMENTARY ON CLAUSE 5.6

It is not proposed to deal in this book with the provisions of the
statutory tax deduction scheme. Reference should be made to the
Finance (No. 2) Act 1975 as amended by the Finance Act 1980,
together with the regulations made thereunder, and to text books on
the subject. Further assistance can be obtained from publications
issued by the Inland Revenue and also from JCT Practice Note 8.

Chapter 9

Injury, damage and insurance

CONTENT

This chapter considers section 6 which deals with indemnities and insurance. It covers personal injury and injury or damage to property arising out of the carrying out of the contract works, and indemnities and insurances related thereto. It also deals with loss or damage to the contract works themselves and insurances in respect of this.

SUMMARY OF GENERAL LAW

(A) Loss or damage to the contract works

As a general principle, subject to relatively few exceptions, a contractor's obligation to complete the works means that he will be obliged, at his own cost, to repair or reinstate the contract works should they be damaged or destroyed before completion. The main exception to this is in relation to damage or destruction which is sufficiently fundamental to cause the contract to become frustrated in law, in which case the contractor will be excused further performance. For a brief discussion of the doctrine of frustration see page 22. In practice, a building contract will often specify expressly who is to bear the risks in the event of loss or damage to the works before completion. If the words used are sufficiently clear, the contract can enable a party whose negligence has caused the loss or damage to nevertheless impose the responsibility for such loss or damage on the other contracting party. A case in point is *Farr* v. *The Admiralty* (1953).

Facts:

The plaintiffs agreed to build a destroyer wharf on behalf of the defendants. The contract provided that the plaintiff should be responsible for, and should make good, any loss or damage arising from any cause whatsoever. A vessel belonging to the defendants collided with and damaged the works as a result of negligent navigation.

Held:

The words 'any cause whatsoever' included negligent navigation of a ship by the defendant's employee and the plaintiffs were accordingly liable under the contract to make good the damage at their own cost.

A frequent means by which the risk of loss or damage to the contract works is imposed upon one of the contracting parties, even if caused by the other, is by the use of the phrase 'sole risk'. Used in the right circumstances this will not only relieve a negligent party from responsibility for the loss or damage caused by his own negligence, but will impose the risk upon the innocent party. There can be very good reasons for this related to insurance of the works. A good example of this is to be found in the House of Lords' case of *Scottish Special Housing Association* v. *Wimpey Construction (UK) Ltd* (1986).

Facts:

This case concerned a JCT 63 Contract (1977 Revision). Clause 20(c) dealing with insurance of the works applied. That clause provided that the existing structures together with the contract work itself 'shall be at the sole risk of the Employer as regards loss or damage by fire . . .'. In the course of carrying out work on the modernisation of 128 houses in Edinburgh, one of the houses was damaged by fire. The question which came before the House of Lords was, whether, assuming for the point of argument that the contractor had been negligent, he was liable to the employer for the damage resulting from the fire.

Held:

The employer was to bear the sole risk of damage by fire including fire caused by the negligence of the contractor or his sub-contractors. This view of clause 20(c) was reinforced by a consideration of clause 18(2) which provided:

'Except for such loss or damage as is at the risk of the Employer under clause . . . 20(c) . . . the Contractor shall be liable for and shall indemnify the Employer against, any expense, liability, loss, claim . . .'

In holding that the contractor was not responsible for his own negligence in damaging the works, the House of Lords followed the earlier English case of *James Archdale & Co. Ltd* v. *Comservices Ltd* (1954).

While it may seem odd that a contractor should be excused from liability even in respect of his own negligence, this is understandable when the insurance background to many of these contractual provisions is considered. The contractual insurance provisions in JCT 63 and indeed in IFC 84 prior to Amendment 1 of November 1986 gave three options, namely:

(1) The contractor to insure for the benefit of both contractor and employer. This applied to new building works and if this option was adopted, even if the risk which materialised was due to negligence on the part of either contractor or employer, as they would both be covered by the policy of insurance, there was no need to provide for one party accepting the sole risk of loss or damage. The main purpose behind including a provision that one party should take the sole risk of loss or damage to the works was so that the insurers, in respect of such loss or damage, having paid out under an insurance claim could not seek recompense from the negligent party. In other words, any right of subrogation which the insurers had was removed by the use of such a phrase. Clearly this elimination of the right of subrogation was generally beneficial as the existence of any subrogation rights would have lead both parties to take out insurance cover in respect of negligent damage to the works which would in insurance terms be both inefficient and uneconomical. If a works policy is taken in joint names then as both contractor and employer are protected by the policy there will be no opportunity for subrogation rights to apply;

(2) The employer to insure (except if a local authority) for the benefit of both employer and contractor. Again, as both parties were covered it would not matter if there was negligence and no rights of subrogation could arise;

(3) In relation to existing structures and work done to them, no one was required to insure and the employer was to take the sole risk. Here, the employer might well have taken out insurance cover and if the employer then claimed under his own insurance policy, the purpose of placing the 'sole risk' with the employer was to prevent the insurer paying out the employer and seeking by way of subrogation to claim against the negligent contractor.

The amended IFC 84 insurance provisions are dealt with in the COMMENTARY ON CLAUSES 6.3 to 6.3C (see page 316).

The insertion of a 'sole risk' provision in relation to loss or damage to contract works in a main contract, expressed to cover sub-contractors also, has had the effect of removing any duty of care owed by the sub-contractor to the employer, thus relieving the sub-contractor from liability for negligently damaging the contract works: *Norwich City Council* v. *Paul Clarke Harvey and Another* (1988): also *Welsh Health Technical Services Organisation* v. *Haden Young Ltd and IDC* (1987).

However, if a party is required to bear the risk rather than the sole risk, this will not absolve the other party from liability in respect of that other party's negligence. Such a clause, being akin to an exclusion clause, will be interpreted strictly against the party relying upon it – *Dorset County Council* v. *Southern Felt Roofing Co. Ltd* (1989).

The main contract may also expressly provide that sub-contractors are to be covered by the works policy or alternatively that any subrogation rights against them in favour of the insurers are waived, for example, clause 6.3.3 of IFC 84.

The existence of insurance cover against loss or damage to the works is of potential benefit to both parties. For the contractor, it guarantees a fund out of which he can meet his primary obligation to complete the contract works notwithstanding their damage or destruction. For the employer, it ensures that his right to have the works completed at no additional cost to him, despite their damage or destruction, is not merely a legal remedy with no substance, as it could be if the contractor was both uninsured and impecunious.

The extent of the insurance cover required will depend on the express terms of the contract. For example, it may require what is often termed 'all risks' cover or alternatively it may be more limited covering only the most frequent and potentially serious risks e.g. fire and flood.

Very often the contract will contain exceptions to the contractor's obligation to insure in respect of certain risks. Generally these arise either because insurance cover is unobtainable e.g. loss or damage due to nuclear explosions; or because the loss or damage has been caused by the employer or by someone for whom he is, for this purpose, responsible e.g. the architect's design failure where design does not form part of the contractor's obligation.

(B)   Third party claims: indemnities and insurance

Most standard forms of building contract will contain express clauses dealing with third party claims, i.e. third party claims from the employer's point of view, arising out of the carrying out of the contract works. In other words, the situation in which a stranger to the contract claims against the employer in respect of a wrong done to the stranger in the form of personal injury or physical damage to his property arising out of the carrying out of the works by the contractor. Typically the express provision will require the contractor in certain circumstances to indemnify the employer against any loss, damage, claims, proceedings, costs etc. for which the employer becomes responsible to the third party. This will generally be coupled with an obligation on the contractor to insure himself against liability in respect of such claims. That liability – and accordingly the insurance to cover it – may well extend to sub-contractors employed by a main contractor.

These indemnity provisions have tended to be construed by the courts very strictly against the party seeking to rely on them, especially where the provision would have the effect of indemnifying one party to the contract in respect of his own negligence. Furthermore, a person dealing as consumer – i.e. if he does not make the contract in the course of a business whereas the other party does – cannot be made to indemnify the other party in respect of that other party's liability for negligence or breach of contract, except in so far as such indemnity satisfies the requirement of reasonableness – see sections 4 and 12 of the Unfair Contract Terms Act 1977. The Act does not purport to control indemnity provisions in non-consumer transactions so that the case law on this topic prior to the Act is still very relevant.

In *Walters* v. *Whessoe Ltd and Shell Refining Co. Ltd* (1960) the Court of Appeal had to consider an indemnity clause in a contract between the first and second defendants following a finding that the second defendants had been negligent. The second defendants sought to recover the damages payable by them to the plaintiffs from the first defendants under an indemnity clause, despite the fact that the second defendants had themselves been at fault when an industrial accident caused the death of an employee of the first defendants. The clause required that the first defendants:

'shall indemnify and hold Shell [the second defendants] their servants and agents free and harmless against all claims arising out of the operations being undertaken by [the first defendants] in pursuance of this contract or order or incidental thereto . . .'.

The court held that the reference to 'all claims' did not indemnify Shell against the results of their own negligence. Lord Justice Sellars said:

'It is well established that indemnity will not lie in respect of loss due to a person's own negligence or that of his servants unless adequate and clear words are used or unless the indemnity could have no reasonable meaning or application unless so applied.'

Lord Justice Devlin said:

'It is now well established that if a person obtains an indemnity against the consequences of certain acts, the indemnity is not to be construed so as to include the consequences of his own negligence unless those consequences are covered either expressly or by necessary implication. They are covered by necessary implication if there is no other subject matter upon which the indemnity could operate.'

In *Canada Steam Ship Lines Ltd* v. *The King* (1952), a Privy Council case, Lord Morton of Henryton set out the approach of the courts to indemnity (and exclusion) clauses as follows:

'(1)   If the clause contains language which expressly exempts the person in whose favour it is made (hereafter called "the *proferens*") from the consequence of the negligence of his own servants, effect must be given to that provision . . . .

(2)   If there is no express reference to negligence, the court must consider whether the words used are wide enough, in their ordinary meaning, to cover negligence on the part of the servants of the *proferens* . . . .

(3)   If the words used are wide enough for the above person, the court must then consider whether "the head of damage may be based on some ground other than negligence", to quote again Lord Greene in the *Alderslade* case. The "other ground" must not be so fanciful or remote that the *proferens* cannot be supposed to have desired protection against it, but subject to this qualification, which is no doubt to be implied from Lord Greene's words,

the existence of a possible head of damage other than that of negligence is fatal to the *proferens* even if the words are *prima facie* wide enough to cover negligence on the part of his servants.'

This test was subsequently applied in the case of *Smith* v. *South Wales Switchgear Co. Ltd* (1978) and again more recently by the Court of Appeal in *Dorset County Council* v. *Southern Felt Roofing Co. Ltd* (1989). However, the wording may of course be such as to make it clear that one contracting party is taking the risk of loss or damage even if caused by the other contracting party's negligence: see *Farr* v. *The Admiralty* (1953), page 288.

Where the indemnity provision does not expressly or by implication cover losses consequent on a person's own negligence, if the loss is due in part to the fault of the contracting party against whom an indemnity is being sought, and in part due to the fault of the party seeking to enforce the indemnity, then, unless the indemnity provision expressly provides for an apportionment to be made, the indemnity provision will fail in its entirety – *A.M.F. (International) Ltd* v. *Magnet Bowling Ltd and G. P. Trentham Ltd* (1968).

The application of the statutory limitation periods in connection with actions on indemnities should be noted. The limitation period under the Limitation Act 1980 will operate from the date when the cause of action on the indemnity accrued i.e. most probably when the person claiming the indemnity actually incurred a liability e.g. when the third party obtains judgment against him. In the case of *County and District Properties Ltd* v. *C. Jenner & Sons and Others* (1976) Mr Justice Swanwick said:

'. . . an indemnity against a breach, or an act, or an omission, can only be an indemnity against the harmful consequences that may flow from it, and I take the law to be that the indemnity does not give rise to a cause of action until those consequences are ascertained.'

If this is a correct statement of the law, in a typical third party claim, while it is the date of the damage to the third party plaintiff which will start time running against the defendant, if that defendant in turn seeks to rely on an indemnity against another defendant, time will not start to run until the first defendant's liability has been ascertained. Only then will his cause of action under the indemnity provisions accrue. However, a contrary view, namely that the cause of action accrues, and therefore time begins to run, as soon as the original loss, damage, claim etc. has been incurred, suffered or made, has received some support as a result of the House of Lords' case of *Scott Lithgow Ltd* v. *Secretary of State for Defence* (1989).

As a matter of convenience and expediency, a defendant will be able to join into the proceedings between the plaintiff and the defendant the person from whom an indemnity is being sought so that if any liability does attach to the defendant, the question of the indemnity can be considered immediately thereafter – see Rules of the Supreme Court Order 16.

The use of contractual indemnities have a number of other advantages or disadvantages depending upon whether the party concerned is the beneficiary under an indemnity or the giver of it. For example, if the act in respect of which the indemnity is given is a negligent act, the indemnity may be capable of extending to the recovery of purely economic loss, even though this may not otherwise have been recoverable against the negligent party. Additionally, the ordinary contractual rules as to remoteness of damage may not apply to the like extent in relation to indemnities which may be worded so as to expressly cover all loss, expense etc. resulting from the act complained of, whether or not the nature of that loss or expense is too remote in terms of ordinary contractual principles as to the measure of damages.

CONSIDERATION OF THE RELEVANT CLAUSES OF IFC 84

INTRODUCTION

As originally published, IFC 84 contained indemnity and insurance clauses which were virtually identical to those contained in JCT 63 and subsequently in JCT 80. These clauses, which have become very familiar to those connected with the building industry, were then thoroughly reviewed and substantially revised, the revisions being introduced in almost identical form in most forms of building contract published by the Tribunal. Amendment 1 to IFC 84 covering amendments to the 'insurance and related liability provisions' was published in November 1986.

Unfortunately, many professionals who are involved with building contracts do not, in general, give this subject the attention it deserves. Their reluctance to become too deeply involved probably stems from the fact that they consider insurance to be secondary in importance to the main task of designing and constructing the works. Lack of interest may also stem from the fact that there is insufficient standardisation in the scope of insurance cover and in policy wordings and, although there has been some movement towards simplification in recent years, the drafting of many policies remains somewhat complex. However, the importance of the subject to the employer and, indeed to all those connected with the building industry cannot be overstressed.

Some working knowledge of the subject is essential. Furthermore, the employer or the employer's insurance brokers must be involved if there is any doubt at all and probably involved as a matter of routine, when the project is in any way complex, in order to ensure that there is adequate insurance cover.

Clause 6.1

Injury to persons and property and indemnity to Employer

6.1.1   The Contractor shall be liable for, and shall indemnify the Employer against, any expense, liability, loss, claim or proceedings whatsoever arising under any statute or at common law in respect of personal injury to or the death of any person whomsoever arising out of[1] or in the course of or caused by the carrying out of the Works, except to the extent that the same is due to any act or neglect[2] of the Employer or of any person for whom the Employer is responsible including the persons employed or otherwise engaged by the Employer to whom clause 3.11 refers.

6.1.2   The Contractor shall, subject to clause 6.1.3 and where applicable, clause 6.3C.1, be liable for, and shall indemnify the Employer against, any expense, liability, loss, claim or proceedings in respect of any injury or damage whatsoever to any property real or personal in so far as such injury or damage arises out of or in the course of or by reason of the carrying out of the Works, and to the extent that the same is due to any negligence, breach of statutory duty, omission or default of the Contractor, his servants or agents or of any person employed or engaged upon or in connection with the Works or any part thereof, his servants or agents or of any other person who may properly be on the site upon or in connection with the Works or any part thereof, his servants or agents, other than the Employer or any person employed, engaged or authorised by him or by any local authority or statutory undertaker executing work solely in pursuance of its statutory rights or obligations.

6.1.3   The reference in clause 6.1.2 to 'property real or personal' does not include the Works, work executed and/or Site Materials up to and including the date of issue of the certificate of Practical Completion or up to and including the date of determination of the employment of the Contractor (whether or not the validity of that determination is disputed) under clause 7.1 to 7.3 or clause 7.5 or 7.6 or 7.8 or, where clause 6.3C applies, under clause 6.3C.4.3 or clause 7.1 to 7.3 or clause 7.5 or 7.6 or 7.8 whichever is the earlier.

COMMENTARY ON CLAUSE 6.1

Injury to persons

For a summary of the general law in relation to indemnities see page 292.

Clause 6.1.1 deals with the indemnity of the employer by the contractor where the employer suffers any expense, liability, loss or claim whatsoever arising under any statute or at common law in respect of personal injury or death of any person arising out of or which is caused by the carrying out of the works. If the employer is held liable to a third party either under statute or at common law, the contractor must indemnify the employer in respect of such a claim but not 'to the extent' that the claim may be due to the act or neglect of the employer or any person for whom the employer is responsible.

Clearly, if the loss sustained by the employer is entirely due to his own 'act or neglect' or the act or neglect of those for whom he is responsible, then no liability rests with the contractor. If, however, the employer is, or those for whom he is responsible are, only partly at fault, then the words 'except to the extent that' mean that liability will be apportioned between the parties, the degree of apportionment reflecting their relative liability. The onus of proving that the employer should accept any part of the liability rests with the contractor. Generally, the employer would not be responsible for independent contractors appointed by him providing reasonable care was taken in their selection. However, it is expressly provided that the employer is responsible for such contractors – see also the second paragraph to clause 3.11 of IFC 84.

It might be thought that the architect is a person for whom the employer is responsible in this connection but this is by no means certain. In *Acrecrest* v. *Hattrell and London Borough of Harrow*

(1982), the Court of Appeal held that the negligence of the architect in failing to ensure compliance of the works with building regulations should not be attributable to the building owners. It might appear therefore that in certain circumstances negligence on the part of the architect which leads to a third party claim for damages in respect of personal injury against the employer could fall within clause 6.1.1, enabling the employer to obtain an indemnity from the contractor on the basis that the architect was not a person for whom the employer is responsible. However, such a contention is, it is submitted, open to some doubt.

Injury or damage to property

Clause 6.1.2 deals with the indemnity of the employer by the contractor against any expense, liability, loss, claim or proceedings in respect of any injury or damage to property other than personal injury which is covered under clause 6.1.1. While the word 'damage' has a reasonably precise meaning, the use of the term 'injury' is not so precise and there is consequently a question as to whether it could be held to apply to, say, interference with the reasonable access by third parties to their property or to a similar situation where the use of a third party's building is seriously affected. There is a possibility that it could also extend to such matters as the infringement of a patent. There is no specific clause in IFC 84 dealing with the infringements of patents and like rights as there is in JCT 80 (clause 9).

The basis of indemnity under this clause differs from that provided under clause 6.1.1. The contractor is only liable to indemnify the employer under this clause to the extent that the loss can be shown to be due to the negligence, breach of statutory duty, omission or default of the contractor or those for whom he is responsible as defined by the clause. This will certainly include sub-contractors, sub-sub-contractors and so on, as well as suppliers. It does not include a local authority or statutory undertaker executing work solely pursuant to its statutory rights or obligations, though they may well be included if they have any contractual relationship with the contractor.

If the injury or damage is not due to such negligence, breach of statutory duty, omission or default, this indemnity is of no effect and the loss will fall upon the employer, unless the employer can recover all or some of the loss from the contractor in some other way under the general law, e.g. in contribution proceedings where the employer suffers loss as a result of being sued by someone in respect of some tortious act of the contractor which does not involve any negligence, breach of statutory duty, omission or default such as nuisance or trespass – see section 1 Civil Liability (Contribution) Act 1978.

The use of the words 'to the extent that' means that even if the employer is partly at fault the indemnity will still apply, but the loss will be apportioned between the parties, the degree of apportionment reflecting the relative responsibility of the contractor.

By clause 6.1.3, the injury or damage to 'any property real or personal' covered by clause 6.1 does not apply to the works themselves or to materials on site, both of which are covered by works insurance under clause 6.3A, 6.3B or 6.3C to which reference is made later. Nevertheless, if part of the works have reached practical completion, for instance in the case of partial possession where the optional clause 2.11 is implemented, then the indemnity provided by clause 6.1.2 will be applicable to that part of the works. However, it must be remembered that the indemnity is limited to loss resulting from 'negligence, breach of statutory duty, omission or default of the contractor, or those for whom he is responsible'. The employer will therefore still need to consider insuring that part of the works which has reached practical completion even if he has no contractual obligation to do so since any loss may well fall outside the indemnity provisions of this clause.

Clause 6.1.2 is expressed to be subject not only to clause 6.1.3, but also to clause 6.3C.1 which imposes upon the employer an obligation to insure existing structures for which he is responsible in the joint names of himself and the contractor against loss or damage from specified perils. The indemnity by the contractor in favour of the employer will not therefore extend to these existing structures if the loss or damage to them has been caused by one or more of the specified perils. This is dealt with in more detail later (page 316).

It is interesting to note with regard to clause 6.1.1 (injury to persons), that the contractor only indemnifies the employer to the extent that the injury or death is not due to any act or neglect of the employer or those for whom he is responsible; whereas here, in clause 6.1.2 (injury or damage to property), the position is reversed and the contractor's indemnity only applies to the extent that the injury or damage arises due to the negligence, breach of statutory duty, omission or default of the contractor or those for whom he is responsible. The onus of proving that the contractor should accept any part of the liability therefore rests with the employer.

NOTES TO CLAUSES 6.1.1 TO 6.1.3

[1] '... *arising out of* ...'

These words were considered in the case of *Richardson* v. *Buckinghamshire County Council and Others* (1971) in connection with the ICE Conditions of Contract 4th Edition, clause 22(1) which is an indemnity clause. The plaintiff had been injured when he fell from his motorcycle at the point of some road works which were the subject of the contract. The defendant local authority successfully resisted a claim from the motorcyclist but the costs which were incurred in so doing were irrecoverable from the plaintiff as he was legally aided. The local authority therefore sought to recover these costs from the contractor under the wording quoted above, contending that the costs were incurred as a result of a claim arising out of the construction of the contract works. It was held that the local authority was not entitled to an indemnity from the contractor as the motorcyclist's claim was not one which arose out of, or in consequence of, the construction of the works. This clearly makes good sense, for otherwise a contractor could find himself having to indemnify an employer in respect of the employer's costs in defending all manner of specious or far-fetched claims. As the motorcyclist could not establish that his injuries arose out of the construction of the road works, it could not be said that the costs arose out of the construction of the works.

[2] '... *except to the extent that the same is due to any act or neglect* ...'.

This exception appears to relate to matters where the employer is at fault, in which case it will not apply where the employer is guilty of a breach of statutory duty where such breach is independent of fault. In such a case the contractor will remain liable to indemnify the employer notwithstanding the employer's breach of statutory duty.

Clauses 6.2.1 to 6.2.3

Insurance against injury to persons or property

6.2.1   Without prejudice to his obligation to indemnify the Employer under clause 6.1 the Contractor shall take out and maintain and shall cause any sub-contractor to take out and maintain insurance which shall comply with clause 6.2.1 in respect of claims arising out of his liability referred to in clauses 6.1.1 and 6.1.2.

The insurance in respect of claims for personal injury to, or the death of any person under a contract of service or apprenticeship with the Contractor or a sub-contractor as the case may be, and arising out of and in the course of such person's employment, shall comply with the Employer's Liability (Compulsory Insurance) Act 1969 and any statutory orders made thereunder or any amendment or re-enactment thereof. For all other claims to which clause 6.2.1 applies the insurance cover shall be not less than the sum stated in the Appendix for any one occurrence or series of occurrences arising out of one event.

[j.1]

6.2.2 As and when he is reasonably required to do so by the Employer the Contractor shall send and shall cause any sub-contractor to send to the Architect/the Contract Administrator for inspection by the Employer documentary evidence that the insurances required by clause 6.2.1 have been taken out and are being maintained, but at any time the Employer may (but not unreasonably or vexatiously) require to have sent to the Architect/the Contract Administrator for inspection by the Employer the relevant policy or policies and the premium receipts therefor.

6.2.3 If the Contractor defaults in taking out or in maintaining, or in causing any sub-contractor to take out and maintain, insurance as provided in clause 6.2.1 the Employer may himself insure against any liability or expense which he may incur arising out of such default and a sum or sums equivalent to the amount paid or payable by him in respect of premiums therefor may be deducted by him from any monies due or to become due to the Contractor under this Contract or such amount may be recoverable by the Employer from the Contractor as a debt.

[j.1] The Contractor or any sub-contractor may, if they so wish, insure for a sum greater than that stated in the Appendix.

COMMENTARY ON CLAUSES 6.2.1 TO 6.2.3

While clauses 6.1.1 and 6.1.2 establish an indemnity on the part of the contractor in favour of the employer, clause 6.2.1 requires the contractor also to insure in respect of any claims arising out of the contractor's liability under clause 6.1. Without that insurance the indemnity given by the contractor may well be worthless if he is financially unable to meet his liabilities under the indemnity provisions. However, even when the contractor's liabilities under clause 6.1.1 and 6.1.2 are adequately covered by insurance, employers should remember the limited nature of that indemnity. Claims arising out of any act or neglect of the employer or of any person for whom the employer is responsible are excluded from the indemnity against injury or death to persons under clause 6.1.1 and, in the case of injury or damage to property under clause 6.1.2, the contractor is only liable if the claim arises out of his negligence, breach of statutory duty, omission or default. A separate policy taken out by the employer to protect him against claims not covered by the indemnity may therefore need to be considered.

Even where a claim made against the employer is covered by the indemnity provisions of clause 6.1.1 or 6.1.2, and where the employer has satisfied himself that the contractor's insurance under clause 6.2.1 is adequate, it would be prudent for the employer to ensure that the contractor has immediately notified his insurers if there is a claim or the possibility of a claim against him under clause 6.1.

In addition to the contractor's obligations to insure under this clause, the contractor must ensure that sub-contractors are similarly insured. With regard to the contractor's own employees, the cover provided by the policy must comply with the Employer's Liability (Compulsory Insurance) Act 1969 or any statutory orders made thereunder. For all other claims which may arise from injury or death to third parties under clause 6.1.1 or injury or damage to property under clause 6.1.2, the insurance cover must be not less than the figure stated in the appendix for any one occurrence or series of occurrences arising out of one event. It may be prudent to obtain the advice of the employer's insurers or insurance brokers before deciding upon the amount to be inserted.

Under clause 6.2.2, the contractor has an obligation to send and to cause any sub-contractor to send to the architect, documentary evidence, for inspection by the employer, that the insurance provisions of clause 6.2.1 have been complied with, and that the policies are being maintained. Care should be taken when checking the evidence submitted by the contractors since policies frequently include limitation or exclusion clauses and there is always a possibility that the policies may not comply with the obligations of the contractor under the indemnity provisions. Where there is any doubt it would be advisable for the employer to seek the advice of his insurers or insurance brokers.

While the indemnity provided by the contractor under clauses 6.1.1 and 6.1.2, and the insurance provisions under clause 6.2.1 covering those indemnities are in respect of claims arising out of the carrying out of the works, the Tribunal recommend contractors and sub-contractors not to terminate their insurance cover at the date of practical completion. The Tribunal suggests that the cover remains until expiry of any defects liability period or the date of making good of defects and possibly until the issue of a final certificate (see the Tribunal's Practice Note 22 page 34/35).

The right given to the employer under clause 6.2.3 to take out insurance in default of the contractor or a sub-contractor insuring is a useful one. The employer needs to be sure that the contractor or sub-contractor as the case may be, will be able to financially withstand any claim against him. A failure by the contractor to insure or cause a sub-contractor to insure, whilst being a breach of contract and entitling the employer to sue, will be a worthless right against an impecunious contractor without the express right to take out the necessary insurance cover. Should that become necessary, the employer can recover the cost by a deduction from any monies due to the contractor or, failing that, such amount may be recovered from the contractor as a debt.

Finally, it should be remembered that, by clause 6.2.5, the contractor is not liable either to indemnify the employer or to insure against personal injury or death or any damage loss or injury caused to any property which is attributable to what is called an 'Excepted Risk' which relates generally to pressure waves from aircraft and nuclear hazards – see clause 8.3 for the definition.

Clauses 6.2.4 and 6.2.5

Insurance – liability etc. of Employer

6.2.4   Where it is stated in the Appendix that the insurance to which clause 6.2.4 refers may be required by the Employer the Contractor shall, if so instructed by the Architect/the Contract Administrator, take out and maintain a Joint Names Policy for such amount of indemnity as is stated in the Appendix in respect of any expense, liability, loss, claim or proceedings which the Employer may incur or sustain by reason of injury or damage to any property other than the Works and Site Materials caused

by collapse, subsidence, heave, vibration, weakening or removal of support or lowering of ground water arising out of or in the course of or by reason of the carrying out of the Works excepting injury or damage:
- for which the Contractor is liable under clause 6.1.2;
- attributable to errors or omissions in the designing of the Works;
- which can reasonably be foreseen to be inevitable having regard to the nature of the work to be executed or the manner of its execution;
- which it is the responsibility of the Employer to insure under clause 6.3C.1 (if applicable);
- arising from war risks or the Excepted Risks.

Any such insurance as is referred to in clause 6.2.4 shall be placed with insurers to be approved by the Employer, and the Contractor shall send to the Architect/the Contract Administrator for deposit with the Employer the policy or policies and the receipts in respect of premiums paid.

The amounts expended by the Contractor to take out and maintain the insurance referred to in clause 6.2.4 shall be added to the Contract Sum.

If the Contractor defaults in taking out or in maintaining the Joint Names Policy as provided in clause 6.2.4 the Employer may himself insure against any risk in respect of which the default shall have occurred.

Excepted Risks
6.2.5   Notwithstanding the provisions of clauses 6.1.1, 6.1.2 and 6.2.1, the Contractor shall not be liable either to indemnify the Employer or to insure against any personal injury to or the death of any person or any damage, loss or injury caused to the Works or Site Materials, work executed, the site, or any property, by the effect of an Excepted Risk.

COMMENTARY ON CLAUSES 6.2.4 AND 6.2.5

Clause 6.2.4 covers the situation where the employer may incur or sustain any expense, liability, loss, claim or proceedings due to injury or damage to property other than the works and site materials, arising out of or in the course of or by reason of the carrying out of the works and which is caused by collapse, subsidence, heave, vibration, weakening or removal of support or lowering of ground water.

A typical example would be a claim made by the owner of an adjoining property due to damage to his property arising from piling, demolition or from basement excavation in connection with the works.

Whether insurance is necessary in order to protect the employer against such risks will clearly depend upon the situation of the site and the works. Close proximity to other buildings will represent a greater risk than building on a green field site where it is unlikely that insurance would be required. Obviously the extent to which injury or damage is likely to be caused to adjoining property will also depend on the nature of the works themselves.

It is important to remember that the indemnity provided by the contractor to the employer under clause 6.1.2 against injury or damage to property other than the works and the insurance under clause 6.2.1 to cover that indemnity, only extends to injury or damage due to the negligence, breach of statutory duty, omission or default of the contractor. The indemnity to the employer does not therefore cover injury or damage arising from any other cause. In some construction situations, that risk can be considerable and the purpose of clause 6.2.4 is therefore to provide insurance cover for both employer and contractor where such cover is deemed to be necessary.

Where such insurance cover is required by the employer, this is achieved by an entry in the appendix to IFC 84 which also provides for the level of indemnity to be stated in respect of any one occurrence or series of occurrences arising out of one event, or alternatively, for an aggregate amount. Clearly, appropriate instructions from the employer and advice from his insurers or insurance brokers should be sought.

The need for such insurance cover was highlighted by the case of *Gold* v. *Patman and Fotheringham Ltd* (1958) which led to a provision similar to clause 6.2.4 being included in JCT 63 and subsequently in JCT 80.

Facts:

A clause in an earlier RIBA form of building contract required the contractor to indemnify the employer against claims in respect of damage to property, provided that the contractor or his sub-contractor was guilty of negligence or default. The contractor was also required to insure as might be required in the bills of quantities. The bills of quantities required the contractor to insure adjoining properties against subsidence or collapse. A specialist piling sub-contractor, without any negligence or default, caused damage to adjoining properties during the piling operation. The contractor had insured himself, but not the employer, against such damage. The employer contended that the contractor was in breach of contract in failing to insure for the employer's benefit.

Held:

The wording of that clause required the contractor to insure himself and not the employer.

Where it is considered by the employer that there is a risk, then clause 6.2.4 provides for the architect to instruct the contractor to take out a policy of insurance which includes the contractor and the employer as the insured (known as a 'Joint Names Policy' – see clause 8.3) to indemnify the employer against claims and losses as defined by the clause.

No action is required from the contractor until he receives an instruction from the architect to obtain a quotation for the insurance cover from insurers approved by the employer. When the employer has approved the quotation the architect should then instruct the contractor to take out the insurance.

It is important that the cover provided conforms to the requirements of this clause and that the policy and premium receipt are sent to the architect for deposit with the employer. Because the greatest risk may well stem from the early work, it is important that the architect ensures that the policy is in place before the contractor commences work on site. If the contractor fails to insure then the employer may do so in which case the premium is clearly not then added to the contract sum which would otherwise be the case.

The insurance cover provided by this clause is by no means all-embracing and the considerable number of specific exceptions should be noted quite apart from the 'Excepted Risks'. While all the exceptions will clearly add to the employer's vulnerability, that which excludes damage 'which can reasonably be foreseen to be inevitable having regard to the nature of the work to be executed or the manner of its execution' could clearly be the cause of considerable dispute. For that reason it may well be prudent for the employer to cover the risk as part of his own insurances.

Clauses 6.3 to 6.3C

[k] **Insurance of the Works – alternative clauses**

6.3.1 Clause 6.3A or clause 6.3B or clause 6.3C shall apply whichever clause is stated to apply in the Appendix.

Definitions

6.3.2 In clauses 6.3A, 6.3B, 6.3C and, so far as relevant, in other clauses of the Conditions the following phrases shall have the meaning given below:

[l] All Risks Insurance:
means insurance which provides cover against any physical loss or damage to work executed and Site Materials excluding the cost necessary to repair, replace or rectify

1 property which is defective due to
   .1 wear and tear,
   .2 obsolescence,
   .3 deterioration, rust or mildew;

---

[k] Clause 6.3A is applicable to the erection of a new building where the Contractor is required to take out a Joint Names Policy for All Risks Insurance for the Works and clause 6.3B is applicable where the Employer has elected to take out such Joint Names Policy. Clause 6.3C is to be used for alterations of or extensions to existing structures under which the Employer is required to take out a Joint Names Policy for All Risks Insurance for the Works and also a Joint Names Policy to insure the existing structures and their contents owned by him or for which he is responsible against loss or damage thereto by the Specified Perils.

[l] The definition of 'All Risks Insurance' in clause 6.3.2 is intended to define the risks for which insurance is required. Policies issued by insurers are not standardised and there will be some variation in the way the insurance for those risks is expressed. See also Practice Note 22 and Guide, Part A.

[k.1] 2 any work executed or any Site Materials lost or damaged as a result of its own defect in design, plan, specification, material or workmanship or any other work executed which is lost or damaged in consequence thereof where such work relied for its support or stability on such work which was defective;

3 loss or damage caused by or arising from

.1 any consequence of war, invasion, act of foreign enemy, hostilities (whether war be declared or not), civil war, rebellion, revolution, insurrection, military or usurped power, confiscation, commandeering, nationalisation or requisition or loss or destruction of or damage to any property by or under the order of any government *de jure* or *de facto* or public, municipal or local authority;

.2 disappearance or shortage if such disappearance or shortage is only revealed when an inventory is made or is not traceable to an identifiable event;

.3 an Excepted Risk (as defined in clause 8.3);

and if the Contract is carried out in Northern Ireland:

.4 civil commotion;

.5 any unlawful, wanton or malicious act committed maliciously by a person or persons acting on behalf of or in connection with any unlawful association; 'unlawful association' shall mean any organisation which is engaged in terrorism and includes an organisation which at any relevant time is a proscribed organisation within the meaning of the Northern

---

[k.1] In any policy for 'All Risks Insurance' taken out under clauses 6.3A, 6.3B or 6.3C.2 cover should not be reduced by the terms of any exclusion written in the policy beyond the terms of paragraph 2; thus an exclusion in terms 'This Policy excludes all loss of or damage to the property insured due to defective design, plan, specification, materials or workmanship' would not be in accordance with the terms of those clauses and of the definition of 'All Risks Insurance'. Cover which goes beyond the terms of the exclusion in paragraph 2 may be available though not standard in all policies taken out to meet the obligation in clauses 6.3A, 6.3B or 6.3C.2; and leading insurers who underwrite All Risks cover for the Works have confirmed that where such improved cover is being given it will not be withdrawn as a consequence of the publication of the terms of the definition in clause 6.3.2 of 'All Risks Insurance'.

Ireland (Emergency Provisions) Act 1973; 'terrorism' means the use of violence for political ends and includes any use of violence for the purpose of putting the public or any section of the public in fear.

Site Materials:
means all unfixed materials and goods delivered to, placed on or adjacent to the Works and intended for incorporation therein.

Sub-contractors – benefit of Joint Names Policies – Specified Perils

6.3.3   The Contractor where clause 6.3A applies, and the Employer where either clause 6.3B or clause 6.3C applies shall ensure that the Joint Names Policy referred to in clause 6.3A.1 or clause 6.3A.3 or the Joint Names Policy referred to in clause 6.3B.1 or in clause 6.3C.1 and 6.3C.2 shall
- either provide for recognition of each sub-contractor referred to in clause 3.3 as an insured under the relevant Joint Names Policy
- or include a waiver by the relevant insurers of any right of subrogation which they may have against any such sub-contractor

in respect of loss or damage by the Specified Perils to the Works and Site Materials where clause 6.3A or clause 6.3B or clause 6.3C.2 applies and, where clause 6.3C.1 applies, in respect of loss or damage by the Specified Perils to the existing structures together with the contents thereof owned by the Employer or for which he is responsible; and that this recognition or waiver shall continue up to and including the date of issue of any certificate or other document which states that the Sub-Contract Works of such a sub-contractor are practically complete or the date of determination of the employment of the Contractor (whether or not the validity of that determination is contested) under clause 7.1 to 7.3 or clause 7.5 or 7.6 or 7.8 or, where clause 6.3C applies, under clause 6.3C.4.3 or clause 7.1 to 7.3 or clause 7.5 or 7.6 or 7.8, whichever is the earlier. The provisions of clause 6.3.3 shall apply also in respect of any Joint Names Policy taken out by the Employer under clause 6.3A.2 or by the Contractor under clause 6.3B.2 or under clause 6.3C.3.

Except in respect of the Joint Names Policy referred to in clause 6.3C.1 (or the Joint Names Policy referred to in clause 6.3C.3 taken out by the Contractor in respect of a default by the Employer under clause 6.3C.1) the provisions of clause 6.3.3 in regard to recognition or waiver shall apply to sub-contractors as referred to in clause 3.2. Such recognition or waiver for such sub-contractors shall continue up to and including the date of issue of any certificate or other document which states that the sub-contract works of such a sub-contractor are practically complete or the date of determination of the employment of the Contractor as referred to in clause 6.3.3 whichever is the earlier.

[k] *Erection of new buildings – All Risks Insurance of the Works by the Contractor*
6.3A.1 The Contractor shall take out and maintain a Joint Names
[l] Policy for All Risks Insurance for cover no less than that
[m.1] defined in clause 6.3.2 for the full reinstatement value of the Works (plus the percentage, if any, to cover professional fees stated in the Appendix) and shall maintain such Joint Names Policy up to and including the date of issue[1] of the certificate of Practical Completion or up to and including the date of determination of the employment of the Contractor under clause 7.1 to 7.3 or clause 7.5 or 7.6 or 7.8 (whether or not the validity of that determination is contested) whichever is the earlier.

Single policy – insurers approved by Employer – failure by Contractor to insure
6.3A.2 The Joint Names Policy referred to in clause 6.3A.1 shall be taken out with insurers approved by the Employer and the Contractor shall send to the Architect/the Contract Administrator for deposit with the Employer that Policy and the premium receipt therefor and also any relevant endorsement or endorsements thereof as may be required to comply with the obligation to maintain that Policy set out in clause 6.3A.1 and the premium receipts therefor. If the Contractor defaults in taking out or in maintaining the Joint Names Policy as required by clauses 6.3A.1 and 6.3A.2 the Employer may himself take out and maintain a Joint Names Policy against any risk in respect of which the default shall have occurred and a sum or sums equivalent to the amount paid or payable by him in respect of premiums therefor may be deducted by him from any monies due or to become due to the Contractor under this Contract or such amount

---

[m.1] In some cases it may not be possible for insurance to be taken out against certain of the risks covered by the definition of 'All Risks Insurance'. This matter should be arranged between the parties prior to entering into the Contract and either the definition of 'All Risks Insurance' given in clause 6.3.2 amended or the risks actually covered should replace this definition; in the latter case clause 6.3A.1, clause 6.3A.3 or clause 6.3B.1, whichever is applicable, and other relevant clauses in which the definition 'All Risks Insurance' is used should be amended to include the words used to replace the definition.

may be recoverable by the Employer from the Contractor as a debt.

Use of annual policy maintained by Contactor – alternative to use of clause 6.3A.2

6.3A.3 .1 If the Contractor independently of his obligations under this Contract maintains a policy of insurance which provides *(inter alia)* All Risks Insurance for cover no less than that defined in clause 6.3.2 for the full reinstatement value of the Works (plus the percentage, if any, to cover professional fees stated in the Appendix) then the maintenance by the Contractor of such policy shall, if the policy is a Joint Names Policy in respect of the aforesaid Works, be a discharge of the Contractor's obligation to take out and maintain a Joint Names Policy under clause 6.3A.1. If and so long as the Contractor is able to send to the Architect/the Contract Administrator for inspection by the Employer as and when he is reasonably required to do so by the Employer documentary evidence that such a policy is being maintained then the Contractor shall be discharged from his obligation under clause 6.3A.2 to deposit the policy and the premium receipt with the Employer but on any occasion the Employer may (but not unreasonably or vexatiously) require to have sent to the Architect/the Contract Administrator for inspection by the Employer the policy to which clause 6.3A.3.1 refers and the premium receipts therefor. The annual renewal date, as supplied by the Contractor, of the insurance referred to in clause 6.3A.3.1 is stated in the Appendix.

6.3A.3 .2 The provisions of clause 6.3A.2 shall apply in regard to any default in taking out or in maintaining insurance under clause 6.3A.3.1.

Loss or damage to Works – insurance claims – Contractor's obligations – use of insurance monies

6.3A.4 .1 If any loss or damage affecting work executed or any part thereof or any Site Materials[2] is occasioned by any one or more of the risks covered by the Joint Names Policy referred to in clause 6.3A.1 or clause 6.3A.2 or clause 6.3A.3 then, upon discovering the said loss or damage, the Contractor shall forthwith give notice in writing both to the Architect/the Contract Administrator and to the Employer of the extent, nature and location thereof.

6.3A.4 .2 The occurrence of such loss or damage shall be disregarded in computing any amounts payable to the Contractor under or by virtue of this Contract.

6.3A.4 .3 After any inspection[3] required by the insurers in respect of a claim under the Joint Names Policy referred to in clause 6.3A.1 or clause 6.3A.2 or clause 6.3A.3 has been completed the Contractor with due diligence shall restore such work damaged, replace or repair any such

Site Materials which have been lost or damaged, remove
and dispose of any debris and proceed with the carrying
out and completion of the Works.

6.3A.4   .4 The Contractor, for himself and for all sub-contractors
referred to in clauses 3.2 and 3.3 who are, pursuant to
clause 6.3.3, recognised as an insured under the Joint
Names Policy referred to in clause 6.3A.1 or clause
6.3A.2 or clause 6.3A.3, shall authorise the insurers to
pay all monies from such insurance in respect of the loss
or damage referred to in clause 6.3A.4.1 to the Employer.
The Employer shall pay all such monies (less only the
percentage, if any, to cover professional fees stated in the
Appendix) to the Contractor by instalments under certificates of the Architect/the Contract Administrator issued
at the intervals to which clause 4.2 refers.

6.3A.4   .5 The Contractor shall not be entitled to any payment in
respect of the restoration, replacement or repair of such
loss or damage, and (when required) the removal and disposal of debris other than the monies received under the
aforesaid insurance.

[k]   *Erection of new buildings – All Risks Insurance of the
Works by the Employer*
6.3B.1   The Employer shall take out and maintain a Joint Names
[l]   Policy for All Risks Insurance for cover no less than that
[m.1]   defined in clause 6.3.2 for the full reinstatement value of
the Works (plus the percentage, if any to cover professional fees stated in the Appendix) and shall maintain
such Joint Names Policy up to and including the date of
issue of the certificate of Practical Completion or up to
and including the date of determination of the employment of the Contractor under clause 7.1 to 7.3 or clause
7.5 or 7.6 or 7.8 (whether or not the validity of that determination is contested) whichever is the earlier.

Failure of Employer to insure – rights of Contractor
6.3B.2   Except where the Employer is a local authority the
Employer shall, as and when reasonably required to do so
by the Contractor, produce documentary evidence and
receipts showing that the Joint Names Policy required
under clause 6.3B.1 has been taken out and is being
maintained. If the Employer defaults in taking out or in
maintaining the Joint Names Policy required under
clause 6.3B.1 then the Contractor may himself take out
and maintain a Joint Names Policy against any risk in
respect of which a default shall have occurred and a sum
or sums equivalent to the amount paid or payable by him
in respect of the premiums therefor shall be added to the
Contract Sum.

Loss or damage to Works – insurance claims –
Contractor's obligations – payment by Employer

6.3B.3   .1 If any loss or damage affecting work executed or any part thereof or any Site Materials[4] is occasioned by any one or more of the risks covered by the Joint Names Policy referred to in clause 6.3B.1 or clause 6.3B.2 then, upon discovering the said loss or damage, the Contractor shall forthwith give notice in writing both to the Architect/the Contract Administrator and to the Employer of the extent, nature and location thereof.

6.3B.3   .2 The occurrence of such loss or damage shall be disregarded in computing any amounts payable to the Contractor under or by virtue of this Contract.

6.3B.3   .3 After any inspection required by the insurers in respect of a claim under the Joint Names Policy referred to in clause 6.3B.1 or clause 6.3B.2 has been completed the Contractor with due diligence shall restore work damaged, replace or repair any Site Materials which have been lost or damaged, remove and dispose of any debris and proceed with the carrying out and completion of the Works.

6.3B.3   .4 The Contractor, for himself and for all sub-contractors referred to in clauses 3.2 and 3.3 who are, pursuant to clause 6.3.3, recognised as an insured under the Joint Names Policy referred to in clause 6.3B.1 or clause 6.3B.2, shall authorise the insurers to pay all monies from such insurance in respect of the loss or damage referred to in clause 6.3B.3.1 to the Employer.

6.3B.3   .5 The restoration, replacement or repair of such loss of damage and (when required) the removal and disposal of debris shall be treated as if they were a Variation required by an instruction of the Architect/the Contract Administrator under clause 3.6.

  [k] *Insurance of existing structures – Insurance of Works in*
     or extensions to existing structures

6.3C.1 The Employer shall take out and maintain a Joint Names Policy in respect of the existing structures together with the contents thereof owned by him or for which he is responsible, for the full cost of reinstatement, repair or replacement of loss or damage due to one or more of the

  [m.2] Specified Perils up to and including the date of issue of the certificate of Practical Completion or up to and including the date of determination of the employment of the Contractor under clause 6.3C.4.3 or clause 7.1 to 7.3 or clause 7.5 or 7.6 or 7.8 (whether or not the validity of that determination is contested) whichever is the earlier. The Contractor, for himself and for all sub-contractors referred to in clause 3.3 who are, pursuant to clause

6.3.3, recognised as an insured under the Joint Names Policy referred to in clause 6.3C.1 or clause 6.3C.3, shall authorise the insurers to pay all monies from such insurance in respect of loss or damage to the Employer.

Works in or extensions to existing structures – All Risks Insurance – Employer to take out and maintain Joint Names Policy

6.3C.2

[l]

[m.2]

The Employer shall take out and maintain a Joint Names Policy for All Risks Insurance (as defined in clause 6.3.2) for the full reinstatement value of the Works (plus the percentage, if any, to cover professional fees stated in the Appendix) and shall maintain such Joint Names Policy up to and including the date of issue of the certificate of Practical Completion or up to and including the date of determination of the employment of the Contractor under clause 6.3C.4.3 or clause 7.1 to 7.3 or clause 7.5 or 7.6 or 7.8 (whether or not the validity of such determination is contested) whichever is the earlier.

Failure of Employer to insure – rights of Contractor

6.3C.3

Except where the Employer is a local authority the Employer shall, as and when reasonably required to do so by the Contractor, produce documentary evidence and receipts showing that the Joint Names Policy required under clause 6.3C.1 or clause 6.3C.2 has been taken out and is being maintained. If the Employer defaults in taking out or in maintaining the Joint Names Policy required under clause 6.3C.1 the Contractor may himself take out and maintain a Joint Names Policy against any risk in respect of which the default shall have occurred and for that purpose shall have such right of entry and inspection as may be required to make a survey and inventory of the existing structures and the relevant contents. If the Employer defaults in taking out or in maintaining the Joint Names Policy required under clause 6.3C.2 the Contractor may take out and maintain a Joint Names Policy against any risk in respect of which the defaults shall

---

[m.2] In some cases it may not be possible for insurance to be taken out against certain of the Specified Perils or the risks covered by the definition of 'All Risks Insurance'. This matter should be arranged between the parties prior to entering into the Contract and either the definition of Specified Perils and/or All Risks Insurance given in clauses 6.3 and 8.3 amended or the risks actually covered should replace the definition; in the latter case clause 6.3C.1 and/or clause 6.3C.2 and other relevant clauses in which the definitions 'All Risks Insurance' and/or 'Specified Perils' are used should be amended to include the words used to replace the definition.

have occurred. A sum or sums equivalent to the premiums paid or payable by the Contractor pursuant to clause 6.3C.3 shall be added to the Contract Sum.

Loss or damage to Works – insurance claims –
Contractor's obligations – payment by Employer

6.3C.4 If any loss or damage affecting work executed or any part thereof or any Site Materials is occasioned by any one or more of the risks covered by the Joint Names Policy referred to in clause 6.3C.2 or clause 6.3C.3 then, upon discovering the said loss or damage, the Contractor shall forthwith give notice in writing both to the Architect/the Contract Administrator and to the Employer of the extent, nature and location thereof and

6.3C.4 .1 the occurrence of such loss or damage shall be disregarded in computing any amounts payable to the Contractor under or by virtue of this Contract;

6.3C.4 .2 the Contractor, for himself and for all sub-contractors referred to in clauses 3.2 and 3.3 who are, pursuant to clause 6.3.3, recognised as an insured under the Joint Names Policy referred to in clause 6.3C.2 or clause 6.3C.3, shall authorise the insurers to pay all monies from such insurance in respect of the loss or damage referred to in clause 6.3C.4 to the Employer;

6.3C.4 .3 if it is just and equitable so to do the employment of the Contractor under this Contract may within 28 days of the occurrence of such loss or damage be determined at the option of either party by notice by registered post or recorded delivery from either party to the other. Within 7 days of receiving such a notice (but not thereafter) either party may give to the other a written request to concur in the appointment of an Arbitrator under article 5 in order that it may be determined whether such determination will be just and equitable;

upon the giving or receiving by the Employer of such a notice of determination or, where a reference to arbitration is made as aforesaid, upon the Arbitrator upholding the notice of determination, the provisions of clause 7.7 shall apply except the words at the end of clause 7.7 'and any direct loss and/or damage caused to the Contractor by the determination'.

6.3C.4 .4 If no notice of determination is served under clause 6.3C.4.3, or, where a reference to arbitration is made as aforesaid, if the Arbitrator decides against the notice of determination, then

after any inspection required by the insurers in respect of a claim under the Joint Names Policy referred to in

clause 6.3C.2 or clause 6.3C.3 has been completed, the
Contractor with due diligence shall restore such work
damaged, replace or repair any such Site Materials
which have been lost or damaged, remove and dispose
of any debris and proceed with the carrying out and
completion of the Works; and

the restoration, replacement or repair of such loss or
damage and (when required) the removal and disposal
of debris shall be treated as if they were a Variation
required by an instruction of the Architect/the Contract
Administrator under clause 3.6.

COMMENTARY ON CLAUSES 6.3 TO 6.3C

(A)   The nature and extent of the insurance required

Before commenting on the detailed provisions, it is useful to comment on the general principles on which the clauses covering the
insurance of the works is based. The insurance provisions include
three methods by which insurance of the works can be arranged. In
the case of the erection of a new building this can be done by either
the contractor insuring the works and site materials (clause 6.3A) or
by the employer insuring (6.3B). Where the works are for the
alteration or extension to an existing building there is no option for
the contractor to insure and the policy must be taken out by the
employer under clause 6.3C. Clauses 6.3A, 6.3B and 6.3C are
therefore mutually exclusive clauses and an entry in the appendix
must indicate which is to apply. It should be noted that in the case of
an existing building, the employer also has a further obligation as far
as insurance is concerned, and that is to insure the remainder of the
building together with its contents.

In all cases, irrespective of whether the contractor or the employer
is to insure, the works themselves must be covered by what is called
a 'Joint Names Policy' which is a policy which includes both the
contractor and the employer as the insured (see clause 8.3 for
definition). This joint insurance has the advantage of preventing the
insurer from exercising any possible subrogation rights as between
the employer and the contractor. Put in its simplest terms, the
insurer (i.e. the insurance company or underwriters), on paying out
the claim, is entitled to stand in the shoes of the insured and seek
recovery of all or part of the loss from anyone having a legal liability
in respect of the loss or damage to the works, e.g. where the loss or
damage was caused by someone's negligence. If for instance the

insurance policy was taken out in the name of the employer alone, and the loss or damage was suffered by reason of the contractor's wrongful act or omission, the insurance company could settle the claim on behalf of the employer and then, standing in the employer's shoes, could seek recovery from the contractor. Insurance in the joint names of contractor and employer protects both from this type of situation.

It should be noted that the conditions provide for 'All Risks' insurance cover for the works irrespective of whether the contractor or the employer insures and irrespective of whether the works are for new construction or are works of alteration or an extension to an existing building. 'All Risks' cover is defined by clause 6.3.2 as insurance which 'provides cover against any physical loss or damage to work executed and Site Materials'. It must be for the full reinstatement value of the works plus any percentage for professional fees stated in the appropriate appendix entry. The words used therefore do not include consequential loss.

Express exclusions include damage to any part of the works resulting from a defect in that parts design, plan, specification, material or workmanship. This exclusion extends to other parts of the works which are lost or damaged in consequence where such other part relied for its support or stability on the defective work. Resulting damage, for example, by fire, to other parts of the works not so dependent will still therefore be covered. Other exclusions include the following:

—property which is defective due to wear and tear, obsolescence, deterioration, rust or mildew;
—consequences of war, invasion, act of foreign enemy etc;
—disappearance or shortage if only revealed when an inventory is made or is not traceable to an identifiable event;
—civil commotion if the contract is carried out in Northern Ireland;
—any unlawful wanton or malicious act committed maliciously in connection with any unlawful association.

Finally, there are the 'Excepted Risks' which are also excluded – see clause 8.3 for the definition of these.

In the case of clause 6.3A the contractor can fulfil his obligation to insure through a suitable existing annual policy in which case its annual renewal date must be stated in the appendix.

The definition of the words 'Site Materials' referred to above covers all unfixed materials and goods delivered to and placed on or adjacent to the works and 'intended for incorporation therein' – see clause 6.3.2. These words clearly exclude items such as form work, the insurance for which would be entirely a matter for the contractor.

Cover does not extend to off-site materials and goods, even though these may have become the property of the employer – see clause 1.11. In such a case the contractor takes the risk of loss or damage, though the employer may still consider it worthwhile to insure them, at any rate if the contractor has not.

When arranging insurance it is important to ensure that the cover provided by the all risks policy conforms to the definition included in the conditions. Where there are any doubts, the insurer should be asked to confirm that the cover provided is no less than that defined in the conditions. Where the cover provided by a policy clearly falls short of that definition, discussions should be held with the insurers with the aim of amending or supplementing the policy in order to provide that required degree of cover. However, if the parties agree to accept a lesser degree of cover, the conditions should be amended accordingly, although it is clearly desirable that the required degree of cover remains unaltered – see footnote [k.1] to clause 6.3.2 and footnote [m.1] to clause 6.3A.1 and 6.3B.1 and footnote [m.2] to clause 6.3C.1.

It may be that while the employer does not want the contractor to insure under clause 6.3A, neither does he wish to insure himself under clause 6.3B. Similarly if the work is in, or an extension to, an existing structure, the employer may not wish to insure the works and/or the existing structures. If this is so, clearly amendments will have to be made. In the case of JCT 80, the Tribunal has provided in Practice Note 22 appendices C and D – Model Clauses 22E to 22K which cover such a situation, as well as modifying the employer's obligations in relation to excesses. Provided care is taken these model clauses may readily be adapted for use with IFC 84.

Where the works are for the alteration or extension to an existing structure the employer has an obligation to insure the existing structure together with its contents against risks termed 'Specified Perils'. This degree of cover is less than that provided for in an all risks policy and it equates to the 'clause 6.3 Perils' included in the conditions prior to the 1986 amendments. 'Specified Perils' means 'fire, lightning, explosion, storm, tempest, flood, bursting or overflowing of water tanks, apparatus or pipes, earthquake, aircraft and other aerial devices or articles dropped therefrom, riot and civil commotion, but excluding Excepted Risks' – see clause 8.3 for definition.

Differences between all risks cover and specified perils should be noted, the most important of which are that the former does whereas the latter does not cover matters such as vandalism, subsidence, impact, malicious damage and theft.

These differences between the two types of cover can of course be crucial. For example, in the case of *Computer and Systems Engineering plc* v. *John Lelliott (Ilford)* (1989), which was concerned with JCT 80 prior to its insurance provisions for the works changing from specified perils to all risks, the insurance was required to cover 'flood, bursting or overflowing of water tanks, apparatus or pipes . . .'. The sub-contractor negligently dropped a purlin which fractured a sprinkler pipe causing damage by the consequent escape of water. The question for consideration by the court was whether or not the escape of water as a result of the bursting of a pipe caused by a negligently created external impact was covered by the clause. The court held that reference in the required cover to flood involved a natural phenomenon of some kind or at least some form of abnormal occurrence. Bursting and overflowing both contemplated a rupture of the pipe due to internal causes. Accordingly, the damage concerned was not covered within the definition of the perils.

Had the cover in this case been of the all risks type, it would no doubt have provided protection.

The cover provided must be for the 'full reinstatement value of the Works plus the percentage, if any, to cover professional fees stated in the Appendix'. The term 'full reinstatement value' is intended to cover the cost of any necessary demolition and the removal of debris together with the cost of reinstating the works and the replacement of any materials or goods on site which may have been lost or damaged. As it is common practice to provide cover for the amount of the contract sum, this should always be kept under review, particularly for a contract with a relatively long contract period or during periods of high inflation. For existing structures and contents, the cover must be for the full cost of reinstatement.

Excesses

As the conditions stipulate that the all risks policy in respect of the works which is to be taken out by either contractor or employer is to be for the full reinstatement value, the conditions do not expressly permit policy excesses to apply. In practice, however, they are likely to do so. Excesses occur where the terms of a policy stipulate that the first stated amount of any claim will not be met by the insurers. Where excesses apply, payment by the insurers will therefore fall short of the full reinstatement value and that will clearly be a breach of the insurance provisions by the party responsible for obtaining the cover. It is therefore the party who has a contractual obligation to insure who will be responsible for meeting any cost arising from those excesses. It is a matter which should be carefully considered since excesses may be substantial.

Inflation cost of remaining work
If any of the risks against which the insurance cover is taken, materialise, it is highly probable that reinstatement will cause delay to overall completion of the works. This will mean that the outstanding work which had not been completed at the time when the loss or damage occurred, will have to be completed later than would otherwise have been the case. If the contract sum is subject to fluctuation provisions by virtue of clause 4.9(b), the increased cost is likely to be for the most part catered for. However, the situation could be very different if it is a fixed price contract. If clause 6.3A applies the cost of the increase in finishing off the outstanding work will fall upon the contractor. The contractor may therefore wish to take special steps to ensure that this risk is covered by some endorsement to the policy or by a separate policy. If clause 6.3B or 6.3C applies, then, by virtue of clause 6.3B.3.5 or 6.3C.4.4 (last paragraph) the increased cost of completing the outstanding work may be covered by the treatment of the contractor's obligation to reinstate following an inspection of the loss or damage as a variation, in which case it is submitted that clause 3.7.8 may apply so as to entitle the contractor to have the outstanding work treated as though it were itself the subject of a variation instruction. This is considered again later as is the question of extensions of time and loss and expense – see page 323. Accordingly, if clause 6.3B or 6.3C applies, the employer may well wish to cover the increased cost for which he may be responsible in relation to the completion of outstanding work by means of additional cover to the all risks policy or by way of a separate policy.

Duration of cover

The insurance cover must be maintained up to and including the date of the issue of the certificate of practical completion, or up to and including the date of determination of the contractor's employment under the contract (even if the validity of that determination is contested), whichever is the earlier.

(B)   Approvals, inspections of policies and evidence of insurance cover

Where clause 6.3A applies (contractor responsible for insuring), the insurer selected by the contractor is subject to the approval of the employer. However, if the contractor has opted to utilise any existing and adequate all risks policy, the employer's approval to the insurer is not required.

If the policy is obtained specifically for the particular contract by the contractor, he must send to the architect for deposit with the employer, the policy and premium receipts including any relevant endorsements.

If risks are covered by the contractor's annual all risks policy and if the contractor sends to the architect for inspection by the employer as and when reasonably required, documentary evidence that the policy is being maintained, e.g. a broker's certificate may suffice, the contractor is discharged from the need to deposit the policy itself and the premium receipts with the employer, though the employer may still call for inspection of the policy and premium receipts provided this is not done unreasonably or vexatiously. If the contractor does utilise his own all risks policy, its renewal date must be stated in the appendix.

If clause 6.3B or 6.3C apply (employer responsible for insuring), there is no provision for the insurer to be approved by the contractor. Unless the employer is a local authority, however, he must, as and when reasonably required by the contractor, produce documentary evidence and receipts showing that a policy has been taken out and is being maintained as required.

The express exception for local authorities is not justifiable. While no doubt one reason for insuring, that is having a fund available to the employer to finance reinstatement, is not likely to be critical for a local authority, this could apply equally to any other public undertaking. Further, as the employer, even if a local authority, is bound to insure, and as the contractor has a right to get the cover if the employer defaults, it would have been sensible and logical for the evidence to be required even where the employer is a local authority.

If clause 6.3C applies, the same requirements falling upon the employer in connection with evidence of insurance extend also to the employer's insurance policy relating to the existing structures and contents.

(C) Default in insuring

If the party upon whom the obligation to insure, fails to do so, the other party may take out the required cover. If it is the contractor who has defaulted, the employer can recover any premiums paid by deduction from monies due or to become due or can claim it as a debt. If it is the contractor who insures as a result of the employer's failure to do so, the premiums paid by the contractor are to be added to the contract sum. Where existing structures and their contents are concerned, the contractor is given a right of access for the purpose of inspecting, carrying out a survey and making an appropriate inventory of the contents.

(D) Occurrence of loss or damage

Where any loss or damage is occasioned to the works by one or more of the insured risks, the contractor must forthwith upon discovery, give notice in writing to both the architect and the employer of the extent, nature and location of the damage.

After an inspection by the insurers has taken place, the contractor is obliged to reinstate and proceed with the carrying out of the remainder of the works. However, if the loss or damage relates to works in, or extensions to, existing structures, the contractor will not have such an obligation if either the employer or the contractor has, within 28 days of the occurrence of the loss or damage, determined the employment of the contractor on the basis that it is just and equitable to do so – see clause 6.3C.4.3 and 6.3C.4.4, and provided further that no objection is taken by the other party within seven days of the notice of determination (or if one is taken where an arbitrator nevertheless upholds the determination). In such a case the contractor will be entitled to the benefits but subject to the obligations which apply following a determination of the contractor's employment under clause 7.7 (dealt with later page 360), except that any direct loss or damage caused by the determination will not be recoverable.

Clearly, where existing structures are concerned, this is a sensible provision as it could well be inappropriate to expect the contractor to reinstate the works, e.g. where the existing structure has itself been destroyed.

In all cases the loss or damage is disregarded in computing any amounts payable to the contractor under or by virtue of the contract. The contractor must on his and his sub-contractor's behalf, authorise the insurers to pay the insurance money over to the employer. The method by which the contractor gets paid for the reinstatement is then as follows:

—In the case of 6.3A (new buildings – contractor to insure) the employer will pay the contractor the insurance monies under separate certificates of the architect. These sums are not included in the contract sum. If there is any shortfall between the insurance monies and the cost of reinstatement it is the contractor who must meet the difference from his own resources;

—In the case of clause 6.3B (new buildings – employer to insure) or 6.3C (works in or extensions to existing buildings – employer to insure) the reinstatement is treated as a variation required by an instruction of the architect under clause 3.6. The instruction will then be valued in accordance with the valuation rules in clause 3.7, so that any risk of a shortfall in the insurance monies will effectively fall upon the employer.

The cost of completing the outstanding work
As noted earlier (see page 320) there is every chance that work yet to be done at the time of the occurrence of the loss or damage will, due to the resultant delays, cost more to complete than would otherwise have been the case.

In the case of clause 6.3A this cost will fall upon the contractor, though if the contract is subject to fluctuations this will clearly assist him.

In the case of clause 6.3B or 6.3C, it must at least be arguable that any increased cost in finishing off can be recovered by the contractor either under clause 3.7.8 and/or as loss and expense under clause 4.12.7. Further, some or all of the extra cost may in any event be recoverable if the contract includes a fluctuations clause.

If the appropriate means of recovery is by way of seeking reimbursement for direct loss and/or expense, certain points should be noted. The loss and expense must be attributable to the variations. Where clause 6.3B applies, the variation will be treated as having been instructed after completion of inspection by or on behalf of the insurers. Clearly, loss or expense could have been incurred between the loss or damage to the works being occasioned and the variation instruction being treated as having been given. There seems no way for the contractor to recover this. In the case of clause 6.3C, the appropriate date for the variation being treated as having been instructed, seems to be either the date of completion of the inspection by or on behalf of the insurers, or the decision on the issue of whether a determination is just and equitable where either party seeks to determine the contractor's employment under clause 6.3C.4.3 and 6.3C.4.4. Any arbitrator's decision could of course take a considerable time. The risks for the contractor here may therefore be considerable.

Whichever party bears the increased cost of completing the outstanding work it should be possible to include for this either as part of the full reinstatement value under the all risks policy or alternatively by way of some special endorsement or even as a separate policy.

(E) Extensions of time

So far as an extension of time is concerned, it is tentatively submitted that the position is as follows:

—If clause 6.3A applies the contractor will obtain an extension of time for any delay to completion resulting from the occurrence of loss or damage provided the loss or damage was due to a specified peril – clause 2.4.3. Accordingly, if the loss or damage is due to, e.g. impact, subsidence or malicious damage, the contractor will not be entitled to an extension of time;

—Where clause 6.3B or 6.3C applies the contractor will similarly be entitled to an extension of time under clause 2.4.3 in relation to specified perils. However, once the architect's instruction is treated as being issued prompting the contractor's obligation to reinstate, the appropriate extension of time event will switch to clause 2.4.5.

(F) Determination of contractor's employment

Quite apart from the 'just and equitable' ground for determination of the contractor's employment referred to earlier (see page 322), if as a result of the loss or damage there is a suspension of the whole or substantially the whole of the uncompleted works for a period of not less than three months, then, provided the loss or damage was occasioned by a specified peril, either the contractor or the employer may determine the contractor's employment – see clause 7.8.1. The contractor cannot utilise this provision if the loss or damage was brought about by his or his sub-contractor's or agent's negligence – clause 7.8.2. This provision and the consequence of a determination under it are dealt with in greater detail in chapter 7, page 368.

(G) Works insurance and the position of sub-contractors

Under clause 6.3.3, the benefit of the joint names policy covering the works taken out by either the contractor or the employer is extended to both domestic and named sub-contractors, but only in respect of specified perils and not to the extent of cover provided by all risks insurance. This is achieved by the policy either providing for recognition of sub-contractors as an insured party or by the inclusion in the policy of a waiver of any rights of subrogation which the insurers may have against any sub-contractor.

The same benefit must also be extended to named sub-contractors in respect of the insurance of existing buildings and contents taken out by the employer under clause 6.3C. It should be noted that in this case the cover is not extended to domestic sub-contractors.

The recognition of a waiver for sub-contractors under this clause continues up to and including the date of the issue of any certificate or other document which states that the sub-contractor's works are practically complete, or the date of determination of the employment of the contractor, whichever is the earlier.

NOTES TO CLAUSES 6.3 TO 6.3C

[1] '*... up to and including the date of issue ...*'
Note that it is not the date of practical completion but the date of the issue of the certificate that is important. This is clearly necessary as otherwise there could be an uninsured period if the date of practical completion is back-dated, as it often is, to a date before the date of the issue of the certificate itself.

[2] '*... loss or damage ... affecting any Site Materials ...*'
The employer may or may not own the materials or goods on site.

Perhaps only in the world of insurance do we find a situation in which on the one hand the contractor could be liable to replace materials or goods owned by the employer (if the employer has paid for them) and yet on the other, the employer can be responsible for the cost of replacing materials or goods still owned by the contractor – clause 6.3B and 6.3C.

[3]   'After any inspection . . .'
The risk of loss or damage brought about by an insured peril will generally fall on the contracator who is contractually responsible for completing the works and will therefore, in pursuance of this obligation, have to reinstate the damaged part as necessary. It might therefore be thought that the obligation to reinstate ought not to be dependent upon any inspection by or on behalf of the insurers being completed. If for some reason the insurers are able to avoid the claim in circumstances where there has been no breach or default by the contractor, it would appear that the obligation to reinstate never arises. The insurance cover ought primarily to be for the protection of the contractor and effective insurance cover should not be a condition precedent to his obligation to reinstate loss or damage to the works.

On the other hand, it is right that the obligation to reinstate etc. should not operate immediately loss or damage has been incurred. The insurance policy will almost certainly provide for the insurers to have the right to inspect the damage before restoration work begins and it would be unreasonable to require the contractor to begin such work before adequate time for an inspection has elapsed.

[4]   '. . . or any Site Materials . . .'
Presumably, this includes *all* unfixed materials and goods, even if they are prematurely or not properly brought on to the site and indeed even if they are not properly protected by the contractor against weather and other casualties. It might be thought harsh for the employer (or his insurers) to meet the cost of replacing materials or goods which perhaps ought not to have been on the site at all. The contractor still owns them and is likely to have them covered under some appropriate policy in any event. The position is even more startling if the loss or damage is restricted to site materials, e.g. accidental damage or theft when all or part of the cost of replacement will be met by the employer due to the inevitable excess on the policy. However, it is no doubt a matter of convenience or expe-

diency to include all site materials which avoids any argument about premature delivery or inadequate protection.

Clause 6.3D

Insurance for Employer's loss of liquidated damages – clause 2.4.3

6.3D.1 Where it is stated in the Appendix that the insurance to which clause 6.3D refers may be required by the Employer then forthwith after the Contract has been entered into the Architect/the Contract Administrator shall either inform the Contractor that no such insurance is required or shall instruct the Contractor to obtain a quotation for such insurance. This quotation shall be for

[m.3] an insurance on an agreed value basis to be taken out and maintained by the Contractor until the date of Practical Completion and which will provide for payment to the Employer of a sum calculated by reference to clause 6.3D.3 in the event of loss or damage to the Works, work executed, Site Materials, temporary buildings, plant and equipment for use in connection with and on or adjacent to the Works by any one or more of the Specified Perils and which loss or damage results in the Architect/the Contract Administrator giving an extension of time under clause 2.3 in respect of the Event in clause 2.4.3. The Architect/The Contract Administrator shall obtain from the Employer any information which the Contractor reasonably requires to obtain such quotation. The Contractor shall send to the Architect/the Contract Administrator as soon as practicable the quotation which he has obtained and the Architect/the Contract Administrator shall thereafter instruct the Contractor whether or not the Employer wishes the Contractor to accept that quotation and such instruction shall not be unreasonably withheld or delayed. If the Contractor is instructed to accept the quotation the Contractor shall forthwith take out and maintain the relevant policy and send it to the Architect/ the Contract Administrator, for deposit with the Employer, together with the premium receipt therefor and also any

---

[m.3] The adoption of an agreed value is to avoid any dispute over the amount of the payment due under the insurance once the policy is issued. Insurers on receiving a proposal for the insurance to which clause 6.3D refers will normally reserve the right to be satisfied that the sum referred to in clause 6.3D.2 is not more than a genuine pre-estimate of the damages which the Employer considers, at the time he enters into the Contract, he will suffer as a result of any delay.

relevant endorsement or endorsements thereof and the
premium receipts therefor.

6.3D.2   The sum insured by the relevant policy shall be a sum
calculated at the rate stated in the Appendix as liquidated
damages for the period of time stated in the Appendix.

6.3D.3   The payment in respect of this insurance shall be calculated at the rate referred to in clause 6.3D.2 for the period
of any extension of time finally given by the Architect/the
Contract Administrator as referred to in clause 6.3D.1 or
for the period of time stated in the Appendix, whichever is
the less.

6.3D.4   The amounts expended by the Contractor to take out and
maintain the insurance referred to in clause 6.3D.1 shall
be added to the Contract Sum. If the Contractor defaults
in taking out or in maintaining the insurance referred to
in clause 6.3D.1 the Employer may himself insure against
any risk in respect of which the default shall have
occurred.

COMMENTARY ON CLAUSE 6.3D

Introduction

In the event of the progress of the works being delayed due to loss or
damage caused by any one or more of the specified perils, the
contractor will be entitled to an extension of time under clause 2.4.3.
It should be noted that the contractor's entitlement to an extension
of time is limited to the events included under the heading of
specified perils as defined in clause 8.3 and not to the wider scope of
all risks as defined in clause 6.3.2. It is quite likely that the
contractor will be entitled to an extension of time even if the
specified peril was brought about by his own negligence.

The award to the contractor of an extension of time and the fixing
of a later completion date will result in the employer being unable to
claim liquidated damages under clause 2.7 for the period of the
delay.

Some employers amend clause 2.4.3 to qualify its application
where the contractor has been negligent. However, it is the architect
who has the duty to determine whether the contractor is entitled to
an extension of time and it may be unwise and perhaps unfair to
require or expect the architect to decide whether there has been
negligence. Accordingly the Tribunal itself has not gone down this
path. Instead, clause 6.3D has been introduced to compensate the

employer for his lost liquidated damages by way of insurance with the added benefit that the insurance monies are payable irrespective of whether the specified peril was the result of the contractor's negligence.

This clause therefore provides a means whereby the employer can require the contractor to arrange insurance to either fully or partially compensate him for the lost liquidated damages.

The operation of clause 6.3D

An entry in the appendix should indicate whether or not insurance under this clause may be required. Immediately after the contract is entered into the employer must decide whether the contractor is to be required to obtain a quotation and the architect must either inform the contractor that no such insurance is required or must instruct the contractor to obtain a quotation on an agreed valued basis for the employer to approve. The architect must then instruct the contractor whether the quotation is to be accepted. If it is, the contractor must forthwith take out and maintain the policy for the employer's benefit, sending it to the architect for deposit with the employer together with premium receipts.

If the appendix entry provides that insurance under this clause may be required then the period of delay for which the employer will require cover should also be entered in the appendix. While the employer's intention will be to obtain reimbursement for the full amount of liquidated damages, the insurers will wish to be satisfied that the amount covered by the policy is no more than a genuine pre-estimate of the employer's loss resulting from any delay. The number of weeks delay which the insurers are prepared to cover may be less than the employer might have wished.

The nature and extent of cover

Clause 2.4.3 which provides for an extension of time where loss or damage is caused by one of the specified perils, is not limited by its terms to loss or damage to the works themselves. Consequently, the scope of loss or damage for which the clause 6.3D cover against lost liquidated damages is applicable, includes loss or damage to the works, work executed, site materials, temporary buildings, plant and equipment for use in connection with and on or adjacent to the works.

It has been said above that in the event of loss or damage brought about by one of the specified perils, there must be an extension of time pursuant to clause 2.4.3. However, it has already been indicated earlier – page 324, that where clauses 6.3B or 6.3C apply, the obligation to reinstate, once it arises, is deemed to be a variation so that as from that time, 2.4.3 would no longer, it is submitted, be appropriate and clause 2.4.5 would take over. This could therefore very significantly restrict the benefit of any cover under a clause 6.3D policy wherever the employer is responsible for insuring the works.

It is possible by incorporating the optional clause 2.11, set out in the appendix to Practice Note IN/1, to have the facility for partial possession in IFC 84. This has been dealt with earlier, page 101. As was noted there, the effect of the employer taking part of the works into possession is that the rate of liquidated damages is proportionately reduced by the value of the works taken over in proportion to the contract sum. If, therefore, clause 2.11 is incorporated, specific provision should be made for the extent of clause 6.3D cover to be reduced. In these circumstances, presumably some proportionate refund of insurance premiums may be possible.

Finally, before any sum becomes payable under a clause 6.3D policy, it is necessary to know the amount of any extension of time awarded by the architect in clause 2.4.3. Extensions awarded prior to practical completion are based upon estimates. Following practical completion, the architect may, up to 12 weeks thereafter, make a more considered and final decision. Does this mean that the monies under the policy will not be paid out against any estimate and that the employer will have to wait and see if the architect makes any adjustments to extensions of time previously given or makes new extensions, within the 12 weeks following practical completion?

The cover provided by a clause 6.3D policy, covers much of the same ground as that already available, particularly to private employers, if they have business interruption cover. As such clause 6.3D will be of little assistance to them. Further, since its introduction in 1986, this form of insurance has not been popular. On balance it adds nothing to the previous situation where the employer could in any event have obtained this type of cover had he wished to. Expressly incorporating it within the contract does not really add much.

Chapter 10

Determination

CONTENT

This chapter considers section 7 which deals with the determination of the employment of the contractor by the employer or by the contractor himself.

SUMMARY OF GENERAL LAW

(A)   Introduction

Contractual obligations may come to an end in a number of ways. In the ordinary course of things this will most commonly be brought about by the parties performing all their contractual obligations or promises.

However, contractual obligations can be determined in a number of other ways, e.g. by accord and satisfaction (mutual discharge by agreement not to require performance of outstanding obligations), by the unilateral release by one party of the other's obligations, by frustration of the contract, discharge by breach (a sufficiently serious breach of contract by one party may entitle the other party to treat himself as discharged from any further performance) or by contractual terms providing for discharge in certain events.

In this summary we are concerned only with the discharge of contractual obligations in accordance with the express conditions of the contract itself and with the discharge of contractual obligations brought about by a breach of contract by one of the parties.

(B)   Provision for discharge in the contract itself

Building contracts often make provision for one or both of the parties to have the right in certain situations either to bring the contract itself to an end or, alternatively, to bring the employment of the contractor under the contract to an end.

At common law, a contracting party will only be able to regard himself as discharged from any further performance of his contractual obligations by a particularly serious breach of contract by the other party. The inclusion therefore of an express provision for determination has the advantage to the party exercising the right that it can permit such a discharge of further obligations for less serious breaches and even for an event which does not amount to a breach of contract at all e.g. receivership or a composition with creditors. However, in certain circumstances a term of this nature would have to be shown to be fair and reasonable by virtue of the provisions of the Unfair Contract Terms Act 1977 (section 3).

There may be very good reasons for determining the employment of the contractor rather than bringing the contract as a whole to an end. It may be desirable for one or both parties that the contract survives and that express provisions can be called into play in the event of the employment of the contractor being determined. Such express provisions may cover: the right of the contractor to payment for work done but not paid for; rights over goods, materials, equipment and plant on site; and recovery of loss or damage incurred by the employer or the contractor as the case may be, as a result of the determination of the employment of the contractor.

If the contract is brought to an end by the breach by one party which is both sufficiently serious to be regarded as an intention by that party no longer to be bound by the contract i.e. a repudiation, and which is accepted as a repudiation by the other party, then the contract falls to the ground except for its partial survival for the purposes of measuring losses arising out of the breach and, of course, for the operation of any arbitration clause contained in the contract.

The inclusion in the contract of express provision for determination does not, unless the express terms of the contract make it clear, take away either party's right to treat the contract as discharged by reason of a repudiation.

The express right to determine either the contract or the contractor's employment thereunder is often made subject to the service of one or more notices and any such notices must be clear and unambiguous in order to be valid. Further, a failure to comply with any procedural requirements e.g. a timetable or a certain mode of service, may prevent the notice from taking effect. More is said about this topic later in this chapter on page 349.

A wrongful determination of the contractor's employment by the employer will itself amount to a repudiation of contract by the employer giving the contractor the right to treat the contract as at an end and to sue for damages for breach of contract.

A determination of the contractor's employment by the employer which is challenged by the contractor can create great difficulties if the contractor is unwilling to leave the site. The employer/landowner will no doubt want to engage another contractor to finish off the work. The continued presence on site of the original contractor would almost certainly make this impossible. The contractor's right to enter the site is likely to be by reason of an express or implied contractual licence to do so. The important question is whether or not, and in which circumstances, this licence may be revocable. An important case in this regard is *London Borough of Hounslow* v. *Twickenham Garden Developments* (1970).

Facts:

The defendants were employed by the London Borough of Hounslow to carry out the sub-structure works on a site known as the Ivybridge Site at Heston and Isleworth in Middlesex. The contract was based on JCT 63 and by clause 25(1) it was provided that if the contractor should make default, *inter alia*, in failing to proceed regularly and diligently with the work, the architect could give notice by registered post, or recorded delivery, specifying the default, and if the contractor continued or repeated the default the council could by notice by registered post or recorded delivery forthwith determine the employment of the contractor. The council purported to do this. The contractors contested the validity of the council's action and stated that they regarded the service of such a notice as a repudiation of the contract by the council which they, the contractors, elected not to accept as a repudiatory breach but instead decided to proceed with the contract works in accordance with the terms of the contract.

The council unsuccessfully tried to obtain possession of the site and issued a writ claiming damages for trespass and seeking an injunction restraining the contractor from trespassing on the site. The council argued that it was entitled, irrespective of the validity or invalidity of the notices given pursuant to clause 25, to evict the contractor and resume possession of its own property. The

contractor claimed to be able to insist on performing the contract.
Held:
There was an implied term of the contract that the council would not
revoke the contractor's licence to enter on the site while the contract
period was still running otherwise than in accordance with the
contract. The court would not grant the council the injunction
requested to expel the contractors from the site because it had not
been decided at that stage whether the contractor's employment had
been validly determined by the council.

The effect of this decision is to produce a legal stalemate. The
decison has been criticised – see *Hudson*: Supplement to 10th edition
at page 712, and earlier in this book page 53.
   The criticism is well deserved. Furthermore, it is now clear (see
American Cyanamid Company v. *Ethicon Ltd* (1975)) that if the
court considers that damages are an adequate remedy, no injuction
will be granted. The contractor would therefore generally be
protected by the employer's undertaking in damages as a condition
of the injunction being granted. Even so, the decision in the case of
Vonlynn Holdings v. *T. Flaherty* (1988) concerning a JCT 80 contract,
provides limited support for the *Hounslow* case. The position under
the ICE Conditions of Contract 5th Edition where the operation by
the employer of the forfeiture provisions under clause 63 of that
contract are dependent upon the certified opinion of the engineer is
very different, and here the contractor is bound to vacate the site
notwithstanding that the engineer's opinion is challenged by the
contractor – see the case of *Tara Civil Engineering Ltd* v. *Moorfield
Developments Ltd* (1989).

(C)   *Discharge by breach*

One party to a contract may, by reason of the other party's breach,
be entitled to treat himself as discharged from any obligation to
further perform his contractual promises under the contract, and
treat that other party's breach as being a repudiation by him of his
contractual obligations. The innocent party will additionally have a
right to claim damages consequent on the breach.
   However, while any breach of contract can give rise to a claim for
damages, it is not every breach of contract which will entitle the
innocent party to a discharge from further liability to perform his
own contractual promises. The default must be of a particularly

serious or fundamental nature. Broadly speaking, it will be sufficiently serious to justify a discharge in three situations:

(i)   In the case of a renunciation by one party of his contractual liabilities where that party by his words or conduct evinces an intention – whether due to unwillingness or inability – to no longer continue his part of the contract.

(ii) Where by breach or default one party has rendered himself unable to perform his outstanding contractual obligations e.g. if a contract requires the builder to be a member of the National House-Building Council and to obtain an appropriate NHBC certificate and he is removed from the register kept by that body during the course of the contract.

(iii) In the case of a very serious failure of performance by one party which will discharge the other. The failure must be of a fundamental nature going to the root of the contract.

The three situations referred to above may be described as repudiation or repudiatory events (sometimes perhaps less accurately as repudiatory breaches).

Once a repudiation has taken place, the innocent party must elect whether to accept the repudiation and this must be clear and unequivocal and must be communicated to the other contracting party. Once made it cannot be withdrawn.

The innocent party may, if he wishes, treat the contract as continuing if this is still a possibility, despite the existence of repudiatory conduct. This will not prevent him from claiming damages while at the same time continuing with his contractual obligations.

If one party mistakenly treats an event as amounting to a repudiation by the other party and purports to accept it as such, this will in itself create a repudiation which the wrongly accused party will very often be forced to accept. Indeed, in practice it is not unusual for both sides to argue that they have accepted a repudiation by the other.

However, if a party refuses to perform the contract, giving an inadequate or wrong reason, he may be able to justify his conduct if he subsequently discovers that at the time there existed another good reason (unknown to him at the relevant time) entitling him to refuse further performance.

Once a repudiation has been accepted by the innocent party, both parties are excused from further performance of their primary obligations under the contract. Instead, secondary obligations are imposed on the guilty party, namely to pay monetary compensation for non-performance – see *Photo Production Ltd* v. *Securicor Transport Ltd* (1980).

Breach of a term of the contract which is not in itself sufficiently serious to amount to a repudiation may become so if it persists, especially after a notice from the innocent party requesting proper performance. Furthermore, the degree of wilfulness of a breach of contract may be a relevant factor in displaying an intention to no longer be bound by the contract terms.

Examples of serious breaches of contract amounting in the particular circumstances of the case to a repudiation include: the prolonged failure by the employer to hand over the site to the contractor (*Carr* v. *J.A. Berriman (Property) Ltd* (1953)); the failure by the contractor to obtain the required performance bond (*Swartz and Son (Property) Ltd* v. *Wolmaransstad Town Council* (1960)).

It has been suggested that a contractual term for due expedition by the contractor of the construction etc. of the contract works must be implied in building contracts, if not expressed, and that the repeated failure by the contractor to proceed with due expedition after notice by the employer will entitle the employer to treat the contractor's failure as a repudiation – see *Hudson*, 10th edition, page 611 *et seq* and see also the supplement to the 10th edition. However, the decision at first instance of Mr Justice Staughton in the case of *Greater London Council* v. *Cleveland Bridge and Engineering Co. Ltd and Another* (1984) has thrown some doubt on the extent and nature of such a term (see page 65).

(D)   Remedies for breach of contract

(i)   Introduction

By far the most important remedy for either party to a building contract in the event of the other party's breach of that contract is a claim for damages. While it is possible in unusual circumstances to obtain an order of specific performance, whereby one party is compelled to honour his contractual obligations, or alternatively to obtain an injunction prohibiting a party from acting in breach of contract, damages remain nevertheless the prime remedy.

(ii) Damages for breach of contract

The essential purpose of damages is to compensate the innocent party for loss or injury suffered through the other party's breach. The innocent party is, so far as money can do it, to be placed in the same position as if the contract had been performed. If no damage has been suffered or no damage can be proven, nominal damages will be awarded. However, difficulties in the assessment of damages does not prevent recovery. The fact that there are future losses involved make the assessment of damages very difficult but nevertheless an assessment must be made and damages in respect of losses incurred and future likely losses must be awarded at the same time. It is not possible to have a series of actions claiming damages for breach of contract as and when the amount of future losses is ascertained.

(iii) Causation and remoteness of damage

Although causation and remoteness are two different concepts, they are often closely related and are therefore discussed together here. It is a matter of causation if the question which requires an answer is 'was the loss caused by the breach of contract?'. On the other hand, it is a matter of remoteness of damage if the question asked is 'was the particular loss within the contemplation of the parties?'.

The leading case on remoteness of damage is *Hadley* v. *Baxendale* (1854).

Facts:

The plaintiff's mill was brought to a standstill by the breakage of their only crankshaft. The defendant carriers failed to deliver the broken shaft to the manufacturer at the time when they had promised to do so. They were sued by the plaintiff who sought recovery of the profits which they would have made had the mill been started up again without the consequent delay due to the late delivery of the broken shaft.

Held:

The facts known to the defendant carriers were insufficient to show that they were reasonably aware that the profits of the mill would have been affected by an unreasonable delay in the delivery of the broken shaft. The loss was therefore too remote. Alderson B said as follows:

'Where two parties have made a contract which one of them has broken, the damages which the other party ought to receive in respect of such breach of contract should be such as may fairly and reasonably be considered either arising naturally, i.e. according to the usual course of things, from such breach of contract itself, or such as may reasonably be supposed to have been in the contemplation of both parties, at the time they made the contract, as the probable result of the breach of it. Now, if the special circumstances under which the contract was actually made were communicated by the plaintiffs to the defendants, and thus known to both parties, the damages resulting from the breach of such contract, which they would reasonably contemplate, would be the amount of injury which would ordinarily follow from a breach of contract under these special circumstances so known and communicated. But, on the other hand, if these special circumstances were wholly unknown to the party breaking the contract, he, at the most, could only be supposed to have had in his contemplation the amount of injury which would arise generally, and in the great multitude of cases not affected by any special circumstances, from such breach of contract.'

The principles laid down in the passage quoted above have been interpreted and restated in more recent times in *Victoria Laundry (Windsor) Ltd* v. *Newman Industries Ltd* (1949) and in *Koufos* v. *C. Czarnikow Ltd (The Heron II)* (1969).

In the *Victoria Laundry* case it was stated that firstly, the aggrieved party is only entitled to recover such part of the loss actually resulting as was at the time of the contract reasonably foreseeable as liable to result from the breach. Secondly, what was reasonably forseeable depends on the knowledge possessed by the parties at the time of entering into the contract. Thirdly, that for this purpose, knowledge 'possessed' is of two kinds: one imputed, the other actual. Everyone is taken to know the 'ordinary course of things' and consequently what loss is liable to result from a breach of contract in that ordinary course, but to this knowledge which the party in breach of contract is *assumed* to possess (whether he actually possesses it or not) there may have to be added in a particular case knowledge which he *actually* possesses of special circumstances outside the 'ordinary course of things', of such a kind that a breach in those special circumstances would be liable to cause a particular loss.

In the *Heron II* case Lord Upjohn stated the broad rule as follows:

'What was in the assumed contemplation of both parties acting as reasonable men in the light of the general or special facts (as the case may be) known to both parties in regard to damages as a result of a breach of contract.'

(iv) Measure of damages and time of assessment

As a general rule, the assessment of damages will be based on the difference between the value of the subject matter of the contract in its defective or incomplete state, as compared to the value it would have had, had the contract been properly completed and fulfilled. However, in certain circumstances this may cause injustice or prove particularly difficult in terms of assessment e.g. a contractor's failure to complete a building. In such a case the measure of damages is likely to be the difference between the contract price and the amount it actually costs the employer to complete the contract works substantially as it was originally intended.

The general rule is that the time at which the assessment of damages will take place will be the date when the cause of action accrues i.e. in the case of breach of contract at the date of the breach, irrespective of when damage is suffered or known to be suffered as a result of the breach. However, in cases where the appropriate measure of damages is the cost of repair or completion of the outstanding work, damages are likely to be assessed at the time when the repairs or completion work ought reasonably to have been undertaken, and this may in appropriate circumstances be as late as the date of the hearing if the plaintiff was acting reasonably in not having the repairs carried out or the outstanding work completed before then.

The New Zealand case of *Bevan Investments* v. *Blackhall and Struthers* (1978) is an example of the modern, more flexible approach to both the question of the most appropriate method for measuring damages in building contract cases and also the time at which such an assessment should be made. The learned authors of Building Law Reports summarised the court's findings in this case as follows: (11 BLR at page 79):

'1.   That the company was entitled to be put into the position it would have been in had the contract been performed;

2.   That in building cases the loss should, *prima facie*, be measured by ascertaining the amount required to rectify the defects complained of and so give to the building owner the equivalent of a building on the land substantially in accordance with the contract;

3.   That this rule was adopted unless the court was satisfied that some lesser basis of compensation could in all the circumstances be fairly employed;

4.   That since the only practicable action in the present case was to complete according to the modified design, the cost of so doing was the reasonable measure of damages;

5.   That in calculating those damages
      (a)   to avoid an element of betterment credit should be given for the hypothetical additional cost of a proper initial design;
      (b)   the assessment should be computed at the date of trial either because of the principle that damages were to be assessed by reference to the date when the reinstatement works could be reasonably carried out and, on the facts, it was reasonable to postpone the work until the issues of liability and damages were settled, or because such damage was not too remote in that it was foreseeable that the company might be unable to complete the works until a trial if the appellant failed to exercise the skill required of him.'

It can be seen therefore that the general principle of English law which provides that, where the proper measure of damages is the cost of repairs etc., that cost is to be calculated on the basis of prices prevailing at or within a reasonable time after the discovery of the defective work is unlikely to be applied when in all the circumstances it is reasonable for the innocent party to postpone the repairs etc. Indeed, in appropriate circumstances the time of assessment may be as late as the date of the hearing itself. There may be good commercial reasons for delaying repairs and the court is entitled to take these into account. However, if the repair work etc. has not been carried out at the date of trial the courts will not use the cost of repair as the basis of measuring damages for breach of contract unless it is satisfied that the innocent party has formed a genuine intention to have the work carried out or alternatively if he undertakes to the court to have it carried out.

If the reason for building work being abandoned before completion is due to the contractor accepting a repudiation by the employer, this gives the contractor the right to sue for damages including his lost profit, if any. However, it is certainly arguable that in such a case the contractor may be able, instead of suing for damages for breach of contract, to bring an action in *quantum meruit* (as much as it is worth) for the work carried out by him. In other words to ignore the original contract sum and claim a reasonable sum instead – *Lodder* v. *Slowey* (1904). This can of course be of considerable advantage to the contractor where his tender price was uneconomic and can be demonstrated to be less than the value of the work carried out. This possibility of being able to claim on a *quantum meruit* appears to have no logical basis. Indeed, it is possible that this principle would not now be readily followed. There seems no reason why the employer should in such a case be penalised by having to pay to the contractor a sum in excess of the true damages suffered by the contractor as a result of the employer's breach.

(v) Mitigation

It is the duty of the innocent party to take reasonable steps to mitigate the loss suffered as a result of the guilty party's breach of contract. It may be said very briefly that there are three rules. Firstly, a plaintiff cannot recover for losses resulting from a defendant's breach of contract where the plaintiff could have avoided the loss by taking reasonable steps. Secondly, if the plaintiff actually avoids or mitigates his loss, even by taking steps which went beyond what was reasonably necessary in all the circumstances, he cannot then recover for such avoided loss. Thirdly, if the plaintiff incurs loss or expense by taking reasonable steps to mitigate the loss, he can claim such loss and expense incurred in taking the mitigating steps from the defendant, even if the result is that the losses flowing from the breach of contract have been exacerbated.

(E)   Limitation of actions

It is a matter of public policy that there should be an end to litigation and that a time should arrive when stale demands can no longer be pursued. Accordingly the Limitation Act 1980, which came into force on 1 May 1981, (consolidating earlier Limitation Acts) prescribes time limits, after the expiry of which a claim will become 'statute-barred'. However, to obtain the benefit of a defence of limitation, the defendant must plead it specifically in his defence.

So far as contracts not under seal are concerned, by virtue of section 5 of the Act, no action founded on contract can be brought after the expiration of six years from the date on which the cause of action accrued. The cause of action accrues at the date of the breach of contract irrespective of when and if damage is suffered as a result of that breach. Typically, in building contracts, the last date therefore on which a cause of action for breach of contract will accrue will be the date of practical completion or substantial completion. If the contract is under seal then the period is 12 years instead of six.

For actions arising out of tortious acts, the limitation period is six years from the date on which the cause of action accrued except in relation to actions for personal injuries when that period is three years. In the tort of negligence, the cause of action does not accrue until damage is suffered. However, in relation to actions for damages for negligence where there are latent defects or damage other than in respect of personal injuries, the limitation period is now subject to the changes introduced by the Latent Damage Act 1986 which amends the 1980 Act.

As a result of the 1986 Act there is a special time limit for commencing proceedings where facts relevant to the cause of action were not known at the date that the action accrued. It extends the present period of limitation to three years from the date on which the plaintiff knew or ought to have known facts about the damage where these facts became apparent later than the usual six years from the date on which the cause of action accrued, namely, in negligence actions, the date when the damage was suffered whether discovered or not.

However, there is an overriding longstop which operates to bar all negligence claims involving latent defects or damage which are brought more than 15 years from the date of the defendant's breach of duty.

Summarising, the time limit runs out in respect of such actions at whichever is the later of the following:

—six years from the occurrence of the damage;
—three years from the discovery of the damage following the expiry of the basic six year period, but subject to;
—a final time limit of 15 years from the breach of duty.

A right of action is given to anyone who acquires property which is already damaged where the fact of such damage is not known and could not be known to him at the time that he acquired the interest.

The three year period commences on the earliest date on which the plaintiff had both the knowledge required for bringing an action and the right to bring such an action. However, in relation to successive owners of buildings etc. if a predecessor in title knew or ought to have known of the damage then the three year period begins at that time and does not start afresh when the new owner acquires his interest.

There are provisions dealing with what knowledge a plaintiff must have or ought to have had before the three year period begins to run. These cover such matters as the knowledge that the damage was attributable to negligence and, if a third party is involved, the identity of that party.

Any deliberate concealment of defective work will prevent time from running until the defect is or ought to have been discovered. Section 32 of the Limitation Act 1980 provides that in certain cases of fraud, concealment or mistake, time will not begin to run until the plaintiff has discovered or could with reasonable diligence have discovered the fraud, concealment or mistake. Fraud in this sense means that the defendant has behaved or acted in an unconscionable manner, for example, if a building contractor knowingly takes a risk by allowing bad workmanship or materials to be covered up: see *Applegate* v. *Moss* (1971).

This amended time-scale will not apply to any actions which would have been statute-barred before 19 September 1986 under the law at that time. Further, it will have no effect on actions already commenced before 19 September 1986. Subject to this the Act is effective in relation to causes of action accruing before as well as after 19 September 1986, except that the provisions dealing with the rights of successors in title apply only in cases where the interest in the damaged property is acquired by the successor in title after 19 September 1986.

Finally, the Act has no application to contractual negligence i.e. breach of a contractually imposed duty of care – see *Iron Trades Mutual Insurance Co. Ltd* v. *Buckenham* (1989).

(F) Effect of determination on the contractual provision for liquidated damages

Subject to what the express terms of the contract may say, the liability of the contractor to pay liquidated damages ends on the determination of the contractor's employment, whether under an express provision or under the general law. However, liquidated damages which have accrued to that date will probably still be deductible from or payable by the contractor.

CONSIDERATION OF THE RELEVANT CLAUSES OF IFC 84

Clauses 7.1 to 7.4

Determination by Employer

7.1   Without prejudice to any other rights or remedies which the Employer may possess, if the Contractor shall make default in any one or more of the following respects:

(a) if without reasonable cause he wholly[1] suspends the carrying out of the Works before completion thereof; or

(b) if he fails to proceed regularly and diligently with the Works; or

(c) if he refuses or neglects to comply with a written notice from the Architect/the Contract Administrator requiring him to remove defective work or improper materials or goods and by such refusal or neglect the Works are materially affected; or

(d) if he fails to comply with the provisions of either clause
3.2   *(Sub-contracting)* or
3.3   *(Named persons)*;

and if the Contractor shall continue such default for 14 days after receipt of a notice by registered post or recorded delivery specifying the default or shall at any time thereafter repeat such default (whether previously repeated or not), then the Employer may thereupon by notice by registered post or recorded delivery determine the employment of the Contractor under this Contract: provided that such notice shall not be given unreasonably or vexatiously.

Contractor becoming bankrupt etc.

7.2   If the Contractor becomes bankrupt or makes a composition or arrangement with his creditors or has a proposal in respect of his company for a voluntary arrangement for a composition of debts or scheme of arrangement approved in accordance with the Insolvency Act 1986, or has an application made under the Insolvency Act 1986 in respect of his company to the court for the appointment of an administrator, or has a winding up order made

or (except for the purposes of amalgamation or reconstruction) a resolution for voluntary winding up is passed
or a provisional liquidator, receiver or manager of his business or undertaking is duly appointed, or has an administrative receiver, as defined in the Insolvency Act 1986,
appointed, or possession is taken, by or on behalf of the
holders of any debentures secured by a floating charge, of
any property comprised in or subject to the floating
charge, the employment of the Contractor under this
Contract shall be forthwith automatically determined but
the said employment may be reinstated and continued if
the Employer and the Contractor, his trustee in bankruptcy, liquidator, provisional liquidator, receiver or manager as the case may be shall so agree.

Corruption: determination by Employer

7.3   Where the Employer is a local authority[2] he shall be
entitled to determine the employment of the Contractor
under this or any other contract, if the Contractor shall
have offered or given or agreed to give to any person any
gift or consideration of any kind as an inducement or
reward for doing or forbearing to do or for having done or
forborne to do any action in relation to the obtaining or
execution of this or any other contract with the Employer
or for showing or forbearing to show favour or disfavour
to any person in relation to this or any other contract with
the Employer, or if the like act shall have been done by
any person employed by the Contractor or acting on his
behalf (whether with or without the knowledge of the
Contractor), or if in relation to this or any other contract
with the Employer the Contractor or any person employed
by him or acting on his behalf shall have committed any
offence under the Prevention of Corruption Acts 1889 to
1916, or shall have given any fee or reward the receipt of
which is an offence under sub-section (2) of section 117
of the Local Government Act 1972 or any re-enactment
thereof.

Consequences of determination under clause 7.1-7.3

7.4   Without prejudice to any arbitration or proceedings in
which the validity of the determination is in issue, in the
event of the determination of the employment of the Contractor under clause 7.1, 7.2, or 7.3 and so long as that
employment has not been reinstated or continued then:

(a) the Contractor shall give up possession of the site of
the Works subject to the orderly compliance of the Contractor with any instruction of the Architect/the Contract
Administrator under the following paragraph:

(b) as and when so instructed in writing by the
Architect/the Contract Administrator the Contractor shall

remove from the Works any temporary buildings, plant, tools, equipment, goods and materials belonging to or hired by him, and if within a reasonable time after any such requirement has been made the Contractor has not complied therewith then the Employer may (but without being responsible for any loss or damage) remove and sell any such property of the Contractor, holding the proceeds less all costs to the credit of the Contractor;

(c) the Employer may employ and pay other persons to carry out and complete the Works and he or they may (whether or not the Contractor has complied with the requirement to give up possession of the site of the Works) enter upon the Works and use all temporary buildings, plant, tools, equipment, goods and materials intended for, delivered to and placed on or adjacent to the Works, and may purchase all materials and goods necessary for the carrying out and completion of the Works;

(d) until after completion of the Works the Employer shall not be bound to make any further payment to the Contractor but upon such completion and the verification within a reasonable time of the accounts therefor the following matters shall be set out in an account drawn up by the Employer:
– the amount of expenses and direct loss and/or damage caused to the Employer by the determination including the payments made to other persons to carry out and complete the Works;
– the amount paid to the Contractor before the date of determination;
and if the total of such amounts exceeds or is less than the total amount which would have been payable on due completion in accordance with the Contract, the difference shall be a debt payable by the Contractor to the Employer or by the Employer to the Contractor as the case may be.

COMMENTARY ON CLAUSES 7.1 TO 7.4

Clauses 7.1 to 7.4 deal with the determination of the contractor's employment by the employer and the consequences flowing from this. These provisions are similar in most respects to those to be found in JCT 63 and JCT 80, though there are some significant differences to which attention is drawn in this Commentary.

Grounds for determination of contractor's employment by the employer

The following are the grounds on which the employer may determine the contractor's employment under the contract.

(1) If the contractor without reasonable cause wholly suspends the carrying out of the works before completion (clause 7.1(a)).

This ground for determination does not apply where the contractor has reasonable cause for suspending the carrying out of the works. Furthermore, a suspension in carrying out a part, even a large part, of the works does not fall within this provision.

A suspension brought about by delays outside the contractor's control will generally amount to reasonable cause. Indeed, the wording may well implicitly require some form of culpability on the part of the contractor.

The proviso at the end of clause 7.1, namely that the notice of determination shall not be given unreasonably or vexatiously must presumably mean that there are situations in which the contractor has without reasonable cause suspended, but where nevertheless it would be unreasonable or vexatious for the employer to determine the employment of the contractor.

This ground for determining the contractor's employment also appeared in JCT 63 (clause 25(1)(a)) and currently appears in JCT 80 (clause 27.1.1).

(2) If the contractor fails to proceed regularly and diligently with the works (clause 7.1(b)).

By clause 2.1 the contractor is under an express duty to proceed regularly and diligently with the works and a failure to do so is a ground for determination. This ground for determining the contractor's employment also appeared in JCT 63 (clause 25(1)(b)) and currently appears in JCT 80 (clause 27.1.2).

The meaning of the words 'regularly and diligently' in the determination provisions of JCT 63 was considered by Mr Justice Megarry in the case of *London Borough of Hounslow* v. *Twickenham Garden Developments Ltd* (1970) where he said (see 7 BLR page 120) as follows:

'These are illusive words, on which the dictionaries help little. The words convey a sense of activity, of orderly progress, and of industry and perseverance: but such words provide little help on the question of how much activity, progress and so on is to be expected. They are words used in a standard form of building contract in relation to functions to be discharged by the architect, and in those circumstances it may be that there is evidence that could be given, whether of usage among architects, builders and building owners or otherwise, that would be helpful in construing the words.'

Certainly the fact that the work is done defectively does not prevent it being carried out regularly – see *Lintest Builders* v. *Roberts* (1978) (10 BLR at page 129). The case of *Greater London Council* v. *Cleveland Bridge and Engineering Co. Ltd* (1984), particularly at first instance, may be relevant here: see page 65.

(3) If the contractor refuses or neglects to comply with a written notice from the architect to remove defective work or improper materials or goods and such refusal or neglect has a material effect on the works (clause 7.1(c)).

The architect has an express power to issue instructions in regard to the removal of any work, materials or goods which are not in accordance with the contract. The instruction under clause 3.14, to be valid, must expressly require the removal. Simply to condemn the work or material is insufficient – see *Holland Hannen and Cubitts* v. *Welsh Health Technical Services Organisation* (1981).

The wording of this provision suggests that it applies before practical completion but not to a failure to make good (as opposed to removal etc.) during the defects liability period – see clause 2.10.

This ground for determination is also in JCT 63 (clause 25(1)(c)) and JCT 80 (clause 27.1.3).

(4) If the contractor fails to comply with the provisions of clauses 3.2 (sub-contracting), or 3.3 (named persons) – (clause 7.1(d)). The failure may be substantial e.g. sub-contracting the whole of the labour element of the works without consent; or it may be much less substantial. In the latter event the proviso relating to unreasonable or vexatious notices of determination must be borne in mind.

There is a similar though not identical provision in JCT 63 (clause 25(1)(d)) and JCT 80 (clause 27.1.4).

Whereas under JCT 63 and 80 failure to comply with the clauses on nomination is not a ground for the determination of the contractor's employment, in IFC 84 a failure by the contractor to comply with clause 3.3 dealing with named persons does qualify as a ground for determination. Bearing in mind that the mischief aimed at in JCT 63 and JCT 80 was that of sub-contracting without consent, it is not easy to see the reasoning behind this additional ground which appears in IFC 84.

General observations on clause 7.1

The opening words 'Without prejudice' make clear what would probably be the position without these words, namely that the determination by the employer of the contractor's employment in accordance with this clause does not prevent the employer pursuing any other remedies which he may have e.g. a claim for damages for breach of contract. The words may even permit the employer in the alternative to accept a repudiation by the contractor which brings not only the contractor's employment to an end but also almost all of the contract terms themselves to an end.

It should also be noted that the mere fact that the contractor complies with clause 7.4 following a purported determination of his employment by the employer under clauses 7.1, 7.2 or 7.3 e.g. giving up possession of the site or compliance with the architect's instructions to remove his plant and equipment, will not prejudice the contractor's right to challenge the validity of the determination.

Clause 7.1 depends for its operation on the contractor being guilty of some default. Such default may not necessarily amount to a breach of contract by the contractor – see *GLC* v. *Cleveland Bridge and Engineering Co. Ltd and Another* (1984) at first instance.

The service of the notices

For a determination of the contractor's employment under clause 7.1, two notices must be served by registered post or recorded delivery. Unlike JCT 63 and JCT 80, the first notice does not have to come from the architect. It can be sent by the employer or by the architect if instructed so to act on the employer's behalf. It is somewhat surprising that the service of the notice is in the hands of the employer rather than the architect. In the case of *J.M. Hill & Sons Ltd* v. *London Borough of Camden* (1980) 18 BLR Lord Justice Ormrod at page 46 said of the provision in JCT 63 clause 25(1)(d) requiring the architect to serve the notice:

> 'The condition requires that such a notice should be given by the architect. A very important qualification in my view . . . It is easy to understand why, in condition 25, it should be the architect: the person who is independent and expert in these matters should be the person to give the certificate and not the employer, who might be peevish or uninformed in one way or another.'

In the case of *London Borough of Hounslow* v. *Twickenham Garden Developments* (1970) the architect, in giving the notice required under clause 25(1) of JCT 63, said:

> 'I therefore hereby give notice under clause 25(1) of the contract dated . . . that in my opinion you have failed to proceed regularly and diligently with the works.'

Of this notice Mr Justice Megarry said as follows (7 BLR at page 115):

> 'I do not read the condition as requiring the architect, at his peril, to spell out accurately in his notice further and better particulars, as it were, of the particular default in question. All that I think the notice need do is to direct the contractor's mind to what is said to be amiss: and this was I think done by this notice.'

Once the first notice has been received by the contractor, he must discontinue the default within 14 days and must not at any time thereafter repeat it. If he continues the default or subsequently repeats it, the employer may determine the contractor's employment by a further notice sent by registered post or recorded delivery.

The employer or architect, as the case may be, should follow the procedural requirements precisely, including the sending of the notice by registered post or recorded delivery. The courts will, however, try to treat the procedural requirements of the contract in a sensible way and not treat them as being too formalistic, treating the notice provisions as directory rather than mandatory if at all possible – see *J.M. Hill & Sons Ltd* v. *London Borough of Camden* (1980).

The second notice i.e. the one determining the contractor's employment operates from the date it is received by the contractor – *J.M. Hill & Sons Ltd* v. *London Borough of Camden* (1980).

Unreasonable or vexatious notice

The notice purporting to determine the contractor's employment must not be given unreasonably or vexatiously. In regard to this Lord Justice Ormrod in *J.M. Hill & Sons Ltd* v. *London Borough of Camden* (1980) at 18 BLR page 49 said:

> '. . . what the word "unreasonably" means in this context, one does not know. I imagine that it is meant to protect the employer who is a day out of time in payment, or whose cheque is in the post, or perhaps because the bank has closed, or there has been a delay in clearing the cheque, or something – something accidental or purely incidental so that the court could see that the contractor was taking advantage of the other side in circumstances in which, from a business point of view, it would be totally unfair and almost smacking of sharp practice'.

'Automatic' determination of contractor's employment (clause 7.2)

Clause 7.2 of the contract provides for an automatic determination of the contractor's employment if any of the following events occur:

(1) the contractor becomes bankrupt (applies to a single individual or an unincorporated association);
(2) the contractor makes a composition or arrangement with his creditors;
(3) the contractor has a proposal in respect of his company for a voluntary arrangement for a composition of debts or a scheme of arrangement approved in accordance with the Insolvency Act 1986;
(4) the contractor has an application made to the court under the Insolvency Act 1986 in respect of his company for the appointment of an administrator;
(5) the contractor has a winding up order made (except for the purposes of amalgamation or reconstruction);
(6) a resolution for voluntary winding up is passed;
(7) a provisional liquidator is appointed;
(8) a receiver or manager of the contractor's business is appointed;
(9) the contractor has an administrative receiver, as defined in the Insolvency Act 1986, appointed;
(10) possession is taken on behalf of the holders of any debentures, secured by floating charge, of any property, comprised in or subject to the floating charge.

The provision for automatic determination has been criticised for a number of reasons. Firstly, it may well not achieve its objective in the case of bankruptcy or liquidation. The consequences of a determination are dealt with in detail below but it can be said here that the contract purports to give the employer certain rights e.g. to withhold payment to make use of materials, goods and plant belonging to the contractor, which may prejudice other creditors of the contractor and in particular run foul of the statutory provisions as to the mandatory *pari passu* discharge of liabilities which cannot be contracted out of – see *British Eagle International Airlines Ltd* v. *Compagnie Nationale Air France* (1975). It may be therefore that, even if this contractual provision is effective to determine the contractor's employment, the consequential provisions set out in clause

7.4 may not be fully effective. However, it should be remembered that in such a situation, the employer will still have some very important rights of set-off under the mutual dealing provisions which apply on bankruptcy and liquidation.

Secondly, on a bankruptcy or liquidation the trustee in bankruptcy or the liquidator, as the case may be, has a statutory right to disclaim unprofitable contracts – see e.g. section 178 of the Insolvency Act 1986. The trustee's of liquidator's obligation is only to carry on the business so far as may be necessary for its beneficial winding up. The decision of whether to complete the contract or disclaim it would, if the automatic determination provisions are effective, be removed from the trustee or liquidator. The notice of intention to disclaim a contract must generally be served within 28 days following an application in writing made by a person interested in that decision. A failure to respond to such a request in writing will be treated as an adoption of the contract. The contractual provisions for determination may therefore be adjudged to be an attempt to circumvent the bankruptcy laws, in which case they will be void.

The contract expressly states that the determination is to be both automatic and forthwith. Even so, it has been suggested that if a contract in truth confers a discretionary right on the employer, some act invoking the provisions of a clause such as this will still be required from the employer – see *Hudson*, 10th edition page 688. In any event the sensible course for the employer would be to inform the contractor, his trustee or liquidator, as the case may be, that the contractor's employment is determined to avoid any argument that there might have been a reinstatement of that employment by the inactivity of the employer being regarded as evidence of his willingness to treat the contractor's employment as reinstated.

The policy of automatic determination of the employment of the contractor in respect of events which fall short of a winding up, seems hard to justify. For instance, the fact of an administrative receiver or an administrator being appointed does not of itself put the contractor in breach of contract and indeed, if permitted to, the contractor might, under the guidance of the administrative receiver or administrator, complete the contract satisfactorily. A determination in such circumstances will leave the employer with only his express contractual rights as set out in clause 7.4. He will not be able to sue for damages for breach of contract as the contractor will not have been guilty of any breach of contract – see *London Borough of Richmond* v. *Samuel Properties (Developments) Ltd* (1983). It may well not be in the best interests of the employer for the contractor's employment to determine automatically in such circumstances.

While clause 7.2 expressly envisages the possibility of there being a reinstatement of the contractor's employment, this is not available as of right to the employer. Depending on the financial position of the contractor at the time of the purported automatic and immediate determination, the employer may have been better served by keeping the contractor's employment intact e.g. there may be little funds in hand to withhold and the contractor may be able to keep trading and complete the contract for the agreed contract sum. Furthermore, it is likely to be easier for an administrative receiver or administrator to obtain a satisfactory novation to another contractor if the contractor's employment is kept intact. It should be recalled that by clause 3.2.1 the contractor is required to include in any sub-contract a provision whereby the sub-contractor's employment automatically determines upon the determination of the contractor's employment under IFC 84. Accordingly, once the sub-contractor's employment has been determined, the contractor or his representatives may be unwilling or unable to carry on. It might therefore be preferable for the employer instead to have an option to determine the contractor's employment in these circumstances rather than for it to be automatic.

Determination of contractor's employment by employer on grounds of corruption (clause 7.3)

Where the employer is a local authority, the contractor's employment may be determined where what may loosely be called matters of corruption are present. Certain of them would no doubt entitle not only a local authority employer but any employer to determine the contract as a whole. No system of notice is provided although clearly the determination must be communicated to the contractor.

Consequences of the determination of the contractor's employment by the employer and of automatic determination (clause 7.4)

Clause 7.4 provides for the consequences of a determination of the contractor's employment under clauses 7.1 to 7.3 inclusive. These may be summarised as follows:

(a) Contractor to relinquish possession of the site
Unlike JCT 63 and JCT 80, IFC 84 contains an express contractual obligation on the contractor to give up possession of the site of the works. It is submitted that the purported determination of the contractor's employment will therefore bring to an end any licence which the contractor may have to occupy the site. Such licence is given by the contract and can be taken away by it. This is so even if

the validity of the purported determination is challenged by the contractor (see the opening two lines of clause 7.4). The decision in the case of *London Borough of Hounslow* v. *Twickenham Garden Developments* (1970), even if a correct statement of the law, which has been doubted (see page 53 and page 333), can have no application under this contract.

(b) Removal from the works of goods, materials and plant etc.
If the architect so instructs, the contractor must remove from the works any temporary buildings, plant, tools, equipment, goods and materials owned by or hired out to him. If he fails within a reasonable time to comply with the instruction, the employer may remove and sell any property belonging to the contractor, holding the net proceeds of sale to the credit of the contractor.

The contractor's failure to remove the temporary buildings etc. would, quite apart from the provisions of the contract, put the employer in the position of an involuntary bailee and as such he may, though the position is not free from doubt, be able to avail himself of the wide powers of sale conferred by sections 12 and 13 of the Torts (Interference with Goods) Act 1977 (see *Chitty on Contract*, 26th edition, paragraph 2660 and paragraphs 2702 to 2707). However, the rights available under the Act will not in any event apply where the employer is aware that the temporary buildings etc. are not owned by the contractor.

(c) Employer may engage others to complete the works and use the goods materials and plant etc. on or adjacent to the works
The employer is given the right to employ and pay others to complete the unexecuted work. This brings into consideration the general duty of a contracting party to mitigate his loss following a breach of contract by the other party. While a determination of the contractor's employment will not necessarily be the result of a breach of contract by the contractor, nevertheless the steps taken by the employer must be reasonable in all the circumstances. This point is discussed further under the next heading.

The employer has the right to enter the site and so do those engaged by him, to complete the unexecuted work, whether or not the contractor has complied with the requirement to give up possession of the site of the works. Clearly, neither the employer nor another contractor would thereby have the right to physically remove the contractor from the site nor use force to obtain control of the temporary buildings etc. The contractor will however be a trespasser and the employer should have little difficulty in obtaining an interim injunction to remove the contractor. Whether the employer could obtain an injunction preventing the contractor from removing his temporary buildings etc, is much more doubtful.

The benefit to the employer of the right to use temporary buildings etc. is likely in practice to be somewhat limited. Firstly, the right cannot apply to any hired or leased property. Secondly, the goods and materials may be the subject of a retention of title clause or the contractor otherwise may not own them and the employer will generally not therefore be able to use them. The subject of retention of title is dealt with earlier in this book on page 29.

Finally, the employer is expressly empowered to purchase all materials and goods necessary for the carrying out and completion of the works and the cost of these will form part of the account required under clause 7.4(d) (discussed under the next heading).

(d) Financial consequences (clause 7.4(d))
The first point to note is the right of the employer to withhold any further payments, even for a certified but unpaid sum, until after completion of the works. Although clause 7.4(d) does not expressly provide for what is to happen if the employer decides to abandon or foreshorten the project, such decision would be likely to be regarded as a completion of the works for the purpose of this clause.

This right to withhold further payment is of considerable importance to the employer. If the contract itself was determined by the employer's acceptance of a repudiation by the contractor, the common law rights of set-off would exist but some quantification might well be necessary to fend off a claim by the contractor for payment for work done. Under the express contractual provisions the right exists to withhold payment until after completion, however small the employer's loss might appear to be and however much is being withheld against it. It is submitted that the employer must act with reasonable dispatch in arranging for the completion of the unexecuted work and that tardiness on his part could well deprive him of this contractual benefit.

On completion of the works and the verification within a reasonable time of the accounts for them, the employer must draw up an account showing the following details:

(i)   *The amount of expenses and direct loss and/or damage caused to the employer by the determination, including the cost of purchasing necessary materials and goods and payments to others to complete the works.*

This can include losses equivalent to those recoverable as damages for breach of contract at common law – see *Wraight Ltd* v. *P.H. and*

T. (Holdings) Ltd (1968). The nature and extent of common law damages has been dealt with briefly earlier in this chapter on page 336.

Though it may be a drafting oversight, there is no express entitlement for the employer to be paid loss and damage as in JCT 63 (clause 25(4)(d)) and JCT 80 (clause 27.4.4). The only reference to it is in relation to the account to be drawn up by the employer which must show such amount. The reference to it in such an account is probably sufficient authority for the employer to claim it.

Included in the amount will be any payments made in purchasing materials and goods or made to other persons to carry out and complete the works. The amount paid by the employer in this connection will be recoverable only to the extent that such expenditure was reasonably incurred. The employer has a duty to take reasonable steps to mitigate his losses. However, the approach of the courts – and so also therefore that of an arbitrator – is unlikely to be a strict one and what is reasonable in all the circumstances is likely to give the employer a good degree of room for manoeuvre. It will not be sufficient for the contractor simply to establish that the outstanding work could have been completed at less cost. He must show that the employer's actions were positively unreasonable.

(ii) *The amount paid to the contractor before the date of determination.*

The account drawn up by the employer must set this out.

Unlike JCT 63 and JCT 80 where the architect has to certify the two amounts referred to, the architect is given no role whatsoever to play in this regard in IFC 84. Contractors may well be disturbed by the absence of an independent official in what is likely to be an acrimonious situation.

The figure produced by (i) and (ii) above has then to be compared with the total amount which would have been payable on a due completion in accordance with the contract, and any difference will be payable to the employer by the contractor, or to the contractor by the employer, as the case may be.

The calculation of the two sets of figures in order to find the balance due or owing will not necessarily be easy. In the first set of figures there will be included, in addition to the amount paid to the original contractor prior to determination, some or all of the amount of the final account of the contract to complete the works, together with any other expense, loss or damage incurred or suffered by the employer arising from the determination. This could include additional professional fees, any additional insurance and the cost of providing security on the site during the period prior to work commencing on the completion contract.

The final account for the completion contract will be based on the rates and prices included in that contract and will include the cost of variations introduced during the period of that contract. The variations which can be included will be not only those variations which would normally have been introduced had the original contractor completed the works, but also those variations requiring the new contractor to repair or replace defective work of the original contractor discovered during the period of the contract so as to complete the works. Such variations will generally be valued on a daywork basis as and when work is carried out. Difficulty may only arise if it can be held that the character and scope of the variations are such as to make it unreasonable to introduce them into the completion contract. If the need for rectification work is known before the completion contract is entered into, then of course it may well be included as measured work in a separate section of the contract documents for that contract.

Clearly any reimbursement to the second contractor of loss and expense arising from a matter for which the employer is responsible will not be included in the calculation of additional cost, unless the loss and expense is suffered as a result of a breach of contract by the original contractor discovered only after the contract had been placed with the completion contractor e.g. discovery of defective work of the original contractor which necessitates the return to site of a direct contractor in the employ of the employer, which in turn causes disturbance of the second contractor's regular progress.

The second figure which must be compared with the cost as indicated above will be the 'notional' cost of completing the contract assuming that there had been no determination. This calculation will be based on the original contract sum adjusted to include not only those variations introduced during the course of the initial contract but also those introduced in the completion contract, valued at the rates and prices contained in the original contract.

The difference between the two sets of figures will, in accordance with the clause, be a debt payable by the contractor to the employer or by the employer to the contractor as the case may be.

Contractor's bankruptcy or liquidation

As has already been noted on page 352, the employer's rights set out in clause 7.4 may not be valid where there has been a purported automatic determination on bankruptcy or liquidation of the contractor.

NOTES TO CLAUSES 7.1 to 7.4

[1] '. . . *wholly* . . .'

It is submitted that the *de minimis* rule will apply so that if the contractor's suspension is virtually complete, though not absolutely, so that a small and insignificant amount of work is continuing, the employer will still be able to rely on this ground.

[2] '. . . *local authority* . . .'.

The right to determine on grounds of corruption might be thought to be appropriate for other public bodies in addition to local authorities.

Clauses 7.5 to 7.7

Determination by Contractor

7.5   Without prejudice to any other rights and remedies which the Contractor may possess the Contractor may give to the Employer a notice by registered post or recorded delivery specifying a matter referred to in this clause 7.5. If the Employer shall continue to make default in respect of the matter specified for 14 days after receipt of such notice or shall at any time thereafter repeat such default (whether previously repeated or not), then the Contractor may thereupon by notice by registered post or recorded delivery determine the employment of the Contractor under this Contract; provided that such notice shall not be given unreasonably or vexatiously.

The matters referred to in the preceding paragraph are:

7.5.1   the Employer does not pay an amount properly due[1] to the Contractor under clause
4.2 *(Interim Payments)*,
4.3 *(Interim Payment on Practical Completion)*
or
4.6 *(Final certificate)*;

7.5.2   the Employer interferes with or obstructs the issue of any certificate due under the Contract;

7.5.3   the carrying out of the whole or substantially the whole of the uncompleted Works (other than the execution of the work required under clause 2.10 *(Defects liability)* is suspended for a continuous period of one month[2] by reason of:

(a) the Architect's/the Contract Administrator's instructions issued under clauses

1.4 *(Inconsistencies)*;
3.6 *(Variation)* or
3.15 *(Postponement)*

unless caused by reason of some negligence or default of the Contractor, his servants or agents or of any person employed or engaged upon or in connection with the Works or any part thereof, his servants or agents other than the Employer or any person employed, engaged or authorised by the Employer or by any local authority or statutory undertaker executing work solely in pursuance of its statutory obligations; or

(b) the Contractor not having received in due time necessary instructions, drawings, details or levels from the Architect/the Contract Administrator for which he specifically applied in writing provided that such application was made on a date which having regard to the Date for Completion or to any extended time fixed under clause 2.3 *(Extension of time)* was neither unreasonably distant from nor unreasonably close to the date on which it was necessary for him to receive the same; or

(c) delay in the execution of work not forming part of this Contract by the Employer himself or by persons employed or otherwise engaged by the Employer as referred to in clause 3.11 or the failure to execute such work or delay in the supply by the Employer of materials and goods which the Employer has agreed to supply for the Works or the failure so to supply; or

(d) failure of the Employer to give in due time ingress to or egress from the site of the Works or any part thereof through or over any land, buildings, way or passage adjoining of connected with the site and in the possession and control of the Employer, in accordance with the Contract Documents after receipt by the Architect/ the Contract Administrator of such notice, if any, as the Contractor is required to give, or failure of the Employer to give, such ingress or egress as otherwise agreed between the Architect/the Contract Administrator and the Contractor.

Employer becoming bankrupt etc.

7.6 If the Employer becomes bankrupt or makes a composition or arrangement with his creditors or has a proposal in respect of his company for a voluntary arrangement for a composition of debts or scheme of arrangement approved in accordance with the Insolvency Act 1986, or has an application made under the Insolvency Act 1986 in respect of his company to the court for the appointment of an administrator, or has a winding up order made or (except for the purposes of amalgamation or reconstruction) has a resolution for voluntary winding up passed or a provisional liquidator, receiver or manager of his business or undertaking is duly appointed, or has an administrative receiver, as defined in the Insolvency Act 1986, appointed, or possession is taken, by or on behalf of the holders of any debentures secured by a floating charge, of any property comprised in or subject to the floating charge, the Contractor may thereupon by notice by registered post or recorded delivery determine the employment of the Contractor under this Contract.

Consequences of determination under clause 7.5 or 7.6

7.7   Without prejudice to the accrued rights or remedies of either party or to any liability of the classes mentioned in clause 6.1 (*Injury*) which may arise either before the Contractor or any sub-contractors shall have removed his or their temporary buildings, plant, tools, equipment, goods or materials or by reason of his or their so removing the same, upon such determination under clause 7.5 or 7.6:

(a) the Contractor shall with all reasonable dispatch and in such manner and with such precautions as will prevent injury, death or damage of the classes in respect of which before the date of determination he was liable to indemnify the Employer under clause 6.1 remove from the site all his temporary buildings, plant, tools, equipment, goods and materials and shall give facilities to his sub-contractors to do the same;

(b) after taking into account amounts previously paid under this Contract the Contractor shall be paid by the Employer the total value of work at the date of determination, any sum ascertained[3] in respect of direct loss and/or expense under clause 4.11 (*Disturbance of progress*), the cost of materials or goods properly ordered for the Works for which the Contractor shall have paid or for which the Contractor is legally bound to pay (and on such payment any materials and goods so paid for shall become the property of the Employer), the reasonable cost of removal under the preceding paragraph and any direct loss and/or damage caused to the Contractor by the determination.

COMMENTARY TO CLAUSES 7.5 TO 7.7

Clauses 7.5 to 7.7 deal with the determination of the contractor's employment by the contractor and the consequences flowing from this. Although there are some similarities with JCT 63 and JCT 80 so far as the grounds for determination are concerned, significant differences exist and attention will be drawn to some of these in the discussion which follows.

Grounds for determination of the contractor's employment by the contractor

(1) If the employer does not pay an amount properly due under an interim or final certificate (clause 7.5.1).

For an interim certificate the employer has 14 days from its date in which to pay under clause 4.2 and 4.3. In the case of the final certificate the sum, if any, payable to the contractor is not due until the twenty-eighth day after the date of the final certificate under clause 4.6. If the failure to pay continues after the contractor has served notice specifying the default or is repeated (not applicable to the final certificate), the contractor can by notice determine his own employment, though in the case of a failure to pay on the final certificate such a course of action by the contractor would generally be pointless as the contractual remedies available to the contractor in clause 7.7 could hardly be applicable.

(2) If the employer interferes with or obstructs the issue of any certificate due under the contract (clause 7.5.2)

Interference or obstruction by the employer in relation to the issue of any certificate by the architect is a very serious matter indeed. It is certainly a serious breach of contract and may well be seen as undermining the independence of the architect in certain circumstances. This is sometimes not fully appreciated by officers within the employers' organisations, particularly auditors within a local or other public authority. Also, lay councillors may not always understand the architect's independent role especially when he is an employee of the authority.

Unlike JCT 63 and JCT 80, under IFC 84 the contractor must first give notice specifying the default and it is only if the default continues or is subsequently repeated that the notice of determination may be issued. Otherwise the ground itself is as stated in JCT 63 and JCT 80.

(3) If the carrying out of the whole or substantially the whole of the uncompleted works (other than work required under the defects liability clause (clause 2.10)) is suspended for a continuous period of one month by reason of:
(a) instructions in relation to:
  (i) inconsistencies etc. (clause 1.4)

(ii)   variations (clause 3.6)

(iii)   postponement (clause 3.15)

unless caused by negligence or default on the part of the contractor, his servants or agents or any person employed or engaged upon or in connection with the works or any part thereof, his servants or agents other than the employer or any person employed, engaged or authorised by the employer or by any local authority or statutory undertaker executing works solely in pursuance of its statutory obligations (clause 7.5.3(a));

(b)   late instructions, drawings, details or levels which the contractor has requested at the appropriate time (clause 7.5.3(b)) (see the Commentary on identical words used in clause 2.4.7 page 79);

(c)   delays or failure by the employer or others employed or engaged by him under clause 3.11 on work not forming part of the contract works or in the supply by the employer of materials or goods which he has agreed to supply for the works (clause 7.5.3(c)) (see Commentary on similar words used in clause 2.4.8 and clause 2.4.9 page 81);

(d)   failure of the employer to give in due time agreed ingress or egress over adjoining property over which he exercises possession and control (clause 7.5.3(d)) (see Commentary on the similar wording of clause 2.4.12, page 83).

The first point to note is that in relation to all these grounds the employer can be said to have some responsibility and in many cases the matter will or should be within his control. Provisions enabling the contractor to determine his own employment in respect of what might be called the neutral events, namely *force majeure*, loss or damage occasioned by clause 6.3 perils and civil commotion, are dealt with separately in IFC 84 in clause 7.8 (see below).

That this is a major improvement over JCT 63 from the employer's point of view becomes clear when the consequences of a determination under clause 7.5 are considered, particularly the entitlement of the contractor to recover his loss and damage (clause 7.7.(b) – dealt with in detail below) and to which under JCT 63 clause 26(1) he is entitled even in respect of a prolonged suspension due to neutral events completely outside the employer's control.

Bearing in mind that loss and damage has a similar meaning to damages recoverable for breach of contract at common law, the correction of the potential injustice thereby created was long overdue. JCT 80 clause 28.1 was, until Amendment 4 of July 1987, similar to JCT 63. However, since then it has mirrored the IFC 84 position.

It is interesting to note that there is no provision at all in IFC 84 for determination following a prolonged suspension due to an instruction as to opening up and testing issued under clause 3.12 (unlike JCT 63 and JCT 80 in this respect).

Clause 7.5.3(a) of IFC 84 as originally published ended with the words 'unless caused by reason of some negligence or default of the Contractor...'. By Amendment 3 of July 1988 this clause was amended by adding a reference to the contractor's servants, agents, or any person employed or engaged upon or in connection with the works or their servants or agents, other than the employer or those for whom the employer is responsible or any local authority or statutory undertaker executing works solely in pursuance of its statutory obligations. This additional wording is intended to make it clear that the negligence or default of the contractor does not have to be personal and can include that of his sub-contractors etc. In the case of *John Jarvis Ltd* v. *Rockdale Housing Association Ltd* (1987) the Court of Appeal had held in relation to clause 28.1.3.4 of JCT 80 that identical words to those in IFC 84 clause 7.5.3, before Amendment 3, referred to negligence or default of the contractor himself, that is, the management or employees of the contractor, and did not extend to nominated sub-contractors (and presumably did not extend to domestic sub-contractors either). The contractor in that case was thus able to determine his own employment after a one month suspension where the cause of postponement was the need for redesign of the piling system following defective work on the part of a nominated sub-contractor. JCT 80 has since been amended by Amendment 4 of July 1987 in a similar way to the amendment to IFC 84, except that while under IFC 84 named sub-contractors are included in the additional wording, under JCT 80 nominated sub-contractors are expressly excluded so that if the facts of the *Jarvis* case were to recur, the result under JCT 80, even as amended, would be the same.

It may be that in any event the different arrangement of the determination clauses in IFC 84 even as originally published (compare clauses 7.5.3 and 7.8.1 of the unamended IFC 84 with clause 28.1.3 of the unamended JCT 80) would have produced a different result. Part at least of the reasoning of the Court of Appeal in support of the decision, namely the reference in clause 28.1.3.2 of JCT 80 to '. . . unless caused by the negligence of the Contractor, his servants or agents or any sub-contractor, his servants or agents . . .' might not apply with equal force to IFC 84 where the equivalent provision is to be found in an entirely separate clause, namely clause 7.8. Nevertheless, the additional words introduced by Amendment 3 put the matter beyond doubt.

The period of suspension is fixed at one month for the listed matters. Although this contract is not designed for contract periods beyond 12 months (see Practice Note 20), some employers may still regard this period as too short in many situations.

The rights given to the contractor by clause 7.5 are without prejudice to any other rights or remedies which he may possess.

The service of the notices
The notices under clause 7.5 must be sent by registered post or recorded delivery. In JCT 63 and JCT 80, except in the case of non-payment where two notices are required, only one notice is required from the contractor to determine his own employment. Under IFC 84 however, there must always be two notices, except in the case of bankruptcy or liquidation etc. (clause 7.6 below). Some of the general comments made earlier in relation to the service of notices and the proviso that the notice shall not be given unreasonably or vexatiously where the employer determines the contractor's employment are, where appropriate, relevant here also (see page 349).

The requirement for a first notice is of course a major benefit to an employer, particularly perhaps the inattentive or inefficient one, as it is less likely that he will be taken by surprise. For example, if an instruction results in a postponement which suspends the carrying out of the uncompleted work for at least one month, the contractor cannot simply determine his own employment. He must first serve a notice specifying on which ground he is entitled to determine and the employer has at least the opportunity of remedying the situation. A similar provision in JCT 80 just might have enabled the employer in the *Jarvis* case to head off a determination by the contractor of his own employment.

(4)   Employer's bankruptcy, liquidation etc. (clause 7.6)

This ground is similar to that in clause 7.2 in relation to the contractor's bankrutpcy, liquidation etc. except that it is not stated to be automatic as in clause 7.2. The contractor must serve notice in the manner prescribed. There is no requirement for a first notice specifying the default or failure. As it is the contractor's choice to determine his own employment, there is no provision here for the possibility of a reinstatement of the contractor's employment by agreement with the employer's trustee in bankruptcy, administrator, liquidator, administrative receiver etc. (or with the employer himself in appropriate circumstances), though there could of course be such an arrangement made between the parties outside the terms of this contract.

Clause 7.6 does not discriminate, as do JCT 63 and JCT 80, between a private edition of the contract (where the equivalent of this provision is to be found) and the local authorities edition (where there is no such provision). In IFC 84 all employers are covered.

Any bankruptcy or liquidation must by definition have taken effect before a notice under clause 7.6 to determine can be issued on such grounds. Clearly there is a real chance in such an event that either the notice will be invalidated as an interference with the bankruptcy laws, or even if valid to determine the contractor's employment, it will be ineffective so far as many of the consequences set out in clause 7.7 are concerned, where such consequences would disturb the due and fair distribution of the employer's assets following the bankruptcy or liquidation.

Many of the comments made above in relation to the contractor's bankruptcy, liquidation etc. are relevant here (see page 351).

Consequences of a determination by the contractor of his own employment (clause 7.7)
Clause 7.7 provides for the consequences of a determination of the contractor's employment under clauses 7.5 or 7.6. The contractor's rights and entitlement are expressly given without prejudice to the accrued rights or remedies of either party e.g. the contractor's right to accept a repudiation of contract by the employer in appropriate circumstances. Despite the express reference to the contractor's other rights and remedies, care should be taken to expressly preserve them where any action taken under clause 7.5 may appear to be inconsistent with them (see comments under general observations on clause 7.1 on page 349).

The consequences of a determination as stated in clause 7.7 are also without prejudice to any liability of the kind covered in clause 6.1 i.e. injury to persons or damage to property and indemnities in

respect thereof, which arise before the removal by the contractor or
any sub-contractor of plant and equipment etc. or by reason of such
removal.

The consequences of a determination under clauses 7.5 or clause
7.6 are as follows:

(a) The contractor must remove his temporary buildings, plant,
     tools, equipment, goods and materials and give his sub-
     contractor the same facilities.

     In so doing the contractor is expressly required to remove
     his temporary buildings etc. in such a manner as to prevent
     personal injury or damage to property.

(b) The contractor is to be paid in respect of the following:
     (i)   value of work done but not yet paid for;
     (ii) loss and expense ascertained but unpaid;
     (iii) cost of goods and materials properly ordered and paid
           for;
     (iv) reasonable cost of removal under (a) above;
     (v)   direct loss and/or damage caused by the determination.

In relation to (iii) such goods and materials, once paid for by the
employer, are to become the employer's property.

Under (v) above the direct loss and/or damage is equivalent to
that which could be recovered as damages for breach of contract at
common law – see *Wraight Ltd* v. *P.H. and T (Holdings) Ltd*
(1968). This topic has been discussed in detail earlier on page 336.

As mentioned on page 362, the loss and/or damage provision
does not extend to what might be termed the 'neutral events' which
can give rise to a determination and which are now separately
treated in clauses 7.8 and 7.9.

Like JCT 80 but unlike JCT 63, IFC 84 contains no express
provision entitling the contractor to take possession of unfixed
goods or materials which may have become the property of the
employer together with a lien over them in respect of monies due.

NOTES TO CLAUSES 7.5 TO 7.7

[1] '. . . properly due . . .'.
The employer is entitled expressly to make certain deductions from
certified sums, the most obvious being liquidated damages. However, the employer may also have certain common law rights of set-
off (discussed earlier in this book on page 206). If the employer

purports to set off and his claim to do so is genuine such as to raise at least a *prima facie* argument of valid set-off, then it is suggested that the contractor needs to take great care before purporting to determine his own employment on this ground since, if he is mistaken, his act will amount to a repudiation, leaving the employer with little alternative but to accept it.

[2] '. . . *month* . . .'.
This means calendar month – see Law or Property Act 1925, section 61.

[3] '. . . *any sum ascertained* . . .'.
In JCT 63 (clause 26(2)(b)(iii)) and JCT 80 (clause 28.2.2.3) there are expressly included sums in respect of loss and expense whether ascertained before or after the date of determination. IFC 84 refers only to 'any sum ascertained'. Does the architect or the quantity surveyor have a duty after the contractor's employment has been determined to carry out and ascertain loss and expense on receipt (whether before or after the date or determination) of an application from the contractor as required under clause 4.11? If not, there is a gap in the contractual machinery and the contractor would have to rely on his common law remedy of damages for breach of contract, if indeed there had been a breach. It is submitted that the words used are sufficiently wide to require an ascertainment to be carried out in respect of any loss and expense incurred for which the contract entitles the contractor to reimbursement provided the appropriate application is made, particularly bearing in mind that the contractual provisions continue as necessary after a determination of the contractor's employment.

Clauses 7.8 and 7.9

Determination by Employer or Contractor

7.8.1   Without prejudice to any other rights or remedies which the Employer or Contractor may possess if the carrying out of the whole or substantially the whole of the uncompleted Works (other than the execution of work required under clause 2.10 (*Defects liability*)) is suspended for a period of three months by reason of:

(a) force majeure; or

(b) loss or damage to the Works occasioned by any one or more of the Specified Perils; or

(c) civil commotion;

then the Employer or the Contractor may by notice by registered post or recorded delivery to the Contractor or to the Employer forthwith determine the employment of the Contractor under this Contract; provided that such notice shall not be given unreasonably or vexatiously.

7.8.2　The Contractor shall not be entitled to give notice under clause 7.8.1 where the loss or damage to the Works occasioned by one or more of the Specified Perils was caused by the negligence of the Contractor, his servants or agents, or by any sub-contractor, his servants or agents.

Consequences of determination under clause 7.8

7.9　Upon such determination under clause 7.8 (or upon determination under clause 6.3C.4.3) the provisions of clause 7.7 shall apply with the exception of the words at the end of clause 7.7 'and any direct loss and/or damage caused to the Contractor by the determination'.

COMMENTARY ON CLAUSES 7.8 AND 7.9

Clauses 7.8 and 7.9 deal with the determination of the employment of the contractor by the employer or the contractor where the carrying out of the whole, or substantially the whole, of the uncompleted works (other than a suspension caused by a failure to remedy defects under clause 2.10 (defects liability)) is suspended for a period of three months owing to certain events whose common thread is that they are beyond the control of either party.

This exercise by either party of the right to determine the contractor's employment is without prejudice to any other rights or remedies which they may possess. The significance of this has been discussed earlier on page 349.

Grounds for determination by employer or contractor

If there is a suspension of the carrying out of the whole, or substantially the whole, of the uncompleted works for a period of three months by:

(a) *force majeure*
(b) loss or damage to the works from a specified peril
(c) civil commotion

either party may by notice determine the employment of the contractor.

The first ground, namely *force majeure* is of uncertain meaning. What is certain is that its interpretation in this contract will be affected by the context in which it appears, so that while in one contract it may extend to freak weather, national strikes, embargoes etc. in another contract, where these other matters are separately listed or treated, the meaning of *force majeure* will not extend to such matters. In IFC 84 clause 2.4 (events for extensions of time) listed separately in addition to *force majeure* are exceptionally adverse weather conditions, specified perils, civil commotion, lockouts etc. Its meaning in clause 2.4 does not therefore include these events.

In clause 7.8.1 apart from *force majeure* there is also reference to specified perils and civil commotion. Weather is not referred to and neither are strikes, lockouts etc. Does *force majeure* when used in clause 7.8.1 therefore include exceptionally inclement weather or strikes, lockouts etc? Although the same words used in the same contract will generally be construed as carrying the same meaning, in which case the words would exclude these events as they do in clause 2.4, this construction will not necessarily prevail where the context in which the words are used differs in such a way as to make it clear that another meaning is to be given to the words. It is submitted that while the context is somewhat different, it would have been logical to include express reference to weather conditions, strikes, lockouts etc., had this been the intention of the parties, and that therefore the words '*force majeure*' carry the same meaning in clause 2.4 and clause 7.8.1.

See Commentary to clause 2.4.1 (page 76) for a further discussion on the meaning of *force majeure*.

In relation to loss or damage to the works occasioned by specified perils, clause 7.8.2 provides that the contractor shall not be entitled to determine his employment under this head where the loss or damage was caused by the negligence of the contractor or his sub-contractor. It is unfortunate that this proviso to clause 7.8.1 is worded differently to the proviso in clause 7.5.3(a) in relation to suspensions attributable to architect's instructions pursuant to clause 1.4 (inconsistencies); clause 3.6 (variations) or clause 3.15 (postponement). Both provisions are almost certainly intended to cover the same ground. It is probable that when clause 7.5.3(a) was amended by Amendment 3 of July 1988, clause 7.8.2 should have been amended to bring the two provisos into line (c.f. JCT 80 clauses 28.1.3.1 and 28A.2).

The contractor's negligence does not of course prevent the employer from determining the contractor's employment under clause 7.8.1. It is surprising that clause 7.8.2 does not purport to disentitle the employer from determining the contractor's employment where the employer or his servants or agents have negligently caused the loss or damage occasioned by the specified peril. This could of course happen, for example, negligence on the part of the employer's direct contractors. While generally to construe words in a contract so as to enable one party to exercise such a fundamental right as a result of its own negligence or the negligence of those for whom that party is contractually responsible would be unlikely without express clear wording to that effect, the fact that here the contractor's right to do so is expressly taken away, whereas the employer's is not, may well, it is submitted, allow the employer to operate clause 7.8.1 in such circumstances.

Alternatively, if it was regarded as necessary for the employer to have the right to determine the contractor's employment even where the loss or damage to existing structures was brought about by the employer's negligence, it would have at least been more reasonable to allow the contractor any loss or damage brought about by a determination taking place in such circumstances.

The suspension must last at least three months. However, unlike the one month's suspension entitling the contractor to determine his own employment under clause 7.5.3, the suspension is not expressly stated to be continuous. It is certainly arguable therefore that intermittent periods of suspension can be added together, and if they exceed three months then the notice of determination can be given.

It is unlikely that individual periods of less than three months, arising from different heads which when added together exceed three months, can be used as a basis for determination of the contractor's employment. If this had been intended one would expect to have seen this clearly stated.

Service of the notice

Only one notice is required, there being no preliminary or warning notice. The notice is to be sent by registered post or recorded delivery. It must not be given unreasonably or vexatiously: see earlier Commentary under clauses 7.1 and 7.5.

Consequences of determination under clause 7.8 (clause 7.9)

The consequences that flow from a determination under clause 7.8 are the same as those applying where the contractor determines his own employment under clause 7.5 or clause 7.6 except that the contractor is not entitled to recover his direct loss and/or damage caused by the determination. This renders the consequences of a determination of the contractor's employment following a suspension due to such events as are listed which are beyond the control of the employer, more equitable than the determination provisions in JCT 63 where in such circumstances the contractor can recover his loss and damage.

Determination under clause 6.3C.4.3

It is possible for either party to determine the employment of the contractor under clause 6.3C.4.3 (as stated in clause 7.9) where, following loss or damage to the works, it is just and equitable so to do – see earlier Commentary to clause 6.3C on page 322 for the probable reasoning behind this provision. However, the notice must be sent within 28 days of the occurrence of such loss or damage. Within seven days of receipt the recipient contractor or employer, as the case may be, may in effect challenge the determination and require the question of whether it is just and equitable to determine the contractor's employment to be adjudicated on in arbitration.

Chapter 11

Interpretation of the contract

CONTENT

This chapter briefly considers section 8 which deals with matters of contract interpretation, construction and definition.

SUMMARY OF GENERAL LAW

Where a contract contains clauses dealing with the interpretation or definition of phrases or words, then clearly reference will have to be made to that clause in deciding the meaning of words or phrases to be found in the contract. However, general principles of law relating to the construction of contracts will also be highly relevant. It is not possible in this book to do more than mention this topic and, if interested, the reader should refer to textbooks which deal with it in more detail e.g. *Chitty on Contracts*, 26th edition at paragraphs 808 to 844; *Building Contracts* by Donald Keating 4th edition, chapter 3.

CONSIDERATION OF THE RELEVANT CLAUSES OF IFC 84

Clause 8.1

References to clauses, etc.

8.1 Unless otherwise specifically stated a reference in the Articles of Agreement, the Conditions, the Supplemental Conditions or the Appendix to any clause means that clause of the Conditions or the Supplemental Conditions.

COMMENTARY ON CLAUSE 8.1

This clause calls for no comment.

Clause 8.2

Articles etc. to be read as a whole

8.2 The Articles of Agreement, the Conditions, the Supplemental Conditions and the Appendix are to be read as a whole and the effect or operation of any article or clause in the Conditions or the Supplemental Conditions or item in or entry in the Appendix must therefore unless otherwise specifically stated be read subject to any relevant qualification or modification in any other article or any of the other clauses in the Conditions, the Supplemental Conditions or item in or entry in the Appendix.

COMMENTARY ON CLAUSE 8.2

This clause makes it quite clear that the articles, clauses and items in the appendix which make up the contract conditions are to be read together. Any particular article, clause etc. must not therefore be considered in isolation but must be read in the light of other articles or clauses etc. and, unless otherwise expressly stated, must be read subject to any qualification or modification in any other article or clause etc.

Clause 8.3

Definitions

8.3 Unless the context otherwise requires or the Articles or the Conditions or the Supplemental Conditions or an item or entry in the Appendix specifically otherwise provides, the following words and phrases in the Articles of Agreement, the Conditions, the Supplemental Conditions and the Appendix shall have the meanings given below:

All Risks Insurance:
see clause 6.3.2.

Appendix:
means the Appendix to the Conditions as completed by the parties.

Approximate Quantity:
means a quantity in the Contract Documents identified therein as an approximate quantity*.

Articles or Articles or Agreement:
means the Articles of Agreement to which the Conditions are annexed, and references to any recital are to the recitals set out before the Articles.

Base Date:
means the date stated in the Appendix.

Contract Sum Analysis (see 2nd recital):
means an analysis of the Contract Sum provided by the
Contractor in accordance with the stated requirements of
the Employer.

Excepted Risks:
means ionising radiations or contamination by radioactiv-
ity from any nuclear fuel or from any nuclear waste from
the combustion of nuclear fuel, radioactive toxic explosive
or other hazardous properties of any explosive nuclear
assembly or nuclear component thereof, pressure waves
caused by aircraft or other aerial devices travelling at
sonic or supersonic speeds.

Form of Tender and Agreement NAM/T:
means the Form issued under that name by the Joint
Contracts Tribunal for use where a person is to be named
as a sub-contractor under clause 3.3.

Joint Names Policy:
means a policy of insurance which includes the Contractor and the Employer as the insured.

person:
means an individual, firm (partnership) or body corporate.

provisional sum:
where the Contract Documents include bills of quantities,
includes a sum provided in such bills for work whether or
not identified as being for defined or undefined work*.

Schedules of Work:
means an unpriced schedule referring to the Works
which has been provided by the Employer and which if
priced by the Contractor (as mentioned in the 2nd recital)
for the computation of the Contract Sum is included in
the Contract Documents.

Site Materials:
see clause 6.3.2.

Specified Perils:
means fire, lightning, explosion, storm, tempest, flood,
bursting or overflowing of water tanks, apparatus or
pipes, earthquake, aircraft and other aerial devices or
articles dropped therefrom, riot and civil commotion, but
excluding Excepted Risks.

Sub-Contract Conditions NAM/SC:
means the Sub-Contract Conditions NAM/SC incorporated by reference in Article 1.2 of Section III of the
Tender and Agreement NAM/T.

Supplemental Conditions:
means the clauses set out or referred to after the Appendix and referred to in clauses
4.9(a) *(Tax etc. fluctuations),*
4.9(b) *(Formulae fluctuations),*
5.5 *(VAT),* and
5.6 *(Statutory tax deductions).*

---

* General Rules 10.1 to 10.6 of the Standard Method of
Measurement 7th Edition provide:

10.1
Where work can be described and given in items in
accordance with these rules but the quantity of work
required cannot be accurately determined, an estimate of
the quantity shall be given and identified as an approximate quantity.

10.2
Where work cannot be described and given in items in
accordance with these rules it shall be given as a Provisional Sum and identified as for either defined or undefined work as appropriate.

10.3
A Provisional Sum for defined work is a sum provided for
work which is not completely designed but for which the
following information shall be provided:
(a) The nature and construction of the work.
(b) A statement of how and where the work is fixed to the
building and what other work is to be fixed thereto.
(c) A quantity or quantities which indicate the scope and
extent of the work.
(d) Any specific limitations and the like identified in Section A35.

10.4
Where Provisional Sums are given for defined work the
Contractor will be deemed to have made due allowance
in programming, planning and pricing Preliminaries. Any
such allowance will only be subject to adjustment in
those circumstances where a variation in respect of other
work measured in detail in accordance with the rules
would give rise to adjustment.

10.5
A Provisional Sum for undefined work is a sum provided
for work where the information required in accordance
with rule 10.3 cannot be given.

10.6
Where Provisional Sums are given for undefined work
the Contractor will be deemed not to have made any
allowance in programming, planning and pricing
Preliminaries.

COMMENTARY ON CLAUSE 8.3

Certain words and phrases are given specific meanings for the
purpose of construing the contract. Many of them are self-
explanatory and need little comment. Where relevant reference is
made to the meaning of these words or phrases in the discussion of
the various articles or clauses in which they appear throughout this
book.

Clause 8.4

'The Architect'/'The Contract Administrator'

8.4   Where the person named in Article 3 is entitled to the
        name 'Architect' under and in accordance with the Archi-
        tects (Registration) Acts 1931 to 1969 the term 'the Con-
        tract Administrator' shall be deemed to have been deleted
        throughout the contract but where the person named is
        not so entitled, the term 'the Architect' shall be deemed
        to have been deleted.

COMMENTARY ON CLAUSE 8.4

There is no comment to be made on this clause.

Clause 8.5

Price Specification or Priced Schedules of Work

8.5   Where in the Conditions there is a reference to the 'Spec-
        ification' or the 'Schedules of Work' then, where the 2nd
        Recital alternative A applies, such reference is to the
        Specification of the Schedules of Work as priced by the
        Contractor unless the context otherwise requires.

COMMENTARY ON CLAUSE 8.5

There is no comment to be made on this clause.

Chapter 12

Settlement of disputes – Arbitration

CONTENT

This chapter looks at section 9 which together with article 5 of IFC 84 deals with arbitration. These provisions satisfy the requirements of section 32 of the Arbitration Act 1950 so as to be an 'arbitration agreement' within the Act. Section 32 provides that an 'arbitration agreement' means a written agreement to submit present or future differences to arbitration, whether an arbitrator is named therein or not.

Section 9 of IFC 84 was introduced by way of Amendment 3 dated July 1988. It not only replaces much of what was previously part of article 5 of IFC 84, it also contains a number of significant additional provisions.

CONSIDERATION OF THE RELEVANT CLAUSES OF IFC 84

Clauses 9.1 and 9.2

9.1   When the Employer or the Contractor require a dispute or difference as referred to in Article 5 to be referred to arbitration then either the Employer or the Contractor shall give written notice to the other to such effect and such dispute or difference shall be referred to the arbitration and final decision of a person to be agreed between the parties as the Arbitrator, or, upon failure so to agree within 14 days after the date of the aforesaid written notice, of a person to be appointed as the Arbitrator on the request of either the Employer or the Contractor by the person named in the Appendix.

9.2   The proper law of this Contract shall be English law.

COMMENTARY ON CLAUSES 9.1 AND 9.2

Article 5 refers to '. . . any dispute or difference as to the construction of this Contract or any matter or thing of whatsoever nature arising thereunder or in connection therewith'. This phrase also appeared in clause 35 of JCT 63, and was given a liberal interpretation by the Court of Appeal in the case of *Ashville Investments Ltd* v. *Elmer Contractors Ltd* (1987) so as to bestow upon the arbitrator a jurisdiction which includes the power to grant rectification of the contract arising out of alleged innocent or negligent mispresentation.

Scope of disputes or differences referable to arbitration and the proper law

The disputes will include disputes as to any matters left by the contract to the discretion of the architect, the withholding by the architect of a certificate; adjustments of the contract sum; rights and liabilities upon determination and the alleged unreasonable withholding of consent or agreement by the employer or the architect or by the contractor where the contract requires such consent not to be unreasonably withheld.

The proper law of contract which is to apply to the settlement of disputes is English law.

There is no restriction as to the time when a reference to arbitration can be initiated; unlike JCT 63 and JCT 80 where only certain matters can be arbitrated before practical completion.

The notice and commencement of the arbitration

When compared with the previous article 5 of IFC 84 prior to Amendment 3 of July 1988, clause 9.1 makes it clearer that one party must give to the other a written notice requiring the dispute or difference to be referred to arbitration.

In the case of *Emson Contractors Ltd* v. *Protea Estates Ltd* (1987) the court considered the arbitration agreement in an early version of JCT 80 together with the reference in clause 30.9.3 of that contract to the words 'If any arbitration or other proceedings have been commenced . . .'. A somewhat similar phrase is to be found in clause 4.7 of IFC 84. His Honour Judge Fox-Andrews QC was of the view, not having however heard argument on the point from counsel, that 'commenced' meant that one party had invited the other to agree to a named person as arbitrator, or, where one party made a written request to the other to concur in the appointment of an arbitrator, whichever first occurred. The issue is now probably put beyond doubt. The arbitration commences for the purposes of clause 4.7 of IFC 84 and for limitation purposes under the Limitation Act 1980 section 34(3) when one party gives written notice as is now expressly required under clause 9.1. One slight doubt remains as a result of the continued use of the phrase, e.g. in Article 5 that any dispute or difference which arises '. . . shall be *and is hereby* referred to arbitration . . .' whereby it might be argued, somewhat faintly, that the arbitration has commenced automatically upon a dispute or difference arising. The obvious inconvenience of such a construction would tend against its being accepted.

As arbitration depends upon an agreement to arbitrate between the parties and also because the notice referring the dispute or difference to arbitration will generally stop time from running for limitation purposes, it is important that the notice describes the dispute or difference in the broadest possible terms, otherwise the recipient of the notice may argue at a later date that a claim actually being made in the arbitration as identified in the statement of case is outside the scope of the written notice. Indeed, the objecting party may even be able to set aside the arbitrator's award made on matters outside the scope of the written notice where that party has properly reserved its position as to the arbitrator's lack of authority, despite having gone ahead with the arbitration including the claim objected to. Some specimen notices to refer disputes or differences to arbitration opt for the broadest possible wording such as 'I require you to concur with me in the appointment of an arbitrator for the settlement of the disputes which have arisen between us . . .' (see *Russel on Arbitration* 20th Edition published by Stephens & Sons). While this may be the safest course it may not be the most helpful. A possible answer is to be as specific as possible in defining the disputes or differences which have arisen and to conclude with some such phrase as 'and without prejudice to the particularity of the foregoing any other disputes or differences which have arisen between us'.

Agreeing the arbitrator

Clause 9.1 requires the dispute or difference to be referred to a person to be agreed between the parties. If no agreement is reached within 14 days of the written notice, either party can request an

appointment to be made by the president or a vice president of
whichever appointing body has been selected in the appendix to the
contract, namely the Royal Institute of British Architects, the Royal
Institution of Chartered Surveyors or the Chartered Institute of
Arbitrators. If the appointor is not selected in the appendix, unlike
JCT 80, which provides for the President or a Vice President of the
Royal Institute of British Architects to make the appointment there
is no fall back position in IFC 84. It would be possible for the court
to appoint an arbitrator in such circumstances under section 19 of
the Arbitration Act 1950 but it would have been preferable to have a
fail safe mechanism in place.

The parties should endeavour to agree a named arbitrator wherever possible. In practice the party serving the written notice (the
claimant) will often include a list of three or four names as possible
arbitrators and invite the other party (the respondent) to agree to
one of them, perhaps adding that in the absence of agreement to one
of those names the respondent may wish to submit further names to
the claimant for consideration. Sometimes, if the parties are hostile
towards one another no names put forward will be agreed so that the
appointing body will have to nominate an arbitrator. The situation
can sometimes become farcical especially where there is only a
limited number of suitable arbitrators and the parties produce long
lists which still result in no agreement. Although in such circumstances the appointing body could select an arbitrator even though
named in the list, nevertheless they are likely to seek to avoid this
with a consequent serious reduction in the choice of arbitrators open
to them. Where a claimant knows the other side to be hostile and
likely to reject any names on the claimant's list, it is not unknown
for the claimant to put on the list those arbitrators he does not want
thereby increasing the chances that those appearing in the respondent's list in response will include one of the names which the
claimant would have liked in the first place. Such a tactical manoeuvre is obviously fraught with a serious risk of its backfiring.

Clause 9.3

9.3   Subject to the provisions of clauses 3.5.2 *(Architect's/*
       Contract Administrator's instructions), 4.7 *(Effect of final*
       certificate) and Supplemental Conditions C4.3 *(Tax etc.*
       fluctuations; or Supplemental Conditions D 8 *(Formulae*
       fluctuations) the Arbitrator shall, without prejudice to the

generality of his powers, have power to rectify the contract so that it accurately reflects the true agreement made by the Employer and the Contractor, to direct such measurements and/or valuations as may in his opinion be desirable in order to determine the rights of the parties and to ascertain and award any sum which ought to have been the subject of or included in any certificate and to open up, review and revise any certificate, opinion, decision, requirement or notice and to determine all matters in dispute which shall be submitted to him in the same manner as if no such certificate, opinion, decision, requirement or notice had been given.

COMMENTARY ON CLAUSE 9.3

Powers of the arbitrator

Clause 9.3 expressly gives the arbitrator extensive powers including the following:

(1) Rectification

The arbitrator has power to rectify the contract so that it accurately reflects the true agreement made between the parties. It is probable that even without this express power to rectify, the scope of the wording of the arbitration clause '. . . any dispute or difference as to the construction of this contract or any matter or thing of whatsoever nature arising thereunder *or in connection therewith . . .*' (Article 5) is sufficient to enable the arbitrator to rectify the contract in many instances – see *Ashville Investments Ltd* v. *Elmer Contractors Ltd* (1987) discussed earlier at page 378.

Despite this express power to rectify, the arbitrator cannot decide on an issue as to his jurisdiction. For example, if a dispute arose as to whether the contract should be rectified so as to carry a deletion of the arbitration agreement, the arbitrator, if he were to decide that the arbitration agreement was intended to be deleted and ordered rectification of the contract accordingly, would be confirming that he had no jurisdiction to act as there would be no arbitration agreement from which his powers could emanate.

(2) Directing measurements or valuations

The arbitrator can direct such measurements or valuations as may in his opinion be desirable in order to determine the rights of the parties.

(3)   Open up review and revise certificates, opinions, decisions,
      requirements or notices

The express reference in clause 9.3 to the opening up etc. of
certificates, opinions, decisions, requirements or notices must relate
to those given in circumstances where the architect uses his professional judgment or discretion under the contract. In other words the
arbitrator's power to open up would not extend to what was in truth
a decision of the employer through the architect, e.g., to vary the
works. In practice it is the role of the architect which will be
reviewed here, for example, decisions as to extensions of time,
certificates relating to interim valuations, opinions as to whether
defects have been made good so as to justify a certificate of making
good defects and so on.

So far as architect's instructions are concerned, while the arbitrator clearly has power to determine whether they are validly issued in
the sense of being empowered by the conditions – he has no power to
open up and review etc. where the architect not only has the power
but also has a discretion as to whether to issue the instruction at all,
e.g. as to the expenditure of a provisional sum.

The opening up and reviewing etc. powers of the arbitrator have
been held to give the arbitrator powers which a court does not
possess. In the Court of Appeal case of *Northern Regional Health
Authority* v. *Derek Crouch Construction Company Ltd and Others*
(1984) the point was argued as to whether the court had power to
open up, review or revise any certificate, opinion or decision of the
architect in the same manner as an arbitrator appointed under clause
35 of JCT 63, which is similar in this respect to clause 9 of IFC 84. It
was argued before the court that where the parties chose to litigate
rather than arbitrate, there was an implied term that the courts
would have the same power as the arbitrator under clause 35. The
Court of Appeal held that it was not necessary to imply such a term.
Lord Justice Browne-Wilkinson said:

> '... in my judgment the court's jurisdiction would be limited to
> deciding whether or not the certificate or opinion was legally
> invalid because given, for example, in bad faith or in excess of his
> [the architect's] powers. In no circumstances would be court have
> power to revise such certificate or opinion solely on the ground
> that the court would have reached a different conclusion since so
> to do would be to interfere with the agreement of the parties...
> The limit of the court's jurisdiction would be to declare inopera-

tive any certificate or opinion given by the architect if the architect had no power to give such certificate or opinion or had otherwise erred in law in giving it.'

It is clear therefore that if either employer or contractor wishes to challenge the opinion or decision of the architect on its merits rather than based upon a contention that the architect was acting outside or in excess of his powers under the contract, the courts will not have the power to review the opinion or decision. The matter will have to proceed by way of arbitration. This provides either party to the contract with very powerful ammunition indeed in seeking to have stayed under section 4 of the Arbitration Act 1950 any court action initiated by the other party. Lord Justice Donaldson, Master of the Rolls, said as follows:

'... our view, if accepted, will virtually give any party a right of veto on any attempt to bypass the arbitration clauses. They will be able to point out that they are thereby being deprived of the benefit of the special powers of the arbitrator under those clauses and they will accordingly have a very strong claim to a stay under section 4 of the Arbitration Act ...'

All of the express powers given to the arbitrator in clause 9.3 are however, made subject to the following provisions:

—clause 3.5.2 (architect's instructions deemed for all purposes to be valid if no written request to concur in the appointment of an arbitrator has been given before compliance with the instruction);
—clause 4.7 (final certificate conclusive evidence as to certain matters);
—supplemental conditions C4.3 and D8 (quantity surveyor and contractor's agreement as to net amount payable to or allowable by the contractor in relation to certain events triggering fluctuations (or in the case of supplemental condition D8 any ascertainment following agreement to altered methods and procedures) is deemed for all purposes of the contract to be the amount payable or allowable thereunder).

Clauses 9.4 and 9.5

> 9.4   Subject to clause 9.5 the award of such Arbitrator shall
>   be final and binding on the parties.
> 9.5   The parties hereby agree and consent pursuant to Sec-
>   tions 1(3)(a) and 2(1)(b) of the Arbitration Act 1979 that
>   either party
>   (a) may appeal to the High Court on any question of law
>   arising out of an award made in an arbitration under this
>   Arbitration Agreement and
>   (b) may apply to the High Court to determine any ques-
>   tion of law arising in the course of the reference;
>   and the parties agree that the High Court should have
>   jurisdiction to determine any such question of law.

COMMENTARY ON CLAUSES 9.4 AND 9.5

Appeals on and applications to determine questions of law in the High Court

Clause 9.5 provides that the parties agree and consent under Sections 1(3)(a) and 2(1)(b) of the 1979 Act that either party:

'(a)   may appeal to the High Court on any question of law arising out of an award made in an arbitration under this Arbitration Agreement; and,

(b)   may apply to the High Court to determine any question of law arising in the course of the reference;

and the parties agree that the High Court should have jurisdiction to determine any such question of law.'

To appreciate the significance of this provision it is necessary briefly to explain the background. The 1979 Act severely restricted the ability of a party to an arbitration bringing before the court questions of law arising either during the course of arbitration or contained in the award of an arbitrator.

By virtue of Section 1 of the 1979 Act an appeal on any question of law arising out of an award can be brought to the courts only with the consent of the parties; or with leave of the court. By virtue of Section 2 of the 1979 Act, in the case of an application being made by any of the parties requesting the court to determine a question of law arising during the course of an arbitration, such an application can be brought only with either the arbitrator's consent and the leave of the court, or the consent of all parties.

So far as appeals on points of law arising out of the award are concerned, Section 1(4) of the 1979 Act provides that the High Court shall not grant leave unless it considers that, having regard to all the circumstances, the determination of the question of law concerned could substantially affect the rights of one or more of the parties to the arbitration agreement. In relation to applications to the High Court for leave to have a point of law arising during the course of the arbitration reference determined, the court is required by Section 2(2) of the 1979 Act to have regard to the same considerations, but it must also be satisfied that the determination of the application might produce substantial savings in costs to the parties.

In practice the High Court also takes other factors into account in deciding whether or not to grant leave and also in deciding whether or not to grant leave to appeal to the Court of Appeal against the High Court's refusal to grant leave under Sections 1 or 2 of the 1979 Act.

The leading case setting out the guidelines to be followed is still that of *Pioneer Shipping Ltd and Others* v. *BTP Tioxide: (The Nema)* (1981) in which Lord Diplock stated the principles in this way:

'My Lords, in view of the cumulative effect of all these indications of Parliament's intention to promote greater finality in arbitral awards than was being achieved under the previous procedure as it was applied in practice, it would, in my view, defeat the main purpose of the first four sections of the 1979 Act if judges, when determining whether a case was one in which the new discretion to grant leave to appeal should be exercised in favour of an applicant against objection by any other party to the reference, did not apply much stricter criteria than those stated in *The Lysland* which used to be applied in exercising the former discretion to require an arbitrator to state such a special case for the opinion of the court.

Where, as in the instant case, the question of law involved is the construction of a one-off clause the application of which to the particular facts of the case is in issue in the arbitration, leave should not normally be given unless it is apparent to the judge, on a mere perusal of the reasoned award itself without the benefit of adversarial argument, that the meaning ascribed to the clause by the arbitrator is obviously wrong; but if on such perusal it appears to the judge that it is possible that argument might persuade him, despite impression to the contrary, that the arbitrator might be right, he should not grant leave; the parties should be left to accept, for better or for worse, the decision of the tribunal that they had chosen to decide the matter in the first instance. The instant case was clearly one in which there was more than one possible view as to the meaning of the one-off clause . . .

For reasons already sufficiently discussed, rather less strict criteria are in my view appropriate where questions of construction of contracts in standard terms are concerned. That there should be as high a degree of legal certainty as it is practicable to obtain as to how such terms apply on the occurrence of events of a kind that it is not unlikely may reproduce themselves in similar transactions between other parties engaged in the same trade is a public interest that is recognised by the 1979 Act, particularly in section 4. So, if the decision of the question of construction in the circumstances of the particular case would add significantly to the clarity and certainty of English commercial law it would be proper to give leave in a case sufficiently substantial to escape the ban imposed by the first part of section 1(4), bearing in mind always that a superabundance of citable judicial decisions arising out of slightly different facts is calculated to hinder rather than to promote clarity in settled principles of commercial law. But leave should not be given, even in such a case, unless the judge considered that a strong *prima facie* case had been made out that the arbitrator had been wrong in his construction; and when the events to which the standard clause fell to be applied in the particular arbitration were themselves one-off events stricter criteria should be applied on the same lines as those I have suggested as appropriate to one-off clauses.

In deciding how to exercise his discretion whether to give leave to appeal under section 1(2) what the judge should normally ask himself in this type of arbitration, particularly where the events relied on are one-off events, is not whether he agrees with the decision reached by the arbitrator, but: does it appear on perusal of the award either that the arbitrator misdirected himself in law or that his decision was such that no reasonable arbitrator could reach? While this should, in my view, be the normal practice, there may be cases where the events relied on as amounting to frustration are not one-off events affecting only the transaction between the particular parties to the arbitration but events of a general character that affect similar transactions between many other persons engaged in the same kind of commercial activity; the closing of the Suez Canal, the United States soya-bean embargo, the war between Iraq and Iran, are instances within the last two decades that spring to mind. Where such is the case it is in the interests of legal certainty that there should be some uniformity in the decisions of arbitrators as to the effect, frustrating or otherwise, of such an event on similar transactions, in order that other traders may be sufficiently certain where they stand as to be able to close their own transactions without recourse to arbitration. In such a case, unless there were prospects of an appeal being brought by consent of all the parties as a test case under section 1(3)(a), it might be a proper exercise of the judge's discretion to give leave to appeal in order to express a conclusion as to the frustrating effect of the event that would afford guidance binding on the arbitrators in other arbitrations arising out of the same event, if the judge thought that in the particular case in which leave to appeal was sought the conclusion reached by the arbitrator, although not deserving to be stigmatised as one which no reasonable person could have reached, was, in the judge's view, not right . . .'

Turning to the specific consideration as to whether the High Court should grant leave to a party to appeal to the Court of Appeal against the judge's refusal to grant leave under the 1979 Act, the principles to have in mind are those outlined in *Antaios Cia Naviera S.A.* v. *Salen Rederierna A.B.; (The Antaios)* (1984) in which Lord Diplock said in the course of his judgment:

'My Lords, I think that your Lordship should take this opportunity of affirming the guidelines given in the Nema . . . that, even in a case that turns on the construction of a standard term, "leave should not be given . . . unless the judge considered that a strong *prima facie* case had been made out that the arbitrator had been wrong in his construction" applies even though there may be dicta in other reported cases at first instance which suggests that on some question of the construction of that standard term there may among commercial judges be two schools of thought. I am confining myself to conflicting dicta, not decisions. If there are conflicting decisions, the judge should give leave to appeal to the High Court, and whatever judge hears the appeal should in accordance with the decision that he favours give leave to appeal from his decision to the Court of Appeal with the appropriate certificate

under section 1(7) as to the general public importance of the
question to which it relates; for only thus can be attained that
desirable degree of certainty in English commercial law which
section 1(4) of the 1979 Act was designed to preserve . . .

It was strenuously urged on your Lordships that, wherever it
could be shown by comparison of judicial dicta that there were
two schools of thought among commercial judges on any question
of construction of a standard term in a commercial contract, leave
to appeal from an arbitral award which involved that question of
construction would depend on which school of thought was the
one to which the judge who heard the application adhered. Maybe
it would; but it is in the very nature of judicial discretion that,
within the bounds of "reasonableness" in the wide Wednesbury
sense of that term, one just may exercise the discretion one way
whereas another judge might have exercised it in another; it is not
peculiar to section 1(3)(b). It follows that I do not agree with Sir
John Donaldson MR where in the instant case he says that leave
should be given under Section 1(3)(b) to appeal to the High Court
on a question of construction of a standard term on which it can
be shown that there are two schools of thought among puisne
judges where the conflict of judicial opinion appears in dicta only.
This would not normally provide a reason for departing from the
Nema guidelines which I have repeated earlier in this speech . . .

This brings me to "the section 1(6A) question" canvassed in
Staughton J's second judgment of 19 November 1982: when
should a judge give leave to appeal to the Court of Appeal from
his own grant or refusal of leave to appeal to the High Court from
an arbitral award? I agree with him that leave to appeal to the
Court of Appeal should be granted by the judge under section
1(6A) only in cases where a decision whether to grant or to refuse
leave to appeal to the High Court under section 1(3)(b) in the
particular case in his view called for some amplification, elucida-
tion or adaption to changing practices of existing guidelines laid
down by appellate courts, and that leave to appeal under section
1(6A) should not be granted in any other type of case. Judges
should have the courage of their own convictions and decide for
themselves whether, applying existing guidelines, leave to appeal
to the High Court under section 1(3)(b) ought to be granted or
not.'

Further refinement of the *Nema* principles has taken place in numerous cases. For example, in the case of *International Sea Tankers, Inc. v. Hemisphere Shipping Co. Ltd; (The Wenjiang)* (1982) Lord Denning said:

> 'In the ordinary way, on the application the judge will have the award before him and counsel to argue it. The judge should look at once to see if it is a one-off case. It may be one-off because the facts are so exceptional that they are singular to this case and not likely to occur again; or because it is a point of construction of a clause singular to this case which is not likely to be repeated. In such a case the judge should not give leave to appeal from the arbitrator if he thinks the arbitrator was right or probably right or may have been right. He should only give leave to appeal if he forms the provisional view that the arbitrator was wrong on a point of law which could substantially affect the rights of one or other of the parties.
>
> When the case is not one-off as so described, but it gives rise to a question of construction of a standard form with facts which may occur repeatedly or from time to time, then the judge should give leave if he thinks that the arbitrator may have gone wrong on the construction of the standard form: but not if he thinks the arbitrator was right.'

And in the case of *Petraco (Bermuda) Ltd* v. *Petromed International S.A.* (1988) in relation to whether leave should be given in respect of a point of law not argued before the arbitrator, Lord Justice Staughton propounded the following additional guidelines for such a case:

(1) The fact that the point which it is proposed to argue is not argued before the arbitrator is not an absolute bar to the grant of leave to appeal.

(2) It is, however, to be taken into account in the exercise of the general discretion provided by Section 1(3).

(3) Where the failure to argue the point below has had the result that all the necessary facts have not been found, this would be a powerful factor against granting leave

(4) Even in such a case, it might be right in very special circumstances to remit the award for further facts to be found with a view to granting leave.

(5) If all necessary facts had been found, the judge should give

such weight as he thinks fit to the failure to argue the point before the arbitrator. In particular he should have regard to whether the new point is similar to points that had been argued or whether it is a totally new and different point.

As can be seen therefore, the provision in advance of the consent of the parties in relation to appeals under Section 1 or applications under Section 2 of the 1979 Act potentially eliminates a very significant obstacle in the way of a party seeking to pursue a point of law before the High Court.

Given the background to the 1979 Act and its obvious intention to severely restrict the ability of a dissatisfied party taking a point of law to the High Court, the obtaining of the consent in advance of any dispute or difference arising by means of a provision in the arbitration agreement may be regarded by some as an attempt to circumvent the spirit if not the letter of the 1979 Act.

The case of *Finelvet A.G.* v. *Vinava Shipping Co. Ltd (The Chrysalis)* (1983) is highly relevant here. In this case Mr Justice Mustill (as he then was) said:

'Counsel for the charterers has, however, contended that the position is different in the present case, because the parties had agreed in advance that there should be a right of appeal on any question of law. This shows, so it is maintained, that the parties wanted an authoritative ruling on the question of frustration and this they would not get from a mode of appeal which precluded the judge from substituting his own opinion for that of the arbitrator on the "judgmental" stage of the reasoning process. I am afraid that I cannot read the agreement as showing any such intention. Its obvious purpose was to save the time and expense involved in a contested application under section 1(3)(b) of the 1979 Act . . .'

This last extract from the judgment of Mr Justice Mustill appears to recognise implicitly that the consent in advance is adequate and effective under section 1 of the 1979 Act.

Even if the consent provisions in clause 9.5 were not to be accepted as a valid consent it may still have some effect upon the exercise by the court of its discretion in deciding whether to grant leave. For example, in the case of *Astro Valiente Compania Naviera*

S.A. v. *Pakistan Ministry of Food and Agriculture (The Emmanuel Colocotronis)* (1982) Mr Justice Staughton (as he then was) said:

'... It was expressly agreed by both sides, at that time, that the arbitrators should give reasons in their award. That, under the 1979 Act, is tacitly an essential condition for there being an appeal to the court from an arbitrator's decision, because unless he gives reasons one cannot say whether it is right or wrong. So it does not look as though, at the time of that agreement, either party contemplated that the arbitrators' decision on this point should inevitably be final. On the contrary, it looks as though they contemplated at least the possibility that it would not be final ...'

Lord Diplock there (i.e. in *The Nema* case) says that leave should not be granted even in such a case unless the judge considers that a strong *prima facie* case had been made out that the arbitrators were wrong. I have said that I am unable to reach that conclusion, because this is a difficult and technical point on which I cannot reach a conclusion either way without benefit of argument and study of the authorities. However, I do not believe that Lord Diplock intended to lay down that as the sole and only test despite all other circumstances. In an earlier passage in his speech, to which counsel for the shipowners referred me, he said that it was legitimate and proper to take into account the circumstances in which the arbitration agreement was made. I have already referred to those circumstances, and to the fact that both parties, through their solicitors, agreed that the arbitrators should give reasons. That does not seem to me to show that the parties, or either of them, were determined, in the words of Lord Diplock, "to accept, for better or for worse, the decision of the tribunal that they had chosen to decide the matter in the first instance". It seems to me to show the contrary, as I have already said ...'

Of course, even if consent given in advance is valid and effective, the courts here are looking at points of law only. There is always the very real risk that the courts may decide that the issue is really one of fact or mixed fact and law thus taking the issue outside of the 1979 Act altogether.

On an optimistic note, the consent provisions making it easier for the parties to take points of law to the High Court coupled with the restrictions on the powers of the court in connection with opening up and reviewing and revising any certificates, opinions, requirements or notices of the architect etc. thus pushing all such matters away from the courts and towards arbitration, could have the effect of leaving the courts, particularly those judges undertaking official referees' business (if appeals under the 1979 Act arising out of construction arbitrations are, as they seem to be, referred to them), somewhat freer to develop a body of case law relating to JCT contracts on a reasonably economic basis.

Clause 9.6

9.6   Where this clause 9.6 is stated in the Appendix to apply,
[m.4]   if the dispute or difference to be referred to arbitration
        under this Contract raises issues which are substantially
        the same as or connected with issues raised in a related
        dispute under a sub-contract entered into in accordance
        with the provisions of clause 3.3 and if the related dis-
        pute has already been referred for determination to an
        Arbitrator, the Employer and Contractor hereby agree that
        the dispute or difference under this Contract shall be
        referred to the Arbitrator appointed to determine the
        related dispute; and the JCT Arbitration Rules applicable
        to the related dispute shall apply to the dispute under this
        Contract; and such Arbitrator shall have power to make
        such directions and all necessary awards in the same
        way as if the procedure of the High Court as to joining
        one or more defendants or joining co-defendants or third
        parties was available to the parties and to him; and the
        agreement and consent referred to in clause 9.5 on
        appeals or applications to the High Court on any question
        of law shall apply to any question of law arising out of the
        awards of such Arbitrator in respect of all related dis-
        putes referred to him or arising in the course of the refer-
        ence of all the related disputes referred to him.

[m.4]   Make appropriate entry in the Appendix.

COMMENTARY ON CLAUSE 9.6

Joinder of parties

Clause 9.6 tackles the potentially complex area of joinder of parties in arbitration where the disputes between the various parties are related. It applies only where stated in the appendix to apply and requires disputes under IFC 84 to be dealt with alongside any related dispute between the main contractor and a sub-contractor appointed pursuant to clause 3.3 (named sub-contractors).

Arbitration is consensual. Therefore a person who has not agreed to arbitrate cannot be compelled to. All of the JCT arbitration agreements are contained in contracts to which there are two parties. It is not unusual, however, for a dispute between the parties under one contract to affect a party under a related contract with one of the disputing parties. For example, an employer may be in dispute with a contractor regarding allegedly defective work. The contractor may likewise be in dispute with a named sub-contractor in respect of the same work. The contractor is a party to both the main and named sub-contract but of course the named sub-contractor is not a party to the main contract and the employer is not a party to the named sub-contract. Both contracts have arbitration agreements but without more if the dispute were to be the subject of arbitration, it would necessarily result in two separate sets of arbitration proceedings in respect of the same issues – one between the employer and the contractor and the other between the contractor and the sub-contractor. There may be different arbitrators and different decisions made even though based upon the same evidence. If this potential problem arises in litigation the court has ample powers to bring all of the affected parties within the same proceedings so that the risk of inconsistent findings as well as the duplication of costs is avoided. The joinder provisions in the IFC 84 arbitration agreement is an attempt to achieve a similar result in arbitration with a resultant saving in time and cost.

Any number of potential difficulties can arise in the operation of these joinder provisions – see for example the attempt, fortunately unsuccessful, in the case of *A. Monk & Co. Ltd* v. *Devon County Council* (1978) (a tripartite arbitration between employer, main contractor and sub-contractor) to tie the hands of the main contractor in his dispute with the employer following a settlement between the main contractor and the sub-contractor. If co-operation and common sense prevail the problems are not insurmountable.

Clause 9.6 provides that if the dispute or difference falling within IFC 84 raises issues which are substantially the same as or connected with issues raised in a related dispute under a named sub-contract with the contractor (known as the 'related dispute') and if the related dispute has already been referred for determination to an arbitrator, then the employer and the contractor agree that the dispute under IFC 84 shall be referred to the arbitrator appointed to determine the related dispute.

It is important to be able to point to the precise moment when the related dispute was referred for determination to an arbitrator. The arbitration agreements in NAM/SC require the giving of a written notice by one party to the other requiring the dispute or difference to be referred to arbitration. Clearly there is a time lag between the notice to refer and the appointment of an arbitrator taking place. Clause 9.6 refers to the related dispute having been referred to 'an Arbitrator' rather than simply referred to arbitration. It seems therefore that unless an arbitrator has actually been appointed in connection with the related dispute, the joinder provisions will not take effect.

Though the current wording of clause 9.6 is in some respects tighter than the earlier wording – see for example JCT 63 and the FASS 'Green Form' of Sub-Contract, some of the problems of the arbitrator's jurisdiction highlighted in cases such as *Higgs & Hill Building Ltd* v. *Campbell Dennis Ltd* (1982); *Multi-Construction (Southern) Ltd* v. *Stent Foundations Ltd* (1988) and *Hyundai Engineering and Construction Co Ltd* v. *Active Building and Civil Construction Pte Ltd* (1988) could still give rise to difficulties.

If the joinder takes place there are a number of ancillary provisions which assist in the process of resolving the enjoined arbitrations including the following:

—The JCT Arbitration Rules apply to the joined dispute. In practice as a result of the joinder there will often be a significant departure from the time-scale set out in the Rules. Read too literally the Rules look, at some points, almost unworkable without modifiction in such circumstances.

—The arbitrator is given power to give such directions and to make such awards as if the High Court procedure on joining defendants and third parties was available to all of the parties and to the arbitrator.

—The consent of the parties to appeals or applications on questions of law (dealt with earlier when considering clause 9.5 – page 384) is extended to all related disputes. However, there appears to be a possible gap in these consent provisions. The consent given to taking a point of law from the arbitrator to the High Court under clause 9.5 is a consent only to the other party to the IFC 84 contract i.e. the employer or the contractor only as the case may be. So the situation could arise where it is, say, a named sub-contractor who wishes to take up a point of law. He will no doubt have the contractor's consent under the arbitration agreement in NAM/SC. However, under IFC 84 there is no provision whereby the employer is bound to consent to the named sub-contractor as opposed to the contractor taking up a point of law. The consent required under Section 1 or Section 2 of the 1979 Act which, if given, avoids the need to obtain the leave of the court to a party taking up a point of law, is that of all the parties to the reference. If the answer is that there are in truth two (or more) separate references to arbitration where joinder takes place so that in the example given only the contractor's consent is required, it makes this insertion of the consent provisions in relation to related disputes irrelevant and unnecessary. Further, the 1979 Act talks of the consent of all of the other parties to the reference rather than simply the other party which is some, albeit limited (as it could be referring to arbitrations arising out of contracts having more than two parties) indication that joinder of related disputes produces a single reference. It may have been sensible to add into clause 9.5 after the words '. . . that either party . . .' some such additional words as 'or any party to a related dispute'. The last seven lines in clause 9.6 could then be removed altogether.

Clause 9.7

> 9.7   If before making his final award the Arbitrator dies or otherwise ceases to act as the Arbitrator, the Employer and the Contractor shall forthwith appoint a further Arbitrator, or, upon failure so to appoint within 14 days of any such death or cessation, then either the Employer or the Contractor may request the person named in the Appendix to appoint such further Arbitrator. Provided that no such further Arbitrator shall be entitled to disregard any direction of the previous Arbitrator or to vary or revise any award of the previous Arbitrator except to the extent that the previous Arbitrator had power so to do under the JCT Arbitration Rules and/or with the agreement of the parties and/or by the operation of law.

COMMENTARY ON CLAUSE 9.7

Appointment of replacement arbitrator

Clause 9.7 contains a very useful provision dealing with the situation where the arbitrator dies or otherwise ceases to act. While in the absence of express provisions the High Court would have power to appoint a replacement arbitrator, nevertheless it is simpler and generally quicker for the original appointing body to re-appoint should the parties fail to agree a named replacement. The replacement arbitrator is bound by any directions or awards of the previous arbitrator to the same extent as a previous arbitrator would have been bound by his own directions or awards.

In arbitrations involving long hearings the parties should give serious consideration to insuring the arbitrator against death or other incapacity from the beginning of the hearing to the publication of the award. Any need for a re-hearing would obviously cause great additional expense.

Clause 9.8

> 9.8   The arbitration shall be conducted in accordance with the
> [m.5]   'JCT Arbitration Rules' current at the Base Date. Provided
> that if any amendments to the Rules so current have
> been issued by the Joint Contracts Tribunal after the
> Base Date the Employer and the Contractor may, by a
> joint notice in writing to the Arbitrator, state that they
> wish the arbitration to be conducted in accordance with
> the JCT Arbitration Rules as so amended.

> [m.5]   The JCT Arbitration Rules contain stricter time limits than
> those prescribed by some arbitration rules or those fre-
> quently observed in practice. The parties should note that
> a failure by a party or the agent of a party to comply with
> the time limits incorporated in these Rules may have
> adverse consequences.

COMMENTARY ON CLAUSE 9.8

The JCT Arbitration Rules

By clause 9.8 the arbitration is to be conducted in accordance with the JCT Arbitration Rules current at the 'Base Date' stated in the appendix to the IFC 84 Conditions, though, as the clause goes on to state, the parties can if they wish agree by a joint notice in writing to the arbitrator that the arbitration be conducted in accordance with any amendment of the rules published after the base date.

The arbitrator should when appointed always seek from the parties the base date in order to see which edition of the JCT Arbitration Rules apply and if there have been any amendments between the base date and his appointment and bring this to the attention of the parties to see if they wish the amended rules to apply and if so to request from them their joint notice in writing.

Interpretation of the Rules

The Rules are a working commercial document regulating the conduct of any arbitration which is subject to them. They are not to be construed or interpreted as though they were part of a statute. This is particularly the case to the extent that the Rules are procedural in character. To the extent that the Rules deal with the powers of the arbitrator in relation to awards or orders rather than simple procedural matters, they may be subject to a somewhat stricter construction. It must always be remembered that the arbitrator is generally the master of the procedure to be adopted and the Rules are designed to regulate the procedures rather than to have any limiting or stultifying effect.

The JCT health warning

In footnote [m.5] to clause 9.8, it is stated that the rules contain stricter time limits than those prescribed by some arbitration rules or those frequently observed in practice and that the parties should note that a failure by a party or the agent of a party to comply with the time limits incorporated in the rules may have adverse consequences. The parties and their representatives need to be aware of them. A failure to meet them could lead to the loss by a party of the opportunity to put his case. If the failure to comply with the time limits is due to ignorance on the part of a party's professional representative it could involve the representative in personal liability.

Appendix 1
Appendix to IFC 84

2·1	Date of Possession	_____
2·1	Date for Completion	_____
2·4·10 and 2·4·11	Extension of time for inability to secure essential labour or goods or materials	[n] Clause 2·4·10 *(Labour)* applies/does not apply [n] Clause 2·4·11 *(Good or materials)* applies/does not apply
2·2 and 2·4·14 and 4·11(a)	Deferment of the Date of Possession	[n] Clause 2·2 applies/does not apply Where clause 2·2 applies, _____ weeks (period not to exceed 6 weeks)
2·7	Liquidated damages	at the rate of £ _____ per
2·10	Defects liability period (if none stated is 6 months from the day named in the certificate of Practical Completion of the Works)	_____
4·2	Period of interim payments if interval is not one month	_____
4·9(a) and C7	Supplemental Condition C: Tax etc. fluctuations	[o] Percentage addition _____ %
4·9(b)	Formulae fluctuations (not applicable unless Bills of Quantities are a Contract Document)	[o] Supplemental Condition D [n] applies/does not apply
D1	Formula Rules (only where Supplemental Condition D applies)	rule 3: Base Month _____ 19 _____ [p] rule 3: Non-Adjustable Element _____ (not to exceed 10%) rules 10 and 30(i) [n] Part I/Part II of Section 2 of the Formula Rules is to apply
5·5	Value Added Tax: Supplemental Condition A	[n] Clause A1·1 of Supplemental Condition A applies/does not apply
6·2·1	Insurance cover for any one occurrence of series of occurrences arising out of one event	£ _____
6·2·4	Insurance-liability of Employer	[n] Insurance may be required/is not required Amount of indemnity for any one occurrence or series of occurrences arising out of one event [p·1] £ _____
6·3·1	Insurance of the Works–alternative clauses	[n] Clause 6·3A/Clause 6·3B/Clause 6·3C applies (See Footnote [k] to clause 6·3)
6·3A·1	Percentage of cover professional fees	_____ %
6·3A·3·1	Annual renewal date of insurance as supplied by Contractor	_____
6·3D	Insurance for Employer's loss of liquidated damages – clause 2·4·3	[n] Insurance may be required/is not required
6·3D·2		Period of time _____
8·3	Base Date	
9·1	Appointor of arbitrator	President or a Vice-President of: [n] Royal Institute of British Architects [n] Royal Institution of Chartered Surveyors [n] Chartered Institute of Arbitrators
9·6	Reference to Arbitrator under NAM/SC	[n] clause 9·6 applies/does not apply

[n] Delete as applicable.

[o] In accordance with clause 4·9, if Supplemental Condition D is not stated to apply then Supplemental Condition C applies.

[p] Only applicable when the Employer is a Local Authority.

[p.1] If the indemnity is to be for an aggregate amount and not for any one occurrence or series of occurrences the entry should make this clear.

Appendix 2
Tender and Agreement NAM/T

Form of Tender and Agreement

for a person to be named by the Employer as a sub-contractor under the
JCT Intermediate Form of Building Contract 1984 Edition (IFC 84), 1st Recital and clause 3·3

*Should there be a separate agreement between the Employer and the Sub-Contractor relating to such matters as are referred to in clause 3·3·7 of the main contract conditions (design etc.), it should **not** be attached to the Form of Tender and Agreement, either where the Form is included in the main contract documents or where it is included in an instruction of the Architect/the Contract Administrator for the expenditure of a provisional sum.*

[a] Section I – Invitation to tender

[b] To_____

_____

_____

_____

_____

You are invited to tender, as a person who is to be employed by the Contractor under a sub-contract in accordance with the Form of Agreement in Section III hereof, a VAT-exclusive Sub-Contract Sum for executing the Sub-Contract Works referred to below by completing [c] SECTION II and returning the whole form to

_____

_____

Your tender must not vary any of the matters set out in SECTION I of this Form.

The documents listed below, hereinafter called 'the Numbered Documents', are enclosed herewith:

_____

_____

_____

_____

_____

Should it be decided to name you as the person to execute the Sub-Contract Works, you will be required to provide the following, hereinafter called 'the Priced Documents':

*a priced copy of the Specification
*a priced copy of the Schedule of Works
*a priced copy of the Bills of Quantities
*a Sub-Contract Sum Analysis
*a Schedule of Rates,

and any such priced copy of a Numbered Document shall be deemed to replace the unpriced copy.

[d] Signed _____

Name _____

Address _____

_____

Date _____19_____

*delete as applicable

Notes on completion

Section I – Invitation to tender

[e] Insert the same
description as in IFC 84,
1st Recital.

[e] Main Contract Works and location_____

_____

_____

_____

_____

_____

_____Job reference_____

Particulars of the Sub-Contract Works_____

_____

_____

Names and address of:

Employer: Tel No:

[f] Main Contract clause 8·4.

[f] *The Architect/The Contract Administrator: Tel No:

Quantity Surveyor: Tel No:

[g] Where item 12(a) applies
the Contractor will not have
been appointed: instead the
names of contractors who
will be invited to tender
should be set out.

[g] Main Contractor: Tel No:

_____

*delete as applicable   © 1988 RIBA Publications Ltd

Notes on completion	

Section I – Invitation to tender

MAIN CONTRACT INFORMATION

1	Form of Main Contract	JCT Intermediate Form of Building Contract 1984 Edition (IFC 84) incorporating Amendments 1 to 4
2	Main Contract alternative or optional provisions (See also item 6)	*Specification/Schedules of Work/Bills of Quantities: **1st Recital** *alternative A/alternative B: **2nd Recital** *The Architect/The Contract Administrator: **Article 3**
3	Any changes from printed Form of Main Contract Conditions identified at item 1:	

[h] See also item 13 page 6.

4   [h]   Execution of Main Contract: *is/is to be *under hand/under seal

5   Inspection of Main Contract: the unpriced *Specification/Schedules of Work/Bills of Quantities and the Contract Drawings may be inspected at:
(where item 12(a) applies the documentation for the Main Contract may not be yet be available)

6   Main Contract: Appendix and entries therein as amended by Amendments 1 to 4
(where item 12(a) applies it is intended that the following entries will be made in the Appendix to the Main Contract. When item 12(b) applies the following are the entries to the Appendix to the Main Contract.)

IFC84 Clause

2·1	[i] Date of Possession	
2·1	[i] Date for Completion	
2·4·10 and 2·4·11	Extension of time for inability to secure essential labour or goods or materials	* Clause 2·4·10 *(Labour)* applies/does not apply * Clause 2·4·11 *(Goods or materials)* applies/does not apply
2·2 and 2·4·14 and 4·11(a)	Deferment of the Date of Possession	* Clause 2·2 applies/does not apply Where clause 2·2 applies, _____weeks (period not to exceed 6 weeks)
2·7	Liquidated damages	at the rate of £ _____ for every week or part of a week

[i] See item 10 page 5.

    *delete as applicable

Section I – Invitation to tender

MAIN CONTRACT INFORMATION continued

---

6   Main Contract: Appendix and entries therein as amended by Amendments 1 to 4

IFC84 Clause

2·10   Defects liability period (if none stated is 6 months from the day named in the certificate of Practical Completion of the Works)   _____

4·2   Period of interim payments if interval is not one month   _____

4·5   Period of Final Measurement and Valuation (if none other stated is 6 months from the day named in the certificate of Practical Completion of the Works)   _____

4·9(a) and C7   Supplemental Condition C: Tax etc. fluctuations   Percentage addition _____ %

4·9(b)   Formulae fluctuations (not applicable unless Bills of Quantities are a Contract Document)   Supplemental Condition D * applies/does not apply

D1   Formula Rules (only where Supplemental Condition D applies)

rule 3: Base Month _____

[A] rule 3: Non-Adjustable Element _____ (not to exceed 10%)

rules 10 and 30(i) * Part I/Part II of Section 2 of the Formula Rules is to apply

6·2·1   Insurance cover for any one occurrence or series of occurences arising out of one event   £ _____

6·2·4   Insurance – liability of Employer   Insurance may be required/is not required

Amount of indemnity for any one occurrence or series of occurrences arising out of one event

£ _____

6·3·1   Insurance of the Works – alternative clauses   * Clause 6·3A/Clause 6·3B/Clause 6·3C applies

6·3A·1   Percentage to cover professional fees   _____ %

6·3B·1

6·3C·2

6·3A·3·1   Annual renewal date of insurance as supplied by Contractor   _____

6·3D   Insurance for Employer's loss of liquidated damages – clause 2·4·3   * Insurance may be required/is not required

6·3D·2   Period of time _____

8·3   Base Date   _____

9·1   Appointor of Arbitrator   President or a Vice-President: *Royal Institution of Chartered Surveyors *Royal Institute of British Architects *Chartered Institute of Arbitrators

9·6   Reference to Arbitrator under NAM/SC   *clause 9·6 applies/does not apply

---

[A] Only applicable where the Employer is a Local Authority.

*delete as applicable   © 1988 RIBA Publications Ltd

Section I – Invitation to tender

MAIN CONTRACT INFORMATION continued

Notes on completion

7 Obligations or restrictions which are or will be imposed by the Employer not covered by the Main Contract Conditions (eg. those which are or will be included in the Specification/the Schedules of Work/the Contract Bills, or are in Variation instructions):

[j]

[j] This information, unless included in the Numbered Documents (see page 1), should be given, eg. by repeating it here or by attaching a copy of the relevant part of the Main Contract Documents.

8 Order of Works: Employer's requirements affecting the order of the Main Contract Works (if any):

9 Location and type of access:

Where item 12(b) or 12(c) applies:

10 New dates for possession or completion where these have been altered from the original dates stated in the Main Contract Appendix reproduced at item 6 pages 3 and 4:

11 Other relevant information (if any) relating to the Main Contract:

Notes on completion

Section I – Invitation to tender

SUB-CONTRACT INFORMATION

12   NAM/T with Section I and II completed, the Priced Documents and such of the Numbered Documents as are not Priced Documents will, if approved, be:

     *(a)   included in the Main Contract Documents for pricing by a contractor (see IFC 84, 1st Recital);

   or *(b)   included in an instruction of the Architect/the Contract Administrator to the Contractor as to the expenditure of a provisional sum (see IFC 84, clause 3·3·2);

   or *(c)   included in an instruction of the Architect/the Contract Administrator to the Contractor naming a replacement sub-contractor (see IFC 84, clause 3·3·3(a)).

13   The Named Sub-Contractor will be required to enter into a sub-contract in accordance with the Form of Agreement in Section III with the Contractor selected by the Employer to execute the Main Contract Works. (Where item 12(b) or (c) applies the Contractor will first have the right to make a reasonable objection to entering into a sub-contract with the Named Sub-Contractor.) Unless otherwise stated below the Form of Agreement shall be entered into in the same manner as the main contract (see item 4 page 3).

[k] See items 6 and 10.

14 **[k]**   The Main Contract Appendix and entries therein will, where relevant, apply to the Sub-Contract unless otherwise specifically stated here:

SPECIMEN

[l] The actual date or dates for commencement of the Sub-Contract Works should be settled by the Contractor and Sub-Contractor taking into account any period of notice required to be given to the Sub-Contractor to commence work on site to be set out in Section II, item 1(3). (See Section II, item 1).

15 **[l]**   The dates between which it is expected that the Sub-Contract Works can be commenced on site:

to be between_____

and_____

Period required by the Architect/the Contract Administrator to approve drawings after

submission_____

*delete as applicable   © 1988 RIBA Publications Ltd

Section I – Invitation to tender

SUB-CONTRACT INFORMATION continued

16 [m] **Sub-Contract Fluctuations**

1   *NAM/SC – clause 33 – Contributions, levy and tax fluctuations will apply

       Duties and taxes on fuels: * included/excluded (NAM/SC 33·2·1)

       Base Date (NAM/SC 33·6·1)_____

       Percentage addition to fluctuation payments or allowances (NAM/SC 33·7)

       _____%

2   *NAM/SC – clause 34 – Formula adjustment will apply

       Sub-Contract/Works Contract Formula Rules are those dated (NAM/SC 34·1)

       _____19_____

       *Part I/Part III of these Rules applies

[n]   Non-Adjustable Element_____% (not to exceed 10%)
       (NAM/SC 34·3·3)

       NAM/SC Formula Rules

       Definition of Balance of Adjustable Work – any measured work not allocated to a Work Category (rule 3)

       _____

       _____

       _____

       _____

[o]   Base Month (rule 3)_____

       Base Date_____

[p]   Method of dealing with 'Fix-only' work (rule 8)

[p]   Part I of Formula Rules only: the Work Categories applicable to the Sub-Contract Works (rule 11)

[p]   Part III of Formula Rules only: Weightings of labour and materials – Electrical Installations or Heating, Ventilating and Air Conditioning Installations (rule 43)

	Labour	Materials
Electrical	_____%	_____%

[q] Heating, Ventilating & Air Conditioning	_____%	_____%
[q] _____	_____%	_____%

[p]   Adjustment shall be affected (rule 61a)

       *upon completion of manufacture of all fabricated components

       *upon delivery to site of all fabricated components

     *delete as applicable

Section I – Invitation to tender

SUB-CONTRACT INFORMATION continued

[r] Any other additional attendance and any other special requirements or any variation to those set out by the Architect/the Contract Administrator here should be detailed by the Sub-Contractor in Section II, item 3.

17 The attendance to be provided by the Contractor free of charge to the Sub-Contractor will be as stated in NAM/SC clause 25·1. The following additional attendance and/or special
[r] requirements will also be provided free of charge to the Sub-Contractor (NAM/SC 25·3):

18 Settlement of disputes – arbitration – appointer (if no appointor is selected the appointor shall be the President or a Vice-President, Royal Institution of Chartered Surveyors)

35·1   President or a Vice-President:
*Royal Institution of Chartered Surveyors
*Royal Institute of British Architects
*Chartered Institute of Arbitrators

*delete as applicable

© 1988 RIBA Publications Ltd

Notes on completion

[s] **Section II – Tender by sub-contractor**

[s] Sub-Contractor to complete the whole of Section II. See page 12 for counter-signature by the Architect/the Contract Administrator.

To: the Employer and Contractor

In response to the invitation in Section I

We_____

of_____

_____Tel No:_____

have duly noted the information therein contained and now OFFER to carry out and complete, as a named person to be employed by the Contractor as a sub-contractor, the Sub-Contract Works identified in the Numbered Documents listed in Section I and in accordance with the entries we have made in items 1-6 of this Section for

the VAT-exclusive Sub-Contract Sum of £_____

and

[t] Where item 12(a) of Section I applies, the Sub-Contractor should note the provisions in clause 3·3·1 of the Main Contract Conditions.

[t] where item 12(a) of Section I applies, to conclude a sub-contract with the Contractor by completing Section III hereof within 21 days of the Contractor entering into the Main Contract with the Employer; or

[u] Where item 12(b) of Section I applies the Sub-Contractor should note the provisions in clause 3·3·2 of the Main Contract Conditions. The Section II items set out below, as completed, are to be counter-signed by the Architect/the Contract Administrator on page 12.

[u] where item 12(b) of Section I applies, to conclude a sub-contract with the Contractor by completing Section III hereof immediately on the issue of the instruction by the Architect/the Contract Administrator under the Main Contract Conditions clauses 3·3·2 or 3·3·3(a) and 3·8 but subject to the right of the Contractor to make a reasonable objection to entering into such a sub-contract within 14 days of the date of issue of the instruction.

AND with: the **daywork percentages** set out below; and

the **fluctuation provisions** set out at Item 4;

any **additional attendance or special requirements** set out at Item 5.

[v] See note [dd] page 12.

[v] **This Tender, subject to any extension of the period for its acceptance, is withdrawn if not accepted by the Contractor within**_____**(weeks) of the date of this Tender.**

The daywork percentages are (NAM/SC 16·2·5):

[w] Where more than one Definition will be relevant set out percentage additions applicable to each such Definition. The four Definitions which may be identified are: those agreed between the Royal Institution of Chartered Surveyors and the Building Employers Confederation; the Royal Institution and the Electrical Contractors Association; the Royal Institution and the Electrical Contractors Association of Scotland; and the Royal Institution and the Heating and Ventilating Contractors Association.

[w] Definition	Labour %	Materials %	Plant %
* RICS/BEC			
* RICS/ECA			
* RICS/ECA (Scotland)			
* RICS/HVCA			

Daywork percentages take into account the 2½% cash discount allowable to the Contractor under Sub-Contract NAM/SC (NAM/SC 19·3·2 and 19·8·2).

*delete as applicable

Section II – Tender by sub-contractor

ITEMS TO BE COMPLETED BY THE SUB-CONTRACTOR

Notes on completion

[x] See also Section I,
item 15.

1   [x]   The periods required (NAM/SC 12·1):

     (1)   for submission of any further sub-contractor's drawings etc. (co-ordination, installation, shop or builder's work or other as appropriate)

         _____weeks from receipt of all necessary information therefor

     (2)   for execution of any Sub-Contract Works off-site

         _____weeks from approval of drawings or establishment of site-dimensions as appropriate

     (3)   for notice to commence work on-site

         _____weeks

     (4)   for execution of Sub-Contract Works on-site

         _____weeks from the date stated in the notice under (3) above.

2   Insurance cover for any one occurrence or series of occurrences arising out of one event (NAM/SC amended by Amendment 1: issued November 1986, clause 8·2)

[y] Must not be less than the amount inserted in Main Contract Appendix: see Section I, item 6.

    [y]   £ _____

[z] See item 17 and sidenote [r] in Section I.

3   [z]   Any additional attendance and/or any other special requirements which vary or add to those referred to in Section I, item 17, and which the Sub-Contractor requires the Contractor to provide free of charge should be set out here in reasonable detail (NAM/SC 25·3):

Notes on completion

Section II – Tender by sub-contractor

ITEMS TO BE COMPLETED BY THE SUB-CONTRACTOR continued

4 **Sub-Contract Fluctuations**

Fluctuations will be in accordance with clauses 33 or 34 of NAM/SC as stated in Section I, item 16.

Where it is stated that clause 33 applies, a list of materials, goods, electricity and fuels (where fuels are to be included: see NAM/SC clause 33·2·1) is attached on a separate sheet.

Where it is stated that clause 34 applies, the following will apply (NAM/SC 34):

Fluctuations – articles manufactured outside the United Kingdom. List of market prices of such articles which the Sub-Contractor is required by the Sub-Contract Documents to purchase and import (see Formula Rules, rule 4 (ii)) – is attached on a separate sheet (NAM/SC 34·4).

[aa] To be completed only to the extent that the Architect/the Contract Administrator has not completed these items in Section I, item 16.

[aa] Method of dealing with 'Fix-only' work (rule 8)

[aa] **Part I only:** the Work Categories applicable to the Sub-Contract Works (rule 11a)

[aa] **Part III only:** Weightings of labour and materials – Electrical Installations or Heating, Ventilating and Air Conditioning Installations (rule 43)

	Labour	Materials
Electrical	_____%	_____%
[bb] Heating, Ventilating & Air Conditioning	_____%	_____%
_____	_____%	_____%

[bb] If both specialist engineering formulae apply to the Sub-Contract the percentages for use with each formula should be clearly identified. The weightings for sprinkler installations may be inserted where different weightings are required.

[aa] Adjustment shall be effected (rule 61a)

*upon completion of manufacture of all fabricated components

*upon delivery to site of all fabricated components

Part III only: Structural Steelwork Installations (rule 64):

(i) Average price per tonne of steel delivered to fabricator's work

£ _____

(ii) Average price per tonne for erection of steelwork

£ _____ _____

Catering Equipment Installations (rule 70a):

apportionment of the value of each item between

[cc] (i) materials and shop fabrication
[cc] (ii) supply of factor items
[cc] (iii) site installations

[cc] Insert values on a separate sheet.

*delete as applicable

Section II – Tender by sub-contractor

ITEMS TO BE COMPLETED BY THE SUB-CONTRACTOR continued

5   Any other matters (eg. special conditions or agreements on employment of labour, limitations on working hours) to be set out here:

Finance (No. 2) Act 1975 – Statutory Tax Deduction Scheme (NAM/SC 18A)

6   The evidence to be produced to the Contractor for the verification of the Sub-Contractor's tax certificate

expiry date_____ 19_____

will be:

SPECIMEN

Signed by or on behalf
of the Sub-Contractor _____

         Dated_____ 19_____

   *(A)   This is the Tender document referred to in the Invitation to Tender issued to the Contractor;

or   *(B)   This is the Tender Document referred to in

       Instruction No _____, dated_____ 19_____

[dd] Signed by or on behalf
of the Architect/the
Contract Administrator _____

         Dated_____ 19_____

*delete as applicable

Section III – Articles of Agreement

This Agreement

is made the_____day of_____19_____

between the Contractor and the Sub-Contractor named or referred to in the foregoing Sections I and II.

Whereas

Recitals

1st The Contractor desires to have executed the works (hereinafter called 'the Sub-Contract Works') referred to in the aforementioned Section I and described in the Numbered Documents identified in that Section;

2nd The Sub-Contractor has submitted the Tender set out in the aforementioned Section II and the Priced Documents identified in Section I;

3rd The Numbered Documents (which as stated in Section I include the Priced Documents), the Schedule of Rates and the Contract Sum Analysis, as appropriate, have been signed by the Contractor and the Sub-Contractor and attached hereto;

4th At the date of this Agreement:

(A) the Sub-Contractor is/is not* the user of a current sub-contractor's tax certificate under the provisions of the Finance (No 2) Act 1975 (hereinafter called 'the Act') in one of the forms specified in Regulation 15 of the Income Tax (Sub-Contractors in the Construction Industry) Regulations, 1975, and the Schedule thereto (hereinafter called 'the Regulations'); where the words 'is not' are deleted, clause 18A of the Sub-Contract Conditions referred to in Article 1·2 shall apply to the Sub-Contract and clause 18B of the said Sub-Contract Conditions shall not apply; where the word 'is' is deleted, clause 18B shall apply to the Sub-Contract and clause 18A·2 to ·8 shall not apply;

(B) The Contractor is/is not* the user of a current sub-contractor's tax certificate under the Act and the Regulations;

(C) The Employer under the Main Contract is/is not* a 'contractor' within the meaning of the Act and the Regulations.

    *delete as applicable

Section III – Articles of Agreement

Now it is hereby agreed as follows

Article 1

Sub-Contractor's
Obligations

1·1 For the consideration mentioned in Article 2 the Sub-Contractor shall, upon and subject to the Sub-Contract Documents, namely this Tender and Agreement NAM/T, the Sub-Contract Conditions and the Numbered Documents, carry out and complete the Sub-Contract Works shown upon and described by or referred to in those documents, and in accordance with the requirements, if any, of the Contractor for regulating the due carrying out of the Works which are agreed by the Sub-Contractor, initialled by the Contractor and Sub-Contractor, attached hereto and incorporated herein, provided that such requirements shall not alter any item set out in Section I or II of NAM/T.

Sub-Contract
Conditions

1·2 The Sub-Contract Conditions are those set out in the 'Sub-Contract Conditions NAM/SC' 1984 Edition (incorporating Amendments 1 to 3) issued by the Joint Contracts Tribunal, which shall be deemed to be incorporated herein.

Article 2

Sub-Contract Sum

The Contractor shall pay to the Sub-Contractor the VAT-exclusive sum of

£_____._____

_____

_____ (words)

(hereinafter referred to as 'the Sub-Contract Sum') or such other sum as shall become payable in accordance with the Sub-Contract.

Article 3

Adjudicator and
Trustee-Stakeholder

3·1 [ee] The name and address of the Adjudicator for the purposes of clause 22·1·2 of NAM/SC is:

_____

_____

_____

3·2 [ee] The name and address of the Trustee-Stakeholder for the purposes of clause 22·3·1·2 of NAM/SC is:

_____

_____

_____

Article 4

Settlement of
disputes – arbitration

[ff] Any dispute or difference between the Contractor and the Sub-Contractor shall be referred to arbitration in accordance with and subject to the provisions of the Sub-Contract.

[ee] *For these provisions to be effective the names and addresses identifying the Adjudicator and the Trustee/Stakeholder should be inserted at the time of entering into this Agreement.*

[ff] *The main provisions dealing with arbitration are to be found in clause 35 of the Sub-Contract Conditions; but references to arbitration may also arise under clauses 18A or 18B, and clause 22.*

Section III – Articles of Agreement

ATTESTATION

[gg] Signed by or on behalf
of the Contractor

in the presence of:

Signed by or on behalf
of the Sub-Contractor

in the presence of:

[gg] The Common Seal of
the Contractor was
hereunto affixed
in the presence of: C.S.

The Common Seal of
the Sub-Contractor was
hereunto affixed
in the presence of: C.S.

[gg] Complete under hand or under seal as applicable: see
items 4 and 13 in Section I.

Appendix 3
RIBA/CASEC Form of Employer/Specialist Agreement (ESA/1)

RIBA/CASEC
Form of Employer/Specialist Agreement

for use between the Employer and a Specialist to be named as a sub-contractor under the
JCT Intermediate Form of Building Contract 1984 Edition (IFC 84)

Between the Employer namely _____

_____

and the Specialist namely _____

The Employer has appointed _____

as the Architect for the purposes hereof and has provided or caused to be provided to the
Specialist the information referred to in the Appendix, part 1, including a brief description of
the Sub-Contract Works hereinafter referred to.

INVITATION

The Specialist is hereby invited:

*Delete either
box A or
box B*

A	To submit herewith a tender for the Sub-Contract Works mentioned in paragraph 6 of the Schedule hereto ('the Tender') and to satisfy the requirements described or referred to in the Schedule hereto ('the Requirements').
B	To submit herewith an approximate estimate in respect of the Sub-Contract Works mentioned in paragraph 6 of the Schedule hereto ('the Approximate Estimate') and to satisfy the requirements described or referred to in the Schedule hereto ('the Requirements'), which include the submission of a tender ('the Tender').

**OFFER AND
AGREEMENT**

Schedule referred to above

1   The Requirements of the Employer described or referred to herein relate to the Sub-Contract
    Works. Save to the extent otherwise agreed, the time requirements are to be satisfied by the
    Specialist subject to the Employer providing or causing to be provided to the Specialist the
    further information (if any) referred to in the Appendix, part 2, at the time or times therein
    prescribed.

2   The information required to be provided by the Specialist to the Architect shall also be
    provided so as to enable the Architect to co-ordinate and integrate the design of the
    Sub-Contract Works into the design for the Main Contract Works as a whole.

3   The Specialist is required to provide to the Architect:

3·1   according to such time requirements as are stated in the Appendix, part 3, the information
      required for the Tender to be used, as the case may be

      either
      – for the purposes of (a) obtaining for the Employer tenders for the Main Contract Works
        and (b) inclusion in the contract documents of the main contract;

      or
      – for the purposes of enabling the Architect to issue an instruction to the main contractor
        as to the expenditure of a provisional sum requiring the Sub-Contract Works to be
        executed by the Specialist as a sub-contractor employed by the main contractor;

Footnote

*This Agreement does not form part of the Agreement between the Employer and the Main
Contractor or of that between the Main Contractor and the Specialist. While it may be sent to the
Specialist with the Form of Tender and Agreement NAM/T for him to complete and return, it should
not be included in the main contract documents or in an instruction of the Architect for the
expenditure of a Provisional Sum; it should be retained for signing/sealing by the Employer.*

OFFER AND AGREEMENT
continued

3·2 according to the time requirements stated or referred to in the Tender and/or the Sub-Contract, such further information relating to the Sub-Contract Works as is reasonably necessary to enable the Architect to provide the main contractor with such information as is reasonably necessary to enable the main contractor:

— to carry out and complete the Main Contract Works in accordance with the conditions of the main contract, including the Sub-Contract Works to be executed by the Specialist as the Sub-Contractor; and

— to provide the Specialist in accordance with the conditions of the Sub-Contract with such information as is reasonably necessary to enable the Specialist to carry out and complete the Sub-Contract Works in accordance with the Sub-Contract.

4 The Employer shall be entitled to use for the purposes of carrying out and completing the Main Contract Works and maintaining or altering the Main Contract Works any drawings or information provided by the Specialist in accordance with this Agreement.

5 The Requirements include the exercise by the Specialist of reasonable care and skill in:

— the design of the Sub-Contract Works insofar as the Sub-Contract Works have been or will be designed by the Specialist; and

— the selection of materials and goods for the Sub-Contract Works insofar as such materials and goods have been or will be selected by the Specialist; and

— the satisfaction of any performance specification or requirement included or referred to in the description of the Sub-Contract Works included in or annexed to the Tender.

Definitions

6 Whether or not the Specialist's tender for the Sub-Contract Works is accepted and a Sub-Contract entered into, in this Offer and Agreement:

— 'the Architect' *means* the Architect named on page 1 as appointed by the Employer or such other person who is named in Section 1 of the Tender as 'the Architect' or as 'the Supervising Officer' for the main contract or such other person as the Employer appoints in place of the person so named;

— 'the Form of Tender NAM/T' *means* the form issued by the Joint Contracts Tribunal for use with the JCT Intermediate Form of Building Contract;

*Delete as appropriate

and 'the Tender' *means* such a form as completed for the purposes of the Specialist's tender (a copy of which * is identified on page 3 and annexed hereto/* will be submitted by the Specialist to the Architect) containing (a) information given to the Specialist about the main contract for which the Sub-Contract Works are required, and (b) information about the proposed conditions of the sub-contract to be entered into between the Specialist and the main contractor for the execution of the Sub-Contract Works;

— 'information' *includes*, wherever appropriate, drawings and the information to be submitted with the Approximate Estimate and/or the Tender;

— 'the Main Contract Works' *means* the works of which the Sub-Contract Works form a part and which are described in Section 1 of the Tender;

— 'the Sub-Contract Works' *means* the work to be executed by the Specialist as a sub-contractor named under the provisions of a building contract between the Employer and a main contractor, after receipt of the Specialist's tender using the Form of Tender NAM/T;

and 'the Sub-Contract' *means* the Sub-Contract entered into by the Specialist as such named person with the main contractor.

Materials and Goods

Delete paragraphs 7·1 and 7·2 if not applicable

7·1 After the date of this Agreement but before the Specialist enters into a sub-contract for the execution of the Sub-Contract Works the Employer may require the Specialist to proceed with the purchase of materials or goods or the fabrication of components for the Sub-Contract Works.

7·2 If the Sub-Contract is not entered into then, subject to any agreement to the contrary, the Employer shall pay the Specialist for any such materials and goods which have been purchased or components properly fabricated whereupon they shall become the property of the Employer.

OFFER AND AGREEMENT
continued

*Delete either
box AA or
box BB*

AA   A copy of the Tender referred to in A on page 1 is **annexed hereto marked**

_____

We offer to satisfy the Requirements described or referred to in the Schedule above in accordance with this Agreement.

This offer is withdrawn if not accepted within_____weeks of the date of our signature below.

BB   The Approximate Estimate referred to in B on page 1 is **annexed hereto marked**

_____

We offer to satisfy the Requirements described or referred to in the Schedule above in accordance with this Agreement:

— **Provided** that if the Sub-Contract is not entered into by the Specialist, the Employer shall pay the Specialist the amount of any expenses reasonably and properly incurred by the Specialist in carrying out work in the designing of the Sub-Contract Works in anticipation of the Sub-Contract and in accordance with the Requirements.

This offer is withdrawn if not accepted within_____weeks of the date of our signature below.

In consideration whereof the sum of £10 (plus VAT as appropriate) shall be payable to the Specialist.

Signed_____ Dated_____

on behalf of the Specialist

This offer is accepted

Signed_____ Dated_____

(on behalf of) the Employer

In witness of this Agreement the parties have hereunto set their seals

this_____day of_____19_____

Signed, sealed and delivered by:

in the presence of:

Or   The common seal of:

was hereunto affixed in the presence of:

Signed, sealed and delivered by:

in the presence of:

Or   The common seal of:

was hereunto affixed in the presence of:

Appendix

Referred to in the Form of Agreement

Part 1   Information provided herewith by the Employer (referred to on page 1)

_____

_____

_____

_____

_____

_____

_____

_____

_____

Part 2   Further information which the Employer is to provide or cause to be provided, referred to in paragraph 1 of the Schedule

Information   Date

_____   _____

_____   _____

_____   _____

_____   _____

_____   _____

Part 3   Time requirements referred to in paragraph 3·1 of the Schedule

For Approximate Estimate (if applicable)

_____

_____

_____

_____

For Tender

_____

_____

_____

_____

_____

_____

Appendix 4
JCT Practice Note IN/1

Introductory Notes on the JCT Intermediate Form of Building Contract IFC 84

General remarks

These notes should be read together with those in Practice Note 20 (revised 1984): 'Deciding on the appropriate form of JCT main contract'.

The Intermediate Form of Building Contract was first issued in 1984 to take its place among the existing forms of building contract issued by the Joint Contracts Tribunal. It is in a single edition for use both by private employers and by local authorities, like the Minor Works Form but unlike the Standard Form, and it can be used either 'With' or 'Without Quantities'.

It was decided to issue the Intermediate Form with the following statement on use endorsed on its cover:

'This Intermediate Form is issued for contracts in the range between those for which the JCT Standard Form of Building Contract With Quantities (1980 Edition) and the JCT Agreement for Minor Building Works (1980 Edition) are issued.

' The Form would be suitable where the proposed Building Works are:

    1. of simple content involving the normally recognised basic trades and skills of the industry; and

    2. without any building service installations of a complex nature, or other specialist work of a similar nature; and

    3. adequately specified, or specified and billed, as appropriate prior to the invitation of tenders.

' Guidance on the intended use of the Form including the length of time and contract value above which the Form should not normally be used is set out in JCT Practice Note 20 (revised 1984).'

The Intermediate Form can be used with Contract Drawings and *either* a Specification *or* Schedules of Work *or* Bills of Quantities; and (unlike the Standard Form) the Intermediate Form provides for a named person to be employed as a sub-contractor by the contractor for work to be priced by the contractor, but does *not* provide for a sub-contractor to be nominated in respect of a prime cost sum item, nor does it provide for nominated suppliers.

The following are also issued by the Tribunal for use with the Intermediate Form:

'JCT Fluctuations Clauses for use with the JCT Intermediate Form of Building Contract IFC 84' (including Formula Rules);

Form of Sub-Contract Tender and Agreement for persons named under clause 3·3: NAM/T;

Sub-Contract Conditions (including fluctuations clauses), referred to in NAM/T: NAM/SC;

Sub-Contract Formula Rules, referred to in NAM/T and NAM/SC; and

Clause for insertion in the Intermediate Form as '2·11 – Partial Possession by Employer', if required: Appendix to this Practice Note.

Outline

The Intermediate Form as a whole is set out generally in accordance with the pattern used for the Standard Form and the Minor Works Form. It will be seen from the table of contents which is printed facing page 1 of the Form that the Conditions are arranged in 8 numbered sections having the same headings for the sections as are used in the Minor Works Form.

In the Recitals, the Articles and the Appendix there are spaces for insertions to be made and alternative wording, or paragraphs, for deletion of whichever does not apply, but in the Conditions no insertions are required (unlike the Minor Works Form), nor deletions (unlike the Standard Form).

The space in Article 3 is for the insertion of the name of the Architect or, as the case may be, the Supervising Officer, as explained in the footnote. References in this Practice Note to 'the Architect' apply also to a person appointed as the Supervising Officer.

On the pages of the Form following the Appendix are printed the Supplemental Conditions A (VAT), B (Statutory tax deduction scheme) and E (Fair Wages), with statements referring to Supplemental Conditions C (Contribution, levy and tax fluctuations) and D (Use of price adjustment formulae), which are available in a separate booklet together with the Formula Rules referred to in Supplemental Condition D.

There is in general less procedural detail in the Intermediate Form than in the Standard Form but otherwise the provisions of these forms have much in common. For example, the payment provisions in Section 4 of the Conditions are broadly similar to those in the Standard Form including the provisions for payment for unfixed materials, which follow those in the Standard Form with the amendment issued in 1984 (after the *Dawber Williamson* case); and provision is made for the percentages not included in the amounts certified for interim payment to have trust status (but not where the Employer is a local authority).

However, it was decided that the provisions relating to insurance should be in the same terms as in the current Standard Form (1984 edition). These are set out at length in Section 6 of the Intermediate Form, with the various provisions which enable the Contractor's insurance to be approved or inspected mentioning that this is to be done by the Employer or by the Architect on behalf of the Employer.

The Intermediate Form differs from the Standard Form in that:

(1) the Intermediate Form provides only for fluctuations in respect of contribution, levy and tax, except that formula price adjustment is also available where the Contract Documents include priced Bills of Quantities;

(2) certain of the events described in the Standard Form as events which may give rise to an extension of time do not appear in the Intermediate Form or are stated to be optional;

(3) the Intermediate Form does not provide for the confirmation of instructions given orally; and

(4) there are some differences in the procedure for, and the consequences of, determination of the Contractor's employment.

Although the Intermediate Form does not include a provision for partial possession before practical completion of the whole of the Works an additional clause for this purpose is set out in the Appendix to this Note.

In addition, the following are some new provisions in the Intermediate Form which have not been included in any of the forms of contract previously issued by the Tribunal:

(a) provision for the Employer to defer giving possession for up to 6 weeks from the date for possession stated in the Appendix to the Conditions;

(b) specific provision for the Architect to make in certain circumstances an extension of time after the due completion date has passed;

(c) provisions relating to the testing of similar work, materials or goods where defects are discovered in work already executed or materials or goods already supplied; and

(d) an optional provision for referring disputes on points of law to the High Court.

More detailed remarks

The rest of this Note relates to the main differences between the provisions of the Intermediate Form and those of the Standard and Minor Works Forms.

The Contract Documents

Under Article 1 and clause 1·1 of the Conditions the Contractor is required to carry out and complete the Works, briefly described in the 1st recital, in accordance with the Contract Documents identified in the 2nd recital, which is in two mutually exclusive alternatives denoted 'A' and 'B'.

Alternative 'A' is for use when the Contract Documents include a Specification *or* Schedules of Work *or* Bills of Quantities, prepared on the instructions of the Employer, which the Contractor has priced in detail and the total of which pricing is the Contract Sum.

Alternative 'B' is for use with Drawings and a Specification where the Contractor has not been required to price the Specification in detail but has been required to provide a Contract Sum Analysis or a Schedule of Rates on which the Contract Sum is based.

In either case, the Contract Documents are defined as including the Agreement and the Conditions and the Contract Drawings, and one other document describing the items of work.

Where the descriptive document consists of priced Bills of Quantities the Intermediate Form follows the Standard Form 'With Quantities' in:

(a) calling these 'the Contract Bills';

(b) requiring them to have been prepared in accordance with the RICS/BEC Standard Method of Measurement, unless otherwise stated; and

(c) providing that the quality and quantity of the work included in the Contract Sum shall be deemed to be that which is set out in the Contract Bills.

It was not considered necessary to provide any further definition of the Contract Bills.

Under Alternative 'B', the descriptive document consists of the Specification, and the Intermediate Form follows the Standard Form 'Without Quantities' by requiring the Contractor to supply another document, which can be used for valuing variations and provisional sum work. This other document may be either a Schedule of Rates (like the Standard Form) *or* a Contract Sum Analysis.

Again, it was not considered necessary to give any further definition of the Specification or Schedule of Rates, but the Contract Sum Analysis is defined in clause 8·3 as:

'an analysis of the Contract Sum provided by the Contractor in accordance with the stated requirements of the Employer'.

It is in the interest of both parties in the effective operation of the Conditions and the avoidance of disputes that the form and degree of the Contract Sum Analysis should be adequate for the purpose required by the Conditions, namely the valuation of variations and provisional sum work in accordance with the rules set out in clause 3·7, particularly rules 3·7·2 and 3·7·3.

'Schedules of Work' are also defined, in clause 8·3, as a Schedule referring to the Works which has been provided by the Employer; and it is to be distinguished from a Schedule of Rates, which the Contractor is to provide where Alternative 'B' applies, unless a Contract Sum Analysis is required instead.

Schedules of Work are included in the Contract Documents only where Alternative 'A' applies, and then only if they are requested to be priced by the Contractor for the computation of the Contract Sum.

The form and content of the Specification/Schedules of Work are not prescribed, but they are mentioned in clause 1·2 as defining the quality of the work included in the Contract Sum where there are no Contract Bills.

The first two paragraphs of clause 1·2, which are irrelevant where there are Contract Bills, allow for the Specification, priced or unpriced, or the priced Schedules of Work to have no quantities for any item; *or* quantities for some but not other items; *or* quantities for all items; and also for the priced Schedules of Work to have quantities for some *or* all items.

To the extent that no quantities are given, the quality and quantity of the work included in the Contract Sum is stated to be 'that in the Contract Documents taken together', but such that the Contract Drawings prevail where there is an inconsistency between them and a description in the Specification/Schedules of Work.

To the extent that quantities are given for items in the Specification/Schedules of Work the quality and quantity of the work included in the Contract Sum for those items is stated to be that which is in the Specification/Schedules of Work, and the Contract Drawings do *not* prevail.

The Architect is required to instruct, under clause 1·4, that an inconsistency, or an error in description, quantity or omission, or a departure from the above mentioned method of preparation of the Contract Bills, shall be removed. If that results in a change from what would have been the case under clause 1·2 described above, the correction is to be valued as a Variation, for adjustment of the Contract Sum by addition or deduction accordingly.

But clause 4·1 expressly states that, subject to such operation of clause 1·4, any error or omission, whether of arithmetic or not, in the computation of the Contract Sum shall be deemed to have been accepted by the parties to the Contract.

Date of Possession

There are provisions in clauses 2·1 (similar to those in the Standard Form) requiring possession of the site to be given to the Contractor on a date stated in the Appendix.

In addition there is a provision in clause 2·2 enabling the Employer to defer possession for a stated period of up to 6 weeks, and provision is included for the Architect to make an extension of time if appropriate (clause 2·4·14) and for payment by the Employer to the Contractor for loss and expense, if any, incurred as a result (clause 4·11).

Extension of Time

The events giving rise to an extension of time are listed in clause 2·4. These are the same as those in the Standard Form except that there is no provision relating specifically to the exercise of statutory powers by the UK Government nor for delay on the part of nominated sub-contractors; the events relating to the non-availability of labour and materials are optional; and, additionally, there is an event relating to deferment of possession, as mentioned above, and an instruction relating to a person named as a sub-contractor may be treated as an event for extension of time in accordance with clauses 3·3·1, 3·3·4 or 3·3·5.

The provisions for notification by the Contractor to the Architect for matters affecting progress are less detailed than the Standard Form.

There is provision for the Architect to extend the time for completion beyond the Date for Completion in the manner described in clause 2·3 in respect of the events listed in clause 2·4.

There is also specific provision in respect of some of those events occurring before Practical Completion when the Date for Completion has passed or after the expiry of any extension previously given. This is make clear that the Architect may extend the time for completion in respect of instructions, the need for which only becomes apparent when the progress of the Works has reached a certain stage. This does not affect the general requirement that the Contractor shall be given information to enable the Contractor to complete the Works by the Date for Completion.

In addition it is provided that the Architect may make an extension of time in respect of *any* events at any time up to 12 weeks after the date of Practical Completion.

The Architect would have to cancel any previously issued certificate of non-completion issued in accordance with clause 2·6; and this in turn could lead to the operation of clause 2·8, requiring the Employer to repay liquidated damages previously deducted under clause 2·7 in respect of the cancelled certificate.

Variations and Provisional Sums

The amount to be added to or deducted from the Contract Sum in respect of instructions requiring a variation or on the expenditure of a provisional sum may be agreed between the Employer and the Contractor prior to the Contractor complying with the instruction.

Where there is no such prior agreement a valuation is to be made by the Quantity Surveyor in accordance with the 9 rules in clause 3·7 (broadly similar to those of the Standard Form) which where appropriate relate the valuation to the values in the 'priced document'.

The 'priced document' is defined in clause 3·7·1 by reference to whichever of the alternatives 'A' or 'B' of the 2nd Recital apply.

Rule 3·7·6 refers to 'items of a preliminary nature'. It should be noted that where the 'priced document' consists of Contract Bills, these are to have been prepared in accordance with the Standard Method of Measurement, which contains a list of such items in 'Section B: Preliminaries'.

Fluctuations

Clause 4·9(a) and Supplemental Condition C contain provisions for tax etc. fluctuations on the same basis as for the Standard Form. Where there are Contract Bills there is the alternative under clause 4·9(b) and Supplemental Condition D for formula adjustment on the same basis as for the Standard Form.

The restriction on fluctuations where the Contractor is in default over completion ('freezing') applies whether the events for extension of time in clauses 2·4·10, 2·4·11 or 2·4·14 are stated in the Appendix to apply or not.

However, fluctuations in respect of work executed by persons named as sub-contractors under clause 3·3 are dealt with separately under clause 4·10 so as to allow for the sub-contract fluctuations to be on a basis different from that for the rest of the Works.

Testing

Clause 3·12 contains provisions (similar to those in the Standard Form) enabling the Architect to issue instructions for inspection or tests of work or materials.

In addition, clause 3·13·1 provides that where a failure of work or materials to be in accordance with the contract is discovered during the carrying out of the Works, the Contractor is required to state in writing to the Architect what action the Contractor will take to establish that there is no similar failure in work already executed or materials supplied.

Nevertheless the Architect may issue instructions in the circumstances described in clause 3·13·1 requiring the Contractor to open up work for inspection or carry out tests. If so, the Contractor must forthwith comply, and, subject to the provisions in clause 3·13·2, at no cost to the Employer; but such compliance is an event which may give rise to an extension of time under clauses 2·3 and 2·4·6.

However, such instructions of the Architect should be reasonable in the circumstances, and if the Contractor states a reasoned objection in writing which is not removed by the Architect withdrawing or modifying the instruction, then any dispute over the reasonableness of the nature or extent of the opening up or testing is stated to be referred for decision under clause 3·13·2 to an arbitrator whose powers include determining the amounts, if any, to be paid by the Employer to the Contractor and any extension of time.

Depending on the failure and the circumstances, progressive sampling or other methods may be considered sufficient, without opening up or testing every similar item of work.

Defects and Errors in Setting Out

It is specifically provided that if the Architect instructs that a defect should not be made good (clause 2·10), or that an error in setting out should not be corrected (clause 3·9), an appropriate deduction should be made from the Contract Sum.

Determination

The Intermediate Form contains in Section 7 detailed provisions relating to determination of the Contractor's employment under the contract. These are similar, but not identical, to those in the Standard Form.

However, unlike the Standard Form, the Contractor is not entitled to payment for direct loss and/or damage resulting from determination consequent upon: force majeure, or loss or damage to the Works occasioned by the specified perils or civil commotion; and the Employer, in addition to the Contractor, is entitled to determine the Contractor's employment after suspension of the Works for three months by reason of such events.

Arbitration

Provision is made in Article 5 for disputes to be settled by arbitration. These provisions are similar as for the Standard Form, but without the restraint on opening references to arbitration while the Works are still in progress; and, in addition, there is an optional provision for the parties to agree at the time of entering into the contract to let questions of law be determined by the High Court under the provisions of the Arbitration Act 1979.

This optional provision is not in any of the forms of contract previously issued by the Tribunal, and its effect has not yet been tested.

There is also an optional provision to refer to the same arbitrator disputes arising under the main contract which are substantially the same as or connected with issues raised in a related dispute which has been referred to an arbitrator under a sub-contract with a person named under the provisions in clause 3·3.

Naming a sub-contractor under clause 3·3 and use of NAM/T

There is no provision for nominated sub-contractors; instead provision is made in clause 3·3 for work which is to be priced by the Contractor to be carried out by a person named in the Specification/Schedules of Work/Contract Bills, or in an instruction for the expenditure of a provisional sum, as a sub-contractor using the prescribed JCT Form of Tender and Agreement NAM/T. The consequences of default of the named person and the provisions for payment relating to the sub-contract works differ from those relating to nominated sub-contractors under the Standard Form.

It is expressly stated in clause 3·3·8 that these provisions do *not* apply to the execution of part of the Works by a local authority or statutory authority executing such work solely in pursuance of its statutory rights or obligations, to whom clause 2·4·13 applies, in the event of delay.

The provisions in clause 3·3 require the Contractor to employ as a sub-contractor for execution of part of the Works a person who may be named in one of the following ways:

(1) in the Specification/Schedules of Work/Contract Bills for pricing by the Contractor in respect of work described therein: **Procedure (1)**; or

(2) in an instruction as to the expenditure of a provisional sum included in the Specification/Schedule of Work/Contract Bills: **Procedure (2)**.

In either case, the Contractor must be given a full description of the work in the Specification/Schedules of Work/Contract Bills under Procedure (1) or in the instruction under Procedure (2), together with the particulars of the tender of the named person in a Form of Tender NAM/T with Sections I and II completed, together with the Numbered Documents referred to therein.

The work of named persons is not made the subject of a prime cost sum in the Contract Documents. Accordingly:

(a) Where the person is named under Procedure (1), the Contractor must consider the description and particulars so given and include in his tender sum an amount for the execution of the work by the named person. The Employer pays that amount to the Contractor; the Contractor pays to the named person the amount stated in Tender and Agreement NAM/T.

(b) Where the person is named under Procedure (2), the Employer pays to the Contractor an amount for the execution of the work by the named person which is *either* agreed between the Employer and Contractor *or* determined by the Quantity Surveyor as being a fair valuation (clause 3·7); the Contractor pays to the named person the amount stated in Tender and Agreement NAM/T.

The use of Tender and Agreement NAM/T is required whichever Procedure is used. It is divided into three sections:

Section I – Invitation to Tender. The whole of this section is to be completed by the Architect (or the Supervising Officer). In Section I, particulars of the Main and Sub-Contract Works are given, together with indications of the sub-contract programme and basis of fluctuations etc.

Section II – Tender by Sub-Contractor. The whole of this Section is to be completed by the tendering sub-contractor. The tender quotes a sub-contract sum together with the percentage additions to the prime cost of daywork which he requires. In addition, he specifies any additional attendances or other special requirements and gives programme information and details of fluctuations arrangements to the extent that they have not already been stated in Section I.

Section III – Articles of Agreement. If the Sub-Contractor's tender is accepted, he and the Contractor complete Section III. This Section consists of the Sub-Contract Articles of Agreement and incorporates by reference the Sub-Contract Conditions NAM/SC, as required by the main Contract Conditions. Upon completion of Section III a sub-contract is formed without any need to execute the Sub-Contract Conditions NAM/SC.

In the case of Procedure (1) clause 3·3·1 provides that:

(a) if the Contractor is unable to enter into a sub-contract with the person named in accordance with the particulars given, then provided the Architect is satisfied with the reasons therefor, he is required to issue further instructions; and

(b) at any time before the Contractor has entered into a formal sub-contract the Architect may, by omitting the work and substituting a provisional sum, instruct that another person shall carry out the work.

In the case of Procedure (2) the circumstances referred to in (a) or (b) above could be dealt with by the Architect issuing a further instruction as to the expenditure of the provisional sum.

Consequences of Determination

The Sub-Contract Conditions NAM/SC include provisions for the determination of the employment of the sub-contractor in the terms of clause 27 (*Sub-Contractor's default*) and clause 28 (*Main Contractor's default*), and it was considered necessary to include, in the main contract, provisions –

(1) for the Architect to issue further instructions under the main contract in the event of the employment of a named person being determined in that or any other way, and

(2) stating the consequential effects for the main contract in respect of extensions of time and adjustment of the Contract Sum.

These provisions are contained in clauses 3·3·3 – 3·3·6.

Design for Sub-Contract Works

The Note endorsed on the cover of the Intermediate Form quoted above states one of the criteria for the use of the Form to be that the Works are 'without any building service installations of a complex nature, or other specialist work of a similar nature'. However, even for such installations or other specialist work not of a complex nature, a named person may be involved in design, and, in the terms of clause 3·3·7, the Contractor is expressly relieved of responsibility to the Employer for defects in the named person's 'design' of the sub-contract work (if any). This relief however is stated not to affect the Contractor's obligation in regard to the supply of goods and materials and workmanship.

The Note on page 1 of NAM/T states: 'Should there be a separate agreement between the Employer and the Sub-Contractor relating to such matters as are referred to in clause 3·3·7 of the main contract conditions (design etc.), it should *not* be attached to the Form of Tender and Agreement, either where the Form is included in the main contract documents or where it is included in an instruction of the Architect/the Supervising Officer for the expenditure of a provisional sum'.

The Tribunal has not issued any such separate agreement, but the Tribunal has been informed by two of the constituent bodies, RIBA and CASEC, that they have prepared such a form of separate agreement, and that this is being published at the same time as the Intermediate Form under the title 'RIBA/CASEC Form of Employer/Specialist Agreement: ESA/1'.

Sub-Contract Conditions NAM/SC

As mentioned above, the sub-contract is formed by executing the Articles of Agreement in Section III of Tender and Agreement NAM/T. The Conditions NAM/SC are incorporated by reference in Article 1·2 and do not require separate execution.

Unlike the Intermediate Form, a section headed format has not been adopted for the NAM/SC Conditions.

Additionally, the matter corresponding to the Supplemental Conditions to the Intermediate Form is set out within the text of the NAM/SC Conditions. A separate set of Sub-Contract Formula Rules has, however, been issued by the Tribunal for use in conjunction with the Conditions.

The principles of the Sub-Contract Conditions NAM/SC generally follow those of the Intermediate Form, subject to the necessary changes to reflect the sub-contractual status of the Conditions.

However, unlike the Intermediate Form, the Sub-Contract Conditions include detailed provisions limiting the right of the Contractor to set-off monies against the named person, in clause 21, and in the event of a dispute between them, a procedure in clause 22 whereby an adjudicator will consider and give a speedy decision as to who shall hold the money pending an arbitration decision on the dispute.

A unique feature of clause 27 of NAM/SC is that where, under clause 3·3·6 (b) of the Intermediate Form the Contractor is obliged to seek to recover from a defaulting named person the additional costs or losses incurred by the Employer in consequence of the issue by the Architect of instructions following the determination of the employment of that named person, then under clause 27·3·3 the named person is to pay such sums to the Contractor.

Further, to avoid any possibility that the named person may raise a legalistic defence to the Contractor's claim, the named person undertakes not to contend that, by virtue of the various provisions of the Intermediate Form, the Contractor has suffered no loss.

**Intermediate Form of Building Contract
1984 Edition**

Practice Note IN/1

Appendix: Revised November 1986

Applicable only where the Intermediate Form includes
'Amendment 1 issued November 1986 – Insurance and related liability provisions'

Partial Possession by Employer before Practical Completion

Clause 2·9 of the Intermediate Form requires the Architect to issue a certificate of Practical Completion when, in his opinion, Practical Completion of the Works is achieved. The defects liability period then begins to run; the Contractor's obligation to keep the Works insured under clause 6·3A (if applicable) ceases; and the Architect is required to issue the certificates for interim payment of 97½%.

The Standard Form of Building Contract includes a provision (clause 18) which is designed for use when the Employer, by agreement with the Contractor, wishes to take possession of part of the Works before practical completion of the whole. When the content of the Intermediate Form was being decided, it was considered that such provision, which would only occasionally be required, should not be included, but that guidance could usefully be issued drawing attention to the various matters in the Conditions which would be relevant, should the Employer wish to take partial possession in the course of the contract.

It was further considered that such guidance should be in the form of a clause which could be used as an addition to Section 2 of the Conditions, and which would, therefore, be numbered 'Clause 2·11' together with consequential amendments.

If it is required to insert the clause and the consequential amendments in the Conditions at the time when the Contract is entered into, this should be shown in the document which formally constitutes the contractual agreement between the parties, that is the Articles of Agreement and the Conditions which are signed or sealed by the parties, and not merely set out or referred to in the Specification/Schedules of Work/Contract Bills. But such provision does *not* provide for phased completion, where the whole of the work is sub-divided into two or more distinct sections, such as is provided for in the Sectional Completion Supplement issued for the Intermediate Form (July 1985).

The additional clause and the consequential amendments are set out on page 2.

Partial Possession by the Employer

2·11   If at any time or times before the date of issue by the Architect/the Supervising Officer of the certificate of Practical Completion the Employer wishes to take possession of any part or parts of the Works and the consent of the Contractor (which consent shall not be unreasonably withheld) has been obtained then, notwithstanding anything expressed or implied elsewhere in this Contract, the Employer may take possession thereof. The Architect/the Supervising Officer shall thereupon issue to the Contractor on behalf of the Employer a written statement identifying the part or parts of the Works taken into possession and giving the date when the Employer took possession (in clauses 2·11, 6·1·4 and 6·3C·1 referred to as 'the relevant part' and 'the relevant date' respectively); and

– for the purpose of clause 2·10 (*Defects liability*) and 4·3 (*Interim payment*) Practical Completion of the relevant part shall be deemed to have occurred and the defects liability period in respect of the relevant part shall be deemed to have commenced on the relevant date;

– when in the opinion of the Architect/the Supervising Officer any defects, shrinkages or other faults in the relevant part which he may have required to be made good under clause 2·10 shall have been made good he shall issue a certificate to that effect;

– as from the relevant date the obligation of the Contractor under clause 6·3A or of the Employer under clause 6·3B·1 or clause 6·3C·2 whichever is applicable, to insure shall terminate in respect of the relevant part but not further or otherwise; and where clause 6·3C applies the obligation of the Employer to insure under clause 6·3C·1 shall from the relevant date include the relevant part;

– in lieu of any sum to be paid or allowed by the Contractor under clause 2·7 *(Liquidated damages)* in respect of any period during which the Works may remain incomplete occurring after the relevant date there shall be paid or allowed such sum as bears the same ratio to the sum which would be paid or allowed apart from the provisions of clause 2·11 as the Contract Sum less the amount contained therein in respect of the said relevant part bears to the Contract Sum.

The consequential amendments are:

Clause	Amendment
6·1·3	Line 1 **delete** 'The' **insert** 'Subject to clause 6·1·4 the'.
	Add as clause 6·1·4: '6·1·4 If clause 2·11 has been operated then, in respect of the relevant part, and as from the relevant date such relevant part shall not be regarded as 'the Works' or 'work executed' for the purpose of clause 6·1·3.'
6·3·3	Line 19, after 'existing structures' **insert** '(which shall include from the relevant date any relevant part to which clause 2·11 refers)'
6·3A·1	Line 7, after 'shall' **insert** '(subject to clause 2·11)'.
6·3B·1	Line 6, after 'and shall' **insert** '(subject to clause 2·11)'.
6·3C·1	Line 3, after 'structures' **insert** '(which shall include from the the relevant date any relevant part to which clause 2·11 refers)'
6·3C·2	Line 6, after 'and shall' **insert** '(subject to clause 2·11)'.
6·3D·3	Line 3, after '6·3D·2' **insert** '(or any revised sum produced by the application of clause 2·11)'

Appendix 5
Intermediate Form of Building Contract (IFC 84)
Sectional Completion Supplement

Intermediate Form of Building Contract (IFC 84)
Sectional Completion Supplement

Issued by the Joint Contracts Tribunal
for the Standard Form of Building Contract July 1985, revised 1989

Guidance notes:

1 The Supplement on pages (2) to (5) provides for the JCT Intermediate Form of Building Contract (IFC 84) to be adapted so as to be suitable for use where the Works are to be completed by sections. It has been issued by the Joint Contracts Tribunal in response to representations that such adaptation was needed where sectional completion was required for Works such as housing rehabilitation or refurbishment, for which the Intermediate Form would otherwise be appropriate having regard to the guidance on use with which that Form is endorsed. A JCT Intermediate Form of Building Contract so adapted is referred to in the following paragraphs as 'the Adapted Contract'.

Practice
2 The Adapted Contract is intended for use only where the Employer requires the Works to be carried out by sections of which the Employer will take possession on Practical Completion of each Section. The Adapted Contract cannot be used for contracts where the Works are not divided into Sections in the Contract Documents.

3 Where tenders are invited, it is essential that the tender documents should identify clearly the Sections which, together, comprise the whole Works. The Sections should be serially numbered and these Section numbers inserted in the Appendix of the Adapted Contract. Care should be taken, in dealing with any part of the Works which is common to all or several Sections (such as a boiler-house serving three separate Sections, each comprising a block of flats), to put this part of the Works into a separate Section.

The Adaptation
4 The principal modifications in the Adapted Contract are the division of the Works into definite Sections in the Contract Documents, and the fixing of separate completion periods for each of the Sections. The Adapted Contract remains, however, a single Contract for which one Final Certificate has to be issued at the end; no provision is made for separate final certificates for each Section.

5 The Intermediate Form of Building Contract is, therefore, modified in the Adapted Contract as follows:

The Contract Documents
5·1 The first recital to the Articles of Agreement is altered so that the Contract Documents (Contract Drawings, and Specification or Schedules of Work or Bills of Quantities as applicable) show and describe the division of the Works into Sections, and in clause 8·3 'Section' is defined by reference to these. Clause 1·1 refers to the obligation of the Contractor 'to carry out and complete the Works by Sections'.

5·2 There are to be entered at the end of Article 6 separately for each Section: Date of Possession, Date for Completion, rate of liquidated damages for delay, and Defects Liability Period: see also Note 2 on page (2).

Carrying out the Works
5·3 The Contractor is to be given possession of the site for each of the Sections of the Works on the date stated. He is then to begin and proceed with each Section concurrently or successively as required by and under the Contract, and to complete each Section within the relevant period stated in the Appendix (clause 2·1) or as extended by the Architect/the Contract Administrator under the Contract (clause 2·3). Liquidated damages for delay relate to each Section (clause 2·7), and are separately calculated where there is delay in completing any Section for which extension of time is not given.

Practical Completion
5·4 On practical completion of any Section the Architect/the Contract Administrator must issue a practical completion certificate for that Section (clause 2·9). In consequence the Contractor is relieved of his corresponding duty to insure for that Section under clause 6·3A·1 (if applicable), and, within 14 days of the date of issue of such certificate, clause 4·3 requires the issue of an Interim Payment Certificate based on 97½% of the value of the work in that Section. A separate Defects Liability Period operates for each Section (clause 2·10).

Final Account
5·5 When all Sections have been carried out the Architect/the Contract Administrator must issue a certificate to that effect (clause 2·10) and the period begins to run within which the final account is to be prepared for the whole of the Works comprising all the Sections (clause 4·5).

Additional clause 2·11 (Partial possession)
5·6 Where the contract includes additional clause 2·11 the term 'Section value' is defined in the adapted clause as 'the value ascribed to the relevant Section by Article 6', and the relevant entry indicates that each Section value is to be the total value of the Section ascertained from the 'priced document' referred to in clause 3·7·1. The Section values must amount in total to the Contract Sum, and should take into account the apportionment of Preliminaries and other like items prices in the 'priced document'. For clause 2·11 refer to the Appendix of Practice Note IN/1, 1984, where guidance and the additional clause 2·11 for inclusion in the IFC 84 Conditions can be found.

Insurance
6 The Architect/The Contract Administrator should particularly note that when making insurance arrangements under clause 6·2·4 agreement must be reached with the insurers as to whether any Section for which a Practical Completion Certificate has been issued is to be treated as continuing to be included in the Works and so not insured OR is to be treated as 'property other than the Works' and so covered by the insurance. Under clause 6·3C, if applicable, agreement should also be reached with the insurers on whether or not to treat such a Section as forming part of the 'existing buildings' which are covered by the Employer's insurance.

IFC 84 Sectional Completion Supplement

This Supplement has been prepared so that the Intermediate Form may be adapted for completion of the Works by sections as follows:

A **Page 2 of IFC 84:**

In 1st Recital:

AFTER 'has caused the following documents'	DELETE 'showing and describing the Works'
and AFTER 'to be prepared'	INSERT 'showing and describing both the work to be done and the division of the Works into Sections for phased completion (hereinafter referred to as 'Sections')'.

B **Pages 4 and 5 of IFC 84:**

Between these pages INSERT **Article 6 – Completion of Works by Sections** and complete the entries to be made therein.

NOTE 1:
Article 6 is set out in the following pages of this Supplement. The adaptation makes it necessary to substitute or add a reference to a 'Section' or 'Sections' in some of the Conditions of the Contract Form where the expression 'the Works' appears. All the adapting modifications to the Conditions are set out seriatim in the table comprised in Article 6.

NOTE 2:
The entries at the end of the table are in substitution for the relevant entries in the Appendix of IFC 84, but whatever is required to suit the circumstances of the Contract should be stated in unambiguous terms.

For example, when commencement of any Section depends on progress with any other, it may be appropriate to insert as the Date of Possession for the second or any subsequent Section a period (e.g. number of days) following Practical Completion (or other identifiable stage) of the earlier Section instead of a fixed date, and to insert a Date for Completion which is fixed in time either by reference to a specific date, or alternatively 'N' weeks from the Date of Possession of the earlier Section the completion of which determines the possession of the later.

If more space than is given in the printed Supplement is needed INSERT against the two entries for clause 2·1 (Date of Possession and Date for Completion) in the Appendix and/or at the end of Article 6:

'See page ... of the Specification/Schedules of Work/Contract Bills'

or otherwise as appropriate.

C **Appendix of IFC 84:**

Against the following items in the Appendix INSERT 'as stated in Article 6':

Clause 2·1   Date of Possession (See NOTE 2 above)
        2·1   Date for Completion (See NOTE 2 above)
        2·7   Liquidated damages
        2·10 Defects liability period

Article 6

Completion of Works
by Sections

The modifications to the Conditions stated in the table set out below and the entries at the end of the table are hereby incorporated in this Contract and the provisions of the Articles of Agreement and the annexed Conditions shall have effect as so modified.

TABLE OF CHANGES TO THE CONDITIONS

The clauses indicated in the first column of this table are modified, at the places therein shown in the second column, by the deletion or insertion of such words as are shown in the third column.

1	2	3
Clause number	Place in text	Words to be deleted / inserted
1·1	AFTER 'complete the Works'	INSERT 'by Sections'
2·1	AFTER 'Possession of'	INSERT 'the relevant part of'
	AFTER 'Date of Possession stated in the Appendix'	INSERT 'in relation to any Section'
	AFTER 'proceed with'	DELETE 'the Works' and INSERT 'that Section'
	AFTER 'Date for Completion stated in the Appendix'	INSERT 'in relation thereto'
2·2	AFTER 'may defer'	INSERT 'in relation to any Section'
2·3 1st paragraph	In line 7 AFTER 'the completion of'	DELETE 'the Works' and INSERT 'any Section or Sections'
	In line 8 AFTER 'Date for Completion stated in the Appendix'	INSERT 'in relation thereto'
	At the end AFTER 'completion of'	DELETE 'the Works' and INSERT 'any such Section or Sections'
2·3 2nd paragraph	AFTER 'length of the delay, if any, to'	DELETE 'the Works' and INSERT 'any Section'
	In last line AFTER 'completion of'	DELETE 'the Works' and INSERT 'that Section'
2·3 3rd paragraph	AFTER 'At any time up to 12 weeks after the'	DELETE 'date of Practical Completion' and INSERT 'date of practical completion of any Section'
	AFTER 'may make an extension of time'	INSERT 'for that Section'
2·4·7	AFTER 'Date for Completion'	INSERT 'for any Section'
2·6 1st paragraph	AFTER 'If the Contractor fails to complete'	DELETE 'the Works' and INSERT 'any Section'
	AFTER 'clause 2·3'	INSERT 'in relation thereto'
2·7	AFTER 'stated in the Appendix'	INSERT 'in relation thereto'
	AFTER 'for the period during which'	DELETE 'the Works' and INSERT 'any Section'
2·9	AFTER 'When in the opinion of the Architect/the Contract Administrator'	DELETE 'Practical Completion' and INSERT 'practical completion of any Section'
	At the beginning of the 2nd sentence: AFTER 'Practical Completion of'	DELETE 'the Works' and INSERT 'that Section'
2.9	AFTER 1st paragraph	INSERT as new 2nd paragraph: 'When in the opinion of the Architect/the Contract Administrator practical completion of all the Sections has been achieved he shall forthwith issue (in addition to any certificates of practical completion of the Sections) a Certificate of Practical Completion of the Works and Practical Completion of the whole of the Works shall for the purpose of clause 4·5 (*Computation of adjusted Contract Sum*) be deemed to have taken place on the day named in such certificate'.

Source   Sectional Completion Supplement issued by the Joint Contracts Tribunal for use with the JCT Intermediate Form of Building Contract (IFC84). See Notes A-C on page (2) of Supplement.

1	2	3
Clause number	Place in text	Words to be deleted / inserted
2·10 1st paragraph	AFTER 'any defects, shrinkages or other faults which appear'	INSERT 'in any Section'
	AFTER 'defects liability period named in the Appendix'	INSERT 'in relation thereto'
2·10 2nd paragraph	AFTER 'discharged'	INSERT 'in respect of the relevant Section'
[1] 2·11	In line 1 AFTER 'If at any time before'	DELETE 'Practical Completion of the Works' and INSERT 'practical completion of any Section'
	In the last two lines (twice)	DELETE 'Contract Sum' and INSERT 'Section value'
	AT END of clause	ADD as new paragraph: 'For the purposes of clause 2·11 the expression 'Section value' shall mean the value ascribed to the relevant Section by Article 6.'
4·2 1st paragraph	AFTER 'Date of Possession stated in the Appendix'	INSERT 'for the first Section'
4·3	At the beginning of the clause	INSERT 'The provisions of this clause apply severally in respect of each Section.'
4·6	AFTER 'issued by the Architect/the Contract Administrator under clause 2·10 (*Defects liability*)'	INSERT 'in respect of all Sections'
6·3A·1	AT END of clause	ADD 'provided that as from the date of issue of the certificate of practical completion of any Section the obligation of the Contractor to insure under this clause shall terminate in respect of that Section, but not further or otherwise'.
8·3	AFTER the definition for 'Schedules of Work'	INSERT as an additional definition: 'Section: *means* one of the Sections into which the Works have been divided for phased completion and shown upon the Contract Drawings and described by or referred to in the Specification/Schedules of Work/Contract Bills and in Article 6.'
C4·7	AFTER 'or C3'	INSERT 'for any Section'
	AFTER 'operative under clause 2·3'	INSERT 'in relation to that Section'
C4·8	AFTER 'not to be applied'	INSERT 'in relation to any Section'
C4·8·1	AFTER 'the printed text of clauses'	DELETE '2·3, 2·4 and 2·5' and INSERT '2·3 and 2·4 as hereinbefore modified, and of clause 2·5'
C4·8·2	AFTER 'if any,'	INSERT 'for that Section'
D12·1	AFTER 'if the Contractor fails to complete'	DELETE 'the Works' and INSERT 'any Section'
	AFTER 'operative under clause 2·3'	INSERT 'in relation thereto'
D13	AFTER 'not to be applied'	INSERT 'in relation to any Section'
D13·1	AFTER 'the printed text of clauses'	DELETE '2·3, 2·4 and 2·5' and INSERT '2·3 and 2·4 as hereinbefore modified, and of clause 2·5'
D13·2	AFTER 'if any'	INSERT 'for that Section'

continued on page (5)

Footnote [1] Delete if clause 2·11 is not included in the Conditions. (Additional clause 2·11 *(Partial possession by Employer)* is set out in the Appendix to Practice Note IN/1.)

[2] The following entries are in substitution for the relevant entries in the Appendix:

Clause		Section Number:	Section Number:	Section Number:
1·1	Section of Works as shown on the Contract Drawings and described in the Specification/Schedules of Work/Contract Bills			
2·1	Date of Possession of Section			
2·1	Date of Completion of Section			
2·7	Rate of liquidated damages for Section	£ _____ per	£ _____ per	£ _____ per
2·10	Defects liability period (if none stated is 6 months from the day named in the certificate of practical completion of the Section)			
[1] 2·11	Section value (total value of Section ascertained from priced document referred to in clause 3·7·1)			

Table of Cases

List of abbreviations in report citations

AC – Appeal Cases
ALJR – Australian Law Journal Reports
All ER – All England Law Reports
BLM – Building Law Monthly
BLR – Building Law Reports
CH – Chancery
CLD – Construction Law Digest
CLJ – Construction Law Journal
CLR – Construction Law Reports
Comm Cas – Commercial Cases
EG – Estates Gazette
Ex – Exchequer
Lloyds Rep – Lloyds Reports
NY – New York State Reports
NZLR – New Zealand Law Reports
QB – Queen's Bench
SALR – South African Law Reports
TLR – Times Law Report
TR – Term Report (by Durnford & East)
WLR – Weekly Law Reports

Table of Statutes

Table of JCT 63 and JCT 80 clauses

JCT 63

Index